IFIP Advances in Information and Communication Technology

437

Editor-in-Chief

A. Joe Turner, Seneca, SC, USA

Editorial Board

IFIP – The International Federation for Information Processing

IFIP was founded in 1960 under the auspices of UNESCO, following the First World Computer Congress held in Paris the previous year. An umbrella organization for societies working in information processing, IFIP's aim is two-fold: to support information processing within its member countries and to encourage technology transfer to developing nations. As its mission statement clearly states,

> IFIP's mission is to be the leading, truly international, apolitical organization which encourages and assists in the development, exploitation and application of information technology for the bene t of all people.

IFIP is a non-profitmaking organization, run almost solely by 2500 volunteers. It operates through a number of technical committees, which organize events and publications. IFIP's events range from an international congress to local seminars, but the most important are:

- The IFIP World Computer Congress, held every second year;
- Open conferences;
- Working conferences.

The flagship event is the IFIP World Computer Congress, at which both invited and contributed papers are presented. Contributed papers are rigorously refereed and the rejection rate is high.

As with the Congress, participation in the open conferences is open to all and papers may be invited or submitted. Again, submitted papers are stringently refereed.

The working conferences are structured differently. They are usually run by a working group and attendance is small and by invitation only. Their purpose is to create an atmosphere conducive to innovation and development. Refereeing is also rigorous and papers are subjected to extensive group discussion.

Publications arising from IFIP events vary. The papers presented at the IFIP World Computer Congress and at open conferences are published as conference proceedings, while the results of the working conferences are often published as collections of selected and edited papers.

Any national society whose primary activity is about information processing may apply to become a full member of IFIP, although full membership is restricted to one society per country. Full members are entitled to vote at the annual General Assembly, National societies preferring a less committed involvement may apply for associate or corresponding membership. Associate members enjoy the same benefits as full members, but without voting rights. Corresponding members are not represented in IFIP bodies. Affiliated membership is open to non-national societies, and individual and honorary membership schemes are also offered.

Lazaros Iliadis   Ilias Maglogiannis
Harris Papadopoulos   Spyros Sioutas
Christos Makris (Eds.)

Artificial Intelligence
Applications
and Innovations

AIAI 2014 Workshops: CoPA, MHDW, IIVC, and MT4BD
Rhodes, Greece, September 19-21, 2014
Proceedings

 Springer

Volume Editors

Lazaros Iliadis
Democritus University of Thrace, Orestiada, Greece
E-mail: liliadis@fmenr.duth.gr

Ilias Maglogiannis
University of Piraeus, Greece
E-mail: imaglo@unipi.gr

Harris Papadopoulos
Frederick University, Nicosia, Cyprus
E-mail: h.papadopoulos@frederick.ac.cy

Spyros Sioutas
Ionian University, Corfu, Greece
E-mail: sioutas@ionio.gr

Christos Makris
University of Patras, Rio, Greece
E-mail: makri@ceid.upatras.gr

ISSN 1868-4238   ISSN 1868-422X (eBook)
ISBN 978-3-662-51595-2   ISBN 978-3-662-44722-2 (eBook)
DOI 10.1007/978-3-662-44722-2
Springer Heidelberg New York Dordrecht London

Typesetting: Camera-ready by author, data conversion by Scientific Publishing Services, Chennai, India

Printed on acid-free paper

Springer is part of Springer Science+Business Media (www.springer.com)

Preface

It has been 58 years since the term artificial intelligence (AI) was coined in 1956 by John McCarthy at the Massachusetts Institute of Technology USA. Since then, after huge efforts of the international scientific community, sophisticated and advanced approaches— e.g., games playing, (computers capable of playing games against human opponents) natural languages, computers able to see, hear, and react to sensory stimuli— that would appear only as science fiction in the past are gradually becoming a reality. Multi-agent systems and autonomous agents, image processing, and biologically inspired neural networks (Spiking ANN) are already a reality. Moreover, AI has offered the international scientific community many mature tools that are easily used, well documented, and applied. These efforts have been continuously technically supported by various scientific organizations like the IFIP.

The International Federation for Information Processing (IFIP) was founded in 1960 under the auspices of UNESCO, following the first historical World Computer Congress held in Paris in 1959. The first AIAI conference (Artificial Intelligence Applications and Innovations) was organized in Toulouse, France, in 2004 by the IFIP. Since then, it has always been technically supported by the Working Group 12.5 "Artificial Intelligence Applications." After 10 years of continuous presence, it has become a well-known and recognized mature event, offering AI scientists from all over the globe the chance to present their research achievements. The 10th AIAI was held in Rhodes, Greece, during September 19–21, 2014.

Following a long-standing tradition, this Springer volume belongs to the IFIP AICT series. It contains the accepted papers that were presented orally at the AIAI 2014 workshops that were held as parallel events. Four workshops were organized, by invitation to prominent and distinguished colleagues, namely:

- The Third CoPA (Conformal Prediction and Its Applications)
- The Third MHDW (Mining Humanistic Data Workshop)
- The Third IIVC (Intelligent Innovative Ways for Video-to-Video Communications in Modern Smart Cities)
- The First MT4BD (New Methods and Tools for Big Data)

It is interesting that three of the above workshops were organized for the third time in the row, which means that they are well established in the AI community.

As the title of the conference denotes, there are two core orientations of interest, basic research AI approaches and also applications in real-world cases. The diverse nature of papers presented demonstrates the vitality of AI computing methods and proves the wide range of AI applications.

All papers went through a peer-review process by at least two independent academic reviewers. Where needed, a third and a fourth reviewer was consulted to resolve any potential conflicts. In the four workshops that were held as parallel events in the framework of the 10th AIAI conference, 36 papers were accepted for oral presentation, which corresponds to an acceptance ratio as high as 44.4%.The authors of accepted papers of the workshops come from several countries, namely: Estonia, Finland, UK, Greece, Cyprus, Portugal, Russia, and Sweden.

Three distinguished keynote speakers were invited to present a lecture at the 10th AIAI conference.

1. Professor Hojjat Adeli, Ohio State University, USA.

   Title: "Multi-Paradigm Computational Intelligence Models for EEG-Based Diagnosis of Neurological and Psychiatric Disorders"

   - Professor in the Departments of Biomedical Informatics, Civil and Environmental Engineering and Geodetic Science and Neuroscience and Centers of Biomedical Engineering and Cognitive Science
   - Director of Knowledge Engineering Lab at the Ohio State University
   - Author of nearly 400 research and scientific publications in various fields of computer science, engineering, and applied mathematics since 1976 when he received his PhD from Stanford University.
   - Author of nine books
   - Founder and Editor-in-Chief of the international research journals *Computer-Aided Civil and Infrastructure Engineering* and *Integrated Computer-Aided Engineering.*
   - Over 100 academic, research, and leadership awards, honors, and recognition
   - Keynote/plenary lecturer at 43 national and international computing conferences held in 28 different countries

2. Professor Plamen Angelov, Lancaster University, UK.

   Title: "Autonomous Learning Systems: Association-Based Learning"

   - Chair in Intelligent Systems and leads the Intelligent Systems Research within the School of Computing and Communications, Lancaster University, UK
   - Founding Chair of the Technical Committee on Evolving Intelligent Systems with Systems, Man and Cybernetics Society, IEEE
   - Co-recipient of several best paper awards at IEEE conferences (2006 and 2009, 2012, 2013)
   - Co-recipient of two prestigious Engineer 2008 Technology + Innovation awards for Aerospace and Defense
   - Co-recipient of the Special Award as well as the Outstanding Contributions Award by IEEE and INNS (2013)
   - Editor-in-Chief of the Springer journal *Evolving Systems,* Associate Editor of the prestigious *IEEE Transactions on Fuzzy Systems* and of Elsevier's *Fuzzy Sets and Systems*

3. Professor Tharam Dillon, La Trobe University, Australia

   Title: "Conjoint Mining of Data and Content with Applications in Business, Bio-medicine, Transport Logistics and Electrical Power Systems"

   - Life Fellow IEEE, FACS, FIE
   - Editor-in-Chief of the *International Journal of Computer Systems Science & Engineering* (UK) 1986–1991 Butterworths, 1992–1996 CRL Publishing
   - Editor-in-Chief of the *International Journal of Engineering Intelligent Systems* (UK) 1993–1996
   - Chief Co-editor of the *International Journal of Electric Power and Energy Systems* (UK) 1978–1991, Butterworths, 1992–1996 Elsevier
   - Associate Editor of *IEEE Transactions on Neural Networks* (USA) 1994–2004

The accepted papers of the 10th AIAI conference are related to the following thematic topics:

- Artificial neural networks
- Bioinformatics
- Feature extraction
- Clustering
- Control systems
- Data mining
- Engineering applications of AI
- Face recognition, pattern recognition
- Filtering
- Fuzzy logic
- Genetic algorithms, evolutionary computing

- Hybrid clustering systems
- Image and video processing
- Multi-agent systems
- Environmental applications
- Multi-attribute DSS
- Ontology, intelligent tutoring systems
- Optimization, genetic algorithms
- Recommendation systems
- Support vector machines, classification
- Text mining

- We wish to thank Professors Harris Papadopoulos (Frederick University, Cyprus), Alex Gammerman, and Vladimir Vovk (Royal Holloway University of London, UK) for their common efforts toward the organization of the third CoPA workshop.
- We are also grateful to Professors Spyros Sioutas, Katia Lida Kermanidis (Ionian University, Greece), Christos Makris (University of Patras, Greece), and Giannis Tzimas (TEI of Western Greece). Thanks to their invaluable contribution and hard work, the third MHDW workshop was held successfully once more and it has already become a well-accepted event running in parallel with AIAI.
- The third IIVC workshop was an important part of the AIAI 2014 event and it was driven by the hard work of Drs. Ioannis P. Chochliouros and Ioannis M. Stephanakis (Hellenic Telecommunications Organization, OTE, Greece) and Professors Vishanth Weerakkody (Brunel University, UK) and Nancy Alonistioti (National and Kapodistrian University of Athens, Greece).

- It was a pleasure to host the MT4BD 2014 in the framework of the AIAI conference. We wish to sincerely thank its organizers for their great efforts. More specifically we wish to thank Professors Spiros Likothanassis (University of Patras, Greece), Dimitrios Tzovaras (CERTH/ITI, Greece), Eero Hyvönen (Aalto University, Finland), and Jörn Kohlhammer (Fraunhofer-Institut für Graphische Datenverarbeitung IGD, Germany).

A Keynote lecture was given by Dr. Dimitrios Tzovaras in the framework of the MT4BD 2014 workshop.

Title: Visual Analytics Technologies for the Efficient Processing and Analysis of Big Data

All workshops had a high attendance from scientists from all parts of Europe and we would like to thank all participants for this.

The organization of the 10th AIAI conference was truly a milestone. After 10 years, it has been established as a mature event with loyal followers and it has plenty of new and qualitative research results to offer the international scientific community. We hope that the readers of these proceedings will be highly motivated and stimulated for further research in the domain of AI in general.

September 2014   Spyros Sioutas
                                                                 Christos Makris

Organization

Executive Committee

General Chair

Tharam Dillon — La Trobe University, Australia

Program Committee Co-chairs

Lazaros Iliadis — Democritus University of Thrace, Greece
Ilias Maglogiannis — University of Piraeus, Greece
Harris Papadopoulos — Frederick University, Cyprus

Workshop Co-chairs

Spyros Sioutas — Ionian University, Greece
Christos Makris — University of Patras, Greece

Organizing Co-chairs

Yannis Manolopoulos — Aristotle University of Thessaloniki, Greece
Andreas Andreou — Cyprus University of Technology, Cyprus

Advisory Co-chairs

Elias Pimenidis — University of East London, UK
Chrisina Jayne — Coventry University, UK
Haralambos Mouratidis — University of Brighton, UK

Website and Advertising Chair

Ioannis Karydis — Ionian University, Greece

Honorary Co-chairs

Nikola Kasabov — KEDRI Auckland University of Technology, New Zealand
Vera Kurkova — Czech Academy of Sciences, Czech Republic
Hojjat Adeli — The Ohio State University, USA

Program Committee

El-Houssaine Aghezzaf — Ghent University, Belgium
Michel Aldanondo — Toulouse University, Mines Albi, France

Vera Kurkova	Czech Academy of Sciences, Czech Republic
Ruggero Donida Labati	Università degli Studi di Milano, Italy
Helge Langseth	Norwegian University of Science and Technology, Norway
Spiridon Likothanassis	University of Patras, Greece
Mario Malcangi	Università degli Studi di Milano, Italy
Manolis Maragoudakis	University of the Aegean, Greece
Francesco Marcelloni	University of Pisa, Italy
Konstantinos Margaritis	University of Macedonia, Greece
Seferina Mavroudi	Technological Education Institute of Western Greece, Greece
Haralambos Mouratidis	University of East London, UK
Nicoletta Nicolaou	University of Cyprus, Cyprus
Vladimir Olej	University of Pardubice, Czech Republic
Eva Onaindia	Universidad Politecnica de Valencia, Spain
Mihaela Oprea	University of Ploiesti, Romania
Stefanos Ougiaroglou	University of Macedonia, Greece
Harris Papadopoulos	Frederick University of Cyprus, Cyprus
Elpiniki I. Papageorgiou	Technological Education Institute of Central Greece, Greece
Efi Papatheocharous	Swedish Institute of Computer Science, Sweden
Miltos Petridis	University of Brighton, UK
Vassilis Plagianakos	University of Central, Greece
Manuel Roveri	Politecnico di Milano, Italy
Alexander Ryjov	Lomonosov Moscow State University, Russia
Alexander B. Sideridis	Agricultural University of Athens, Greece
Ioannis Stephanakis	OTE Hellenic Telecommunications S.A., Greece
Ilias Sakellariou	University of Macedonia, Greece
Christos Schizas	University of Cyprus, Cyprus
Kyriakos Sgarbas	University of Patras, Greece
Alexei Sharpanskykh	Delft University of Technology, The Netherlands
Dragan Simic	University of Novi Sad, Serbia
Spyros Sioutas	Ionian University, Greece
Stefanos Spartalis	Democritus University of Thrace, Greece
Anastasios Tefas	Aristotle University of Thessaloniki, Greece
Konstantinos Theofilatos	University of Patras, Greece
Nicolas Tsapatsoulis	Cyprus University of Technology, Cyprus
Theodore Tsiligiridis	Agricultural University of Athens, Greece
Giannis Tzimas	CTI, Greece
Theodoros Tzouramanis	University of the Aegean, Greece

Table of Contents

IIVC Workshop

MHDW Workshop

COPA Workshop

MT4BD Workshop

Modern Video-to-Video Communications to Enhance Citizens' Quality of Life and Create Opportunities for Growth in "Smart" European Cities

Ioannis P. Chochliouros[1], Anastasia S. Spiliopoulou[2], Ioannis M. Stephanakis[2], Evangelos Sfakianakis[1], Evangelia Georgiadou[1], Maria Belesioti[1], Luís Cordeiro[3], and João Gonçalves[3]

[1] Research Programs Section Fixed,
Hellenic Telecommunications Organization (OTE) SA
99 Kifissias Avenue, 151 24, Athens, Greece
{ichochliouros,esfak,egeorgiadou,mbelesioti}@oteresearch.gr
[2] Hellenic Telecommunications Organization (OTE) SA
99 Kifissias Avenue, 151 24, Athens, Greece
{aspiliopoul,stephan}@ote.gr
[3] OneSource, Consultoria Informatica, Lda
{cordeiro,john}@onesource.pt

Abstract. The LiveCity Project effort structures a city-based "Living-Lab" and associated ecosystem to "pilot" and test live interactive high-definition video-to-video (v2v) on ultrafast wireless and wireline Internet infrastructure for the support of suitably selected public service use cases also involving a number of city user communities in five major European cities (Athens, Dublin, Luxembourg (city), Valladolid and Greifswald). The "core" target is to allow the citizens of a city to interact with each other in a more productive, efficient and socially useful way by using v2v over the Internet in a variety of distinct use cases that discussed and analysed, in detail, in the present work.

Keywords: Future Internet (FI), interoperability, Living-Lab, quality of service (QoS), right of way (RoW), video-to-video (v2v) communication.

1 The Scope of the LiveCity Initiative

In the past decade several video-based communications have entered the market [1], to become usual in some domains including our working environment, our homes and applications on mobile devices ([2], [3]). However, we are still lagging behind from the goal of video calls being as easy and ubiquitous as phone calls are today – across any network and between variable devices. There are a number of existing gaps on the pathway to the realization of the *Future Internet (FI)* ([4], [5]) with live *video-to-video (v2v)*. These "gaps" do appear at a number of different aspects (i.e.: end-user, service provider, network infrastructure, infrastructure design, service delivery platform, applications provider, end-point device). LiveCity aims to fill most -*if not all*- of the above mentioned gaps, by providing detailed and appropriate responses,

L. Iliadis et al. (Eds.): AIAI 2014 Workshops, IFIP AICT 437, pp. 1–12, 2014.

per separate case. To this aim, LiveCity considers standard video encoding already available in "off-the-shelf" devices and seeks to resolve diverse issues relevant to the following matters: (i) Intercarrier interoperability; (ii) Service interoperability between different carriers; (iii) Compatibility with most used laptops, notebooks, smart-phones & tablets; (iv) Promotion of the "First-for-consumers" option, by supporting simplicity of installation and usage; (v) Seamless operation between end-point devices; (vi) *Right of Way (RoW)* to be provided over various network types; (vii) Consideration of a large number of contacts addressable, together with globally reachable solutions; (viii) Integration of video application within vertical applications; (ix) Extension of the FI ecosystem to include mass-market in several types of communities [6].

LiveCity considers a variety of existing underlying network infrastructures and/or related facilities to explicitly serve its defined purposes [7]. This includes, *for example:* 4G wireless technology (i.e.: LTE and WiMAX) coupled with 3.5G wireless technology as well as fixed line xDSL broadband technology in those cases where fixed line is more appropriate than wireless, in urban environments. The "core" aim is to establish a real network with the involved cities together with the inclusion of a RoW without interference from unwanted traffic; this shall permit any potential user, in any of the involved cities, to "experience" live HD v2v. Thus, LiveCity is a pure "technological incorporation trial", that provides modern Internet-based services to over 2,750 users in the involved cities, via the implementation of selected pilot actions. LiveCity creates a wide ecosystem consisting of a variety of market "actors" including public service providers, network infrastructure operators, technology providers and subject matter experts, with the aim of supporting several well-defined public service use cases. LiveCity establishes a *"Living-Lab"-like* infrastructure [8] for the implementation of the selected pilots with diverse user communities, by sharing common service platforms [9]. It actually includes five European cities (i.e.: Athens, Dublin, Luxembourg (city), Valladolid and Greifswald) within an operational public-private-people (PPP) partnership-*like* framework [10] (as shown in Fig.1).

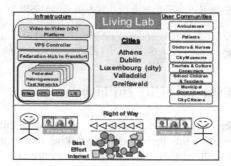

Fig. 1. LiveCity "Living-Lab"-*like* infrastructure

This "Living-Lab"-*like* scheme offers the possibility for a methodical user co-creation approach, which integrates v2v-based technology for the design and the final delivery of modern services, adopted to conform to the requirements of real-life use cases [11].

The original LiveCity concept permits to all involved actors-stakeholders to take into account both the global performance of the v2v service as well as its expected potential adoption by users. LiveCity structures a European network of "smart cities", where its smartness attribute originates from the capability to be interactive and live between (remote) locations between citizens in any appropriately connected city. LiveCity public service use cases are driven by public service operators to ensure a proper orientation towards fulfilling real market-born requirements. The technology and network partners come together to leverage both already existing and emerging network infrastructure combined with emerging "off-the-shelf" Virtual Path Slice (VPS) control software platforms and v2v applications software components, together with display/camera devices to realize video communication. This "unifying" approach leads to an operational and exploitable FI-based platform, which can later be rolled-out to other cities and communities, on a broader mass-market basis [12].

LiveCity supports technology that has the ability to massively scale. During the recent years, previous "barriers" to video calling have been remarkably reduced as a result of the widely penetrating high-bandwidth Internet access in most urban areas, together with lower cost of cameras and high-definition multimedia display devices in the market sector. LiveCity *"joins together"* all necessary components in a low-cost manner and offers a "testing ground" for a v2v mass-market deployment to cities in Europe. Also, it gathers and activates several user groups who can obtain economic and social benefit(s) from the proper deployment-use of live HD interactive v2v service. LiveCity assesses the effect of v2v usage by these "actors" to demonstrate that a "homogenized service" can be available on a mass-market basis, over heterogeneous network infrastructure as actually happens in modern cities. The present public Internet provides a globally reachable "best-effort" network [13]. It is foreseen that FI should comprise a RoW for a user's traffic, lacking interference from unwanted traffic. Such RoWs are already implemented in virtual private networks by using diverse quality of service (QoS) mechanisms; nevertheless, a RoW without interference from unwanted traffic is not offered to the broad spectrum of users in the mass European market [14]. In order to offer maximum utility for all potential users, such RoW has to be globally reachable and must be supported on most -*if not at all*-Internet environments, and at low -*or at least at affordable*- cost so that to be applicable by the users. LiveCity's explicit vision is to "conduct" v2v pilots for a range of user communities initially in the five participating European cities for a range of diverse applications and to support the network infrastructure with a proper RoW, in order that the inelastic traffic can get to the user and support the live experience [15].

2 Developing Scenarios of Societal and Economic Impact

LiveCity's core aim is to permit the involved citizens of the participating cities to really interact with each other in a more creative, effective and simultaneously socially useful way, via the use of high quality v2v over the Internet [16]. Video-to-video communication can be incorporated -*or can be the central element*- in a huge variety of services/applications, as a "means" to offer advanced services so that to: improve city administration; enhance education and learning capabilities for pupils/teachers; improve or extend cultural, educational and artistic applications offered by involved museums/cultural institutions, thus enhancing city experiences; support patient tele-monitoring and related medical applications in the eHealth sector, *and*; save patients' lives via v2v inclusion in ambulances, to assist medical personnel for the realization of more effective emergency services. Furthermore, v2v also implicates significant environmental benefits via the reduction of fuel costs and carbon footprint, as it reduces travels within cities, in the scope of all previous purposes. To perform its promising visions, LiveCity is structured around the development of some selected and well-defined scenarios and/or use-cases, of major business and societal impact, as described below, *per distinctive case:*

Municipal Services: The use of v2v for public administration and/or public information services can drastically enhance the access to and range of services reachable by the Internet. Likewise, this proves to the citizens' communities the effectiveness of a direct face-to-face communication for several municipal services which, *in turn*, leads to a more efficient user-service provider relationship and can so support the penetration of new services as opposed to voice-only or web-based services. Furthermore, this can decrease fuel costs and green-house emissions as a result of the reduction of travel and traffic within a city [17]. LiveCity focuses upon the provision of pilot platforms for use within specific local contexts. It undertakes extensive user requirements capture and development of use-cases drawn from working with real end-users. Besides, v2v can further enhance efficiency and utility of the proposed services-facilities and this is expected to improve accessibility and "comfort in use" of e-government services and physical service to the less ICT savvy, disabled and elderly citizens. As obvious, the above may have significant social and economic impact, resulting in remarkable benefits for individuals. (Effort for this purpose is deployed in the city of Valladolid, in Spain, and is mainly guided by Fundación Cartif and Ayuntamiento de Valladolid). Two specific use cases have been implemented: (1) the *"Advisory service for entrepreneurs and enterprise initiatives"* and; (2) the *"Education for Adults"* (see Fig.2). For the former, the main goal is to offer an integral service for entrepreneurs, having all the required and necessary support during the process of creation for a new enterprise so these initiatives could be properly arranged. In this scope, the phases of the intended activity comprise: (i) A video link for a Valladolid citizen to the interactive municipal online service office; (ii) A proactive advising session; (iii) Purpose/aim to explain -or to support- citizens about their questions concerning *entrepreneur service*, and; (iv) More approachable communications to be offered from the municipal offices to interested citizens.

Regarding the latter use case, most of the libraries in Valladolid are sited into multifunctional buildings called as "civic centres", where many activities are developed (i.e.: theatre, music, adult education). The *"Education for Adult support"* is a high-demand service and LiveCity aims to cover the need for the creation and development of volunteering platforms, through which the older people could "channel" their knowledge-experiences to "benefit" others, and enhance both their quality of living and social welfare of the community, *in general*. The goal here is to encourage the active participation of older people through the volunteer labour in technological and social projects, inter-alia with an innovating will that would imply an active social role and will also generate relationship and coexistence. The work performed -still being in progress- also includes the further refinement of testing methodologies for each use-case and will identify specific elements of the pilots which can bring added social, economic and environmental benefits.

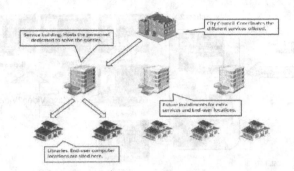

Fig. 2. Physical locations for municipal services in the city of Valladolid

City Education: As ICT turns out to be increasingly widespread within the environment of schools, a corresponding LiveCity pilot action focuses on the development of a framework where a v2v network is established between city secondary schools to improve learning and education, as well as to support/extend interactivity-collaboration between pupils and teachers. This latter option offers the possibility, to the involved schools, to work together to realize several projects via a proper "use and sharing" of any available resources and of teachers ([18], [19]).

Besides, this scope of interactivity can offer benefits of social value such as: (i) Ability, for leading teachers, to organize classes for students across the globe; (ii) Possibility to provide teachers' training facilities between the involved schools; (iii) Option for enhanced cultural exchanges between participating European schools, thus raising awareness of arts, science and social issues; (iv) Increase of parental participation in education, through the provision of teacher-parent training; (v) Capability to generate richer educational experiences to advance pupil attainment; (vi) Upgrading of the learning understanding of pupils; (vii) Better transfer of skills between pupils and teachers from different participating schools, in a cost-efficient way; (viii) Support of political ambitions -both at EU and national level- to get better education standards; (ix) Support for the formation of an EU-wide education system, potentially reachable by all sections of society, able to sustain gain of knowledge through access to expert teachers; (x) Contribution to the generation of new knowledge, new ideas for learning and teaching through the use of video archives. Work to fulfill these expectations takes places in Dublin and in Athens.

User scenarios that are under exploration include using v2v in a literacy education scenario, in the subject area of arts and in a sport education scenario. LiveCity implements two educational use cases by using FI capabilities to realize v2v in the involved schools in Dublin. These are: (1) The "School-to-School" (S2S) use case, where v2v is established between schools for applications such as projects, culture awareness and other topics (as in *Fig.3a*), and; (2) the "Ask-the-Expert Channel" use case, where v2v is used between children/students after school from their school to a central educational support provider. The purpose is to assist and enhance school activities (*see Fig.3b*). (The effort is mainly realized by Skryne School, in cooperation with other LiveCity partners such as RedZinc Services Ltd., QuartzSpark Ltd., Magnet Networks Ltd. and OneSource, Consultoria Informatica Lda).

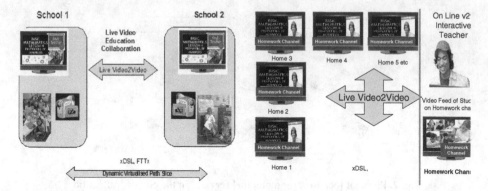

Fig. 3a. The *"School-to-School"* use case, with video interaction

Fig. 3b. The *"Ask-the-Expert Channel"* use case, with video interaction

City Experiences: With the purpose of enhancing tourism, cultural and city marketing information, a dedicated LiveCity pilot action considers v2v use between museums and/or involved cultural institutes. In the same context, local city administrations can also be involved in the related effort *"by choosing the figure and further promoting it"*. This "opens new horizons" in the corresponding domains as it can widen the capability of any involved legal entities -located across Europe- to offer modern and shared (occasionally mobile) content and/or experiences for both real and virtual visitors.

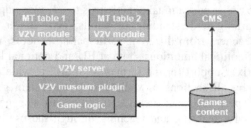

Fig. 4. Diagram of the proposed solution for the LiveCity *Museum v2v City Experience* use case

This use case also implicates for one more beneficial impact as it can improve the interactive and social nature of the (participating) museum's exhibits and can thus "be a magnet" for extra visitors who will stay connected for longer periods of time. Through the use of suitable IT equipment which also includes multi-touch tables (MTs), cameras, TV screens and appropriate content management system (CMS) -*as shown in Fig.4-* this initiative can "make bigger and longer" the cultural experience also with the intention of covering a greater part of the involved city; this can be done by opening-up the possibility for creating exhibits with "mixed" content which is created by the original museum in cooperation with real-world originating content, potentially promoted/created by museum visitors. The core aim is to enable telecom museums visitors to "*interact with one-another in a joint experience*", via the deployment of fitting games [20]. Games should: be of a well-known concept and be familiar to visitors; flexible and adjustable to different content; multiplayer, *but can also be played alone*; enhance knowledge about museum exhibits (local and remote).

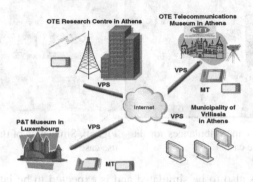

Fig. 5. The LiveCity *Museum v2v City Experience* use case

Developed applications are useful for the participating museums to generate common, shared and multi-user experiences that include visitors both *on-* and *off*-site; this is a remarkable aspect as it implicates potential for "cooperation and content sharing" between geographically diverse locations. Such actions also promote the development of cultural-oriented tourism within the participating cities as well as in a wider EU framework. This can advance the visits and profitability of tourist locations and operate as "a way of creating new revenue streams" [21]. The pilot takes place between the "twin" cities of Athens and Luxembourg (via an active collaboration between *OTE's Telecoms Museum* and the *P&T Museum, correspondingly*) as shown in Fig.5. (The main partners participating to the realization of this use case are OTE, the University of Luxembourg/SnT and the Municipality of Vrilissia-Athens).

City Emergency: This specific use case considers a city hospital emergency department (ED) team who has reduced/limited effectiveness because they cannot "see" -and optimally treat- the patient remotely, during the critical "golden hour" before the patient gets to hospital in the ambulance [22]. The related pilot activity experiments with live high-quality v2v between the ambulance and the hospital and examines options to enhance patient outcomes (e.g., shorter time to administer thrombolysis clot busting drugs in the case of a stroke or survival in a case of a polytrauma) and, *consequently*, to have early availability of expert opinion at the scene [23], when necessary (see Fig.6a). This implicates speedier access to senior decision making and optimal referring to appropriate centre to facilitate further medical management-treatment [24]. The three specific cases which are being implemented with ambulance video link are: (i) Elevation Myocardial infarction fast track; (ii) Cerebrovascular event/stroke, *and;* (iii) Entrapment of injured patient (trauma) [25]. The corresponding pilot action takes place in the city of Dublin (via the participation of RCSI (Royal College of Surgeons in Ireland) personnel (stationed at the Beaumont Hospital) and working collaboratively with Health Service Executive-HSE and Telefónica O2 Ireland Ltd.).

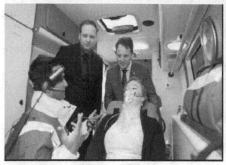

Fig. 6a. Use of v2v in Ambulances for the *City Emergency* use case

Fig. 6b. Simulation of the *Medical Emergency* use case

This use case is also to be simulated and is expected to be later implemented in a live clinical setting. A realistic patient simulator test phase [26] will allow the LiveCity to adopt the needed technical and educational needs for the real world implementation (see Fig.6b). Simulation takes place in the city of Greifswald, in Germany (via the participation of the Ernst-Moritz-Arndt University of Greifswald).

City eHealth: This use case is related to a glaucoma patient who needs monitoring and/or medical treatment (see Fig.7). The use of v2v between the home of a patient and hospital/clinic support group enables medical personnel (GP, doctor specialist or nurse) to provide enhanced support and have regular contact with that patient during his/her medical treatment, at lower cost and higher utility [27], [28]. By being able to provide faster and more reliable video feeds and medical telemetry tests (tele-diagnosis and tele-monitoring) [29], LiveCity facilitates and improves: the quality of the health service as more often than usual contacts may apply; the quality of life, and; ambient-living conditions for the patients.

Fig. 7. City *eHealth* use case, for patient glaucoma treatment

The inclusion of v2v -*that may be also accessible at community level and at mass accesses and least possible price*- can provide better solutions to the health problems and it can result in fewer complications and lower medical costs, in parallel with reduced travels and traffic. This pilot initiative is actually deployed in Athens (via the participation of the National & Kapodistrian University of Athens-UoA, mainly in cooperation with the Attikon University Hospital and Innovators S.A.).

All previous use cases are fairly representative among those "*shaping the wider profile*" of citizen's living in a "smart city" where Internet-based facilities are conceived as the "means" to perform transition to a more inclusive and digitally-based economy and society. Therefore, LiveCity covers a variety of services that are expected to be developed conformant to real users' needs. As v2v planned actions are realized via the joint effort of city authorities and (public) service operators, resulting outcome is important, applicable and beneficial in real-life terms. This can also further strengthen the role of the "user-citizen" in the social and technological innovation lifecycle. For all use cases, stakeholder evaluation and key performance indicators (KPIs) are measured, analyzed and disseminated, *appropriately* [30]. The essential hypothesis is that these indicators -once properly identified- can then be significantly enhanced with the usage of v2v communication [31]. Work, for this purpose, is coordinated for all use cases by the Brunel University, UK.

3   Concluding Remarks

The previously discussed LiveCity use cases are all "*active and indispensable parts*" of the local cities ecosystems and may have major benefits to improve citizens' living standards. The full scope of the related effort focuses upon interoperable actions for testing and validation processes of the proposed facilities, for the uptake of new services and for the benefit of multiple recipients, in real and applicable scenarios [32]. It is worth mentioning that end-users are dynamic "participants" and also "assessors" of the proposed solution(s), while their feedback is taken into account and affects further evolutionary steps.

LiveCity creates a city-based Living-Lab and associated ecosystem to pilot live interactive high-definition v2v on a variety of existing Internet infrastructure, among 5 EU cities. It implements public service use cases with over 2,750 involved users, by focusing on the application of v2v in a number of selected domains (i.e.: Municipal Services, City Experience, Education and Learning, Emergency and eHealth) with a significant validity for the citizen, also intending enhance a range of KPIs. It also aims to support the acceleration of v2v mass-market start across cities in Europe, by showcasing successful use cases, producing "how-to guides" and providing a "get-up and go" solution to public service operators.

LiveCity considers the proper use and the exploitation of already existing infrastructures-technologies that are truly offered by market operators and so its solutions can be applicable without necessitating any additional investment of resources. It implements selective pilots for motivating the development of innovative FI-based platforms and services; this is performed in a "balanced" partnership scheme including "actors" from different sectors (industry and business sector, citizens, public authorities and academia) with the aim of developing possible "synergies" of value [33]. Expected results should be important for market development, for supporting local economies and for deploying new services and products, implicating social, technical and economic impacts [34].

Acknowledgments. The present article has been structured in the context of the *LiveCity* ("*Live Video-to-Video Supporting Interactive City Infrastructure*") European Research Project and has been supported by the Commission of the European Communities - *DG CONNECT* (FP7-ICT-PSP, Grant Agreement No.297291). The present work has been realized by the support of all involved LiveCity partners and their contributions to the various work packages, *per case*.

References

1. Wang, Y., et al.: Video Processing and Communications. Prentice Hall (2002)
2. Wang, H., Kondi, L., Luthra, A., Ci, S.: 4G Wireless Video Communications. John Wiley & Sons (2009)
3. Al-Mualla et al.: 4G Wireless Video Coding for Mobile Communications: Efficiency, Complexity and Resilience. Elsevier Science, Academic Press (2009)
4. Future Internet Assembly (FIA): Position Paper: Real World Internet (2009), http://rwi.future-internet.eu/index.php/Position_Paper
5. Tselentis, G., Domingue, L., Galis, A., Gavras, A., et al.: Towards the Future Internet - A European Research Perspective. IOS Press (2009)
6. Organization for Economic Co-operation and Development (OECD):The Seoul Declaration for the Future of the Internet Economy.Paris, France (2008)
7. Chochliouros, I.P., Stephanakis, I.M., Spiliopoulou, A.S., Sfakianakis, E., Ladid, L.: Developing Innovative Live Video-to-Video Communications for Smarter European Cities. In: Iliadis, L., Maglogiannis, I., Papadopoulos, H., Karatzas, K., Sioutas, S. (eds.) Artificial Intelligence Applications and Innovations, Part II. IFIP AICT, vol. 382, pp. 279–289. Springer, Heidelberg (2012)
8. Chochliouros, I.P., Spiliopoulou, A.S., Sfakianakis, E., Georgiadou, E.M., Rethimiotaki, E.: Living Labs in Smart Cities as Critical Enablers for Making Real the Modern Future Internet. In: Iliadis, L., Papadopoulos, H., Jayne, C. (eds.) EANN 2013, Part II. CCIS, vol. 384, pp. 312–321. Springer, Heidelberg (2013)
9. Mulvenna, M., et al.: Living Labs as Engagement Models for Innovation. In: Cunningham, P., Cunningham, M. (eds.) Proceedings of eChallenges-2010, pp. 1–11. International Information Management Corporation (IIMC) (2010)
10. Commission of the European Communities: Communication on "A Public-Private Partnership on the Future Internet" [COM(2009) 479 final, 28.10.2009]. European Commission, Brussels (2009)
11. Kusiak, A.: Innovation: The Living Laboratory Perspective. Computer-Aided Design and Applications 4(6), 863–876 (2007)
12. Chesbrough, H.: Open Innovation: The New Imperative for Creating and Profiting from Technology. Harvard Business School, Boston (2003)
13. Sood, A., Tellis, G.: Technological Evolution and Radical Innovation. Journal of Marketing 69(3), 152–168 (2005)
14. Commission of the European Communities: Communication on "Future Networks and the Internet" [COM(2008) 594 final, 29.09.2008]. European Commission, Brussels (2008)

15. Klaue, J., Rathke, B., Wolisz, A.: Evalvid - A Framework for Video Transmission and Quality Evaluation. In: Proceedings of the 13th International Conference on Modelling Techniques and Tools for Computer Performance Evaluation, pp. 255–272 (2003)
16. Afuah, A., Tucci, C.L.: Internet Business Models and Strategies: Text and Cases. McGraw-Hill (2000)
17. Parkinson, M., et al.: H.: Competitive European Cities: Where Do The Core Cities Stand? European Institute for Urban Affairs, Office of the Deputy Prime Minister, London (2003)
18. Chochliouros, I.P., Spiliopoulou, A.S., Sfakianakis, E., Stephanakis, I.M., Morris, D., Kennedy, M.: Enhancing Education and Learning Capabilities via the Implementation of Video-to-Video Communications. In: Iliadis, L., Maglogiannis, I., Papadopoulos, H., Karatzas, K., Sioutas, S. (eds.) AIAI 2012, Part II. IFIP AICT, vol. 382, pp. 268–278. Springer, Heidelberg (2012)
19. European Commission, Information Society and Media: Education and Information Society: Linking European Policies. Luxembourg (2006)
20. Montola, M., Stenros, J., Waern, A.: Pervasive Games: Theory and Design. Morgan Kaufmann, San Francisco (2009)
21. Chochliouros, I.P., et al.: (Semi-) Pervasive Gaming Educational and Entertainment Facilities via Interactive Video-to-Video Communication over the Internet, for Museum Exhibits. In: Papadopoulos, H., Andreou, A.S., Iliadis, L., Maglogiannis, I. (eds.) AIAI 2013. IFIP AICT, vol. 412, pp. 474–485. Springer, Heidelberg (2013)
22. Palma, D., Goncalves, J., Cordeiro, L., Simoes, P., Monteiro, E., Magdalinos, P., Chochliouros, I.: Tutamen: An integrated personal mobile and adaptable video platform for health and protection. In: Papadopoulos, H., Andreou, A.S., Iliadis, L., Maglogiannis, I. (eds.) AIAI 2013. IFIP AICT, vol. 412, pp. 442–451. Springer, Heidelberg (2013)
23. Padki, A., et al.: Healthcare Provider Perspective on Telemedicine in Pre-Hospital Care: A LiveCity Survey of Doctors and Nurses in Emergency Department of Beaumont Hospital, Dublin. In: IAEM (ed.), Proceedings of the Annual Irish Association for Emergency Medicine 2013 Conference (2013)
24. Houlihan, A.P., et al.: Perceptions and Views of Use of Telemedicine in Emergency Care: A LiveCity Survey of Patients attending the Emergency Department. In: IAEM (ed.), Proceedings of the Annual IAEM 2013 Conference (2013)
25. Panayides, A., et al.: An Overview of Recent End-to-End Wireless Medical Video Telemedicine Systems Using 3G. In: IEEE (ed.), Proceedings of the Annual Conference on Engineering in Medicine and Biology Society (EMBC), pp. 1045–1048. IEEE (2010)
26. Metelmann, B., Metelmann, C., et al.: Videoübertragung in Echtzeit zur Verbesserung der Notfallmedizin. Accepted Paper to Appear in the Proceedings of the National German Congress VDE 2014, Frankfurt am Main, Messe, Germany, October 20-21 (2014)
27. Diamantopoulos, P., Bompetsis, N., et al.: E-health Applications for Smart Cities Infrastructures based on Live Video-to-Video Solutions. Accepted Paper to Appear in the Proceedings of the EuCNC-2014 Congress in Bologna, Italy, June 23-26 (2014)
28. Stamatelatos, M., Katsikas, G., Makris, P., Alonistioti, N., Antonakis, S., Alonistiotis, D., Theodossiadis, P.: Video-to-Video for e-Health: Use Case, Concept and Pilot Plan. In: Iliadis, L., Maglogiannis, I., Papadopoulos, H., Karatzas, K., Sioutas, S. (eds.) Artificial Intelligence Applications and Innovations, Part II. IFIP AICT, vol. 382, pp. 311–321. Springer, Heidelberg (2012)
29. Corchado, J., et al.: Using Heterogeneous Wireless Sensor Networks in a Telemonitoring System for Healthcare. IEEE Trans. Inf. Technol. Biomed. 14(2), 234–240 (2010)

30. Weerakkody, V., El-Haddadeh, R., Chochliouros, I.P., Morris, D.: Utilizing a high definition live video platform to facilitate public service delivery. In: Iliadis, L., Maglogiannis, I., Papadopoulos, H., Karatzas, K., Sioutas, S. (eds.) AIAI 2012 , Part II. IFIP AICT, vol. 382, pp. 290–299. Springer, Heidelberg (2012)
31. Molnar, A., Weerakkody, V., El-Haddadeh, R., Lee, H., Irani, Z.: A Framework Reference for Evaluating User Experience when Using High Definition Video-to-Video to Facilitate Public Services. In: Dwivedi, Y.K., Henriksen, H.Z., Wastell, D., De', R. (eds.) TDIT 2013. IFIP AICT, vol. 402, pp. 436–450. Springer, Heidelberg (2013)
32. Commission of the European Communities: Communication on Europe 2020: A Strategy for Smart, Sustainable and Inclusive Growth [COM(2010) 2020 final, 03.03.2010].European Commission,Brussels (2010)
33. Blumenthal, M.S., Clark, D.D.: Rethinking the Design of the Internet: The End-to-End Arguments vs. The Brave New World. ACM Trans. on Internet Tech. 1(1), 70–109 (2001)
34. Castells, M.: The Information Age: Economy, Society, and Culture. Blackwell, Oxford (1996)

Optimal Video Delivery in Mobile Networks Using a Cache-Accelerated Multi Area eMBMS Architecture

Ioannis M. Stephanakis[1], Ioannis P. Chochliouros[2],
George L. Lymperopoulos[3], and Kostas Berberidis[4]

[1] Hellenic Telecommunication Organization S.A. (OTE),
99 Kifissias Avenue, GR-151 24, Athens, Greece
stephan@ote.gr
[2] Research Programs Section, Hellenic Telecommunication Organization S.A. (OTE)
99 Kifissias Avenue, GR-151 24, Athens, Greece
ichochliouros@oteresearch.gr
[3] Head of Network Evolution Dept., Fixed & Mobile,
COSMOTE Mobile Telecommunications S.A.
Pelika & Spartis 1 Str., GR-151 22, Maroussi, Athens, Greece
glimperop@cosmote.gr
[4] Head of Signal Processing and Communications Lab.,
Dept. of Com. Engineering & Informatics, University of Patras, 26500 Patras, GR
berberid@ceid.upatras.gr

Abstract. Long-Term Evolution (LTE) evolved into enhanced *Multimedia Broadcast/Multicast Service* (eMBMS) that features improved perfomance, higher and more flexible LTE bit rates, Single Frequency Network (SFN) operations and carrier configuration flexibility. Multiple eMBMS service areas allow for efficient spectrum utilization in the context of mobile *Content-Delivery-Networks* (CDNs) as well as for delivering broadcast and push media over modern broadband networks. This paper intends to highlight novel service architectures for efficient content delivery. Content caching is a widely used technique in the networking industry that brings content closer to end-users improving, thus, servive performance and latency. Next generation wireless networks use edge located cache as well as core located cache in order to reduce traffic between the gateway and the internet and make the response of mobile network faster. This is referred to as hierarchical/distributed caching. Dimensioning of individual scenario-based traffic models for health care, museum virtual tours and interactive educational use cases in the context of the *LiveCity* European Research Project is attempted in this work. ON/OFF source models featuring state dependent active and inactive periods are used in order to describe multi-threaded resource transmissions. A group of chain ON/OFF models holds the parameters of the basic services and defines cache requirements. An architecture that associates distributed cache deployed at eMBMS Gateways per service area is proposed in order to reduce backhaul traffic. A simple algorithm that determines the optimal cache size in a mixture of chain ON/OFF modeled services is presented.

L. Iliadis et al. (Eds.): AIAI 2014 Workshops, IFIP AICT 437, pp. 13–23, 2014.
© IFIP International Federation for Information Processing 2014

Keywords: LTE, Evolved Multimedia Broadcast/Multicast Service (eMBMS), Single Frequency Network transmission, DVB-NGS, cache optimization, ON/OFF Chain Models, traffic modeling.

1   Introduction

Recent advances in the delivery of multimedia content are investigated in this paper. They allow for enhanced user experience and for novel offering from telecom providers. The Third Generation Parnership Project (3GPP) defined multimedia broadcast/multicast service in 2005 in order to optimize the distribution of video traffic in 3GPP specifications release 6 (Rel-6) for Universal Mobile Telecommunications System (UMTS). The standard refers to terminal, radio, core network, and user service aspects. MBMS standard eventually evolved into enhanced MBMS (eMBMS) [1] that features improved perfomance, higher and more flexible LTE bit rates, single frequency network (SFN) operations and carrier configuration flexibility. The introduction of MBMS over Single Frequency Network (SFN) transmission (MBSFN), which is described in Rel-7 of 3GPP specification version, overcomes cell-edge problems of MBMS and allows for the increase of the capacity of broadcast channels by a factor up to 3 or 4 in certain deployment conditions [2]. An identical waveform is transmitted from multiple cells with a tight synchronization in time in SFN operation. Nevertheless, it is not possible to use the same frequency for MBMS and non-MBMS transmissions in UMTS deployments making. Current 3GPP Rel-11 (2012) specifies improvements in the areas of service layer and Coordinated Multipoint Operation (CoMP) and allows for offloading the LTE network and mobile backhaul through eMBMS. Pushed content via user equipment caching as well as machine-to-machine services are distinctive enhancements. The evolution of the MBMS standard and its performance are presented in many scientific and technical papers [3,4,5,6]. Its efficiency in content delivery is addressed in [7].

eMBMS enables the possibility to deliver premium content to many users with secured quality of service in defined areas. It efficiently delivers rich media to mass users over unicast and broadband/multicast transmissions. A user may send and receive data individually over one-to-one transmissions in the context of such applications as Video-On-Demand, e-mail services, web-browsing and media downloads. One-to-many transmissions on the other hand utilize mobile spectrum more efficiently and result in lower cost for common content. They are used for live video and audio streaming, push media, e-publications, application downloading and other services. Several telecommunication companies adopt eMBMS as an efficient and low-cost means to deliver multimedia content. Qualcomm and Ericsson[1] demonstrated their eMBMS solution at Mobile World Congress (MWC) in 2012. Alcatel-Lucent and Huawei have also introduced their end-to-end solution to support broadcast video delivery in LTE networks. Verizon and Telstra announced plans to launch a live video broadcast service for sport events based on eMBMS technology

---

[1] Ericsson at `http://www.ericsson.com/res/docs/whitepapers/` `wp-lte-broadcast.pdf` last accessed 30th of April 2014.

after conducting live tests. Vodafone Germany has conducted live tests with LTE Broadcast in collaboration with Ericsson and Qualcomm as well[2].

Cache-accelerated architectures are suggested for current deployments of broadband and content delivery networks (CDNs). Dynamic site acceleration is a suite of technologies that make websites reliant on dynamically served content. The adoption of such technologies makes internet applications perform better and load faster. Traditional CDNs improved performance of telecommunication networks by caching critical content closer to end users. *Software-as-a-Service* (SaaS) and novel enterprise applications (B2B and B2C) base their functionality upon such notions as personalized recommendations, transactional and secure check-out and shopping carts. This poses stringent requirements on the delivery of dynamic, transactional content, as well as on the demand for e-commerce and web retailers. Traditional cache-accelerated architectures implement most or all of the following technologies (most of which are dealing with optimizing bit delivery across the network):

- **TCP optimization,** which includes algorithms designed to handle network congestion and packet loss.
- **Route optimization**, which is a technology that optimizes the route of the request from the end-user to the origin and ensures the reliability of the connection.
- **Connection management**, that includes persistent connection provision and HTTP multiplexing. Reusable and optimized HTTP connections from the edge servers to the origin servers as well as between the edge servers themselves are maintained rather than initiating a new connection for every request.
- **On-the-fly compression**, that compresses text objects shortly after they leave the origin servers alleviating, thus, computational burden from the origin servers without requiring additional bandwidth or hardware.
- **SSL offload** in order to speed up the critical secure transaction processes such as a check-out at an online store.
- **Pre-fetching**, that will parse through a served HTML file and will prefetch from the origin server the cacheable objects embedded in the file. A CDN can prefetch only cacheable content. Dynamic content by definition is contextual and can only be requested by the user.
- **Whole-site delivery,** that identifies cacheable and dynamic content and whether dynamic site acceleration techniques are applicable to the dynamic content of a site instead of simple fetches from the origin server.

Novel approaches have introduced such techniques as preresolving (performing DNS lookup in advance), preconnecting (opening a TCP connection in advance) and prewarming (sending a dummy HTTP request to the origin server) [8]. Response time for a mobile application is defined as the time between clicking to request information and loading a web page. Specifying cache size requires accurate modeling of application traffic. ON/OFF internet traffic models are frequently used for such a cause [9]. Measuring analysis and modeling of traffic has still been one of the main research

---

[2] GlobeNewswire at http://globenewswire.com/
news-release/2014/02/25/612970/10069866/en/Europe-s-first-
live-trial-of-LTE-Broadcast-revolutionizing-video-delivery-
across-mobile-networks.html last accessed 30th of April 2014.

challenges. Several studies have been carried-out over the last fifteen years on analysis of network traffic in the internet [10,11], traffic measurements in high speed networks [12] as well as measurements in next generation networks [13]. A brief description of the novel features and the architecture of eMBMS is presented in *Section* 2 whereas a comparison with the DVB-NGH standard - which is currently under development - is attempted is *Section* 3. Dimensioning issues for cache acceleration are considered in *Section* 5 and a novel approach for dynamic parameterization of buffers at edge servers based upon optimal storage for chain ON/OFF modeled services in the context of eMBMS is proposed and evaluated.

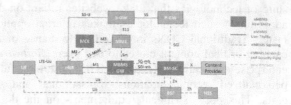

Fig. 1. eMBMS new entities

2   Evolved Multimedia Broadcast/Multicast Service (eMBMS) in LTE-Advanced

2.1   Features and Services

An evolved architecture is required to support eMBMS transmission in LTE network. The novel logical network entities proposed for eMBMS operation are the following [14]:

- **BM-SC (Broadcast Multicast Service Center)**, which is the entity that connects the Content Provider and the Evolved Packet Core. This entity plays the role of traffic shaper and authorizing content provider/terminal request. It is in charge of SYNC protocol to synchronize transmitted data among eNBs. SYNC protocol associates a specific header to IP packets, providing Time Stamps and session information.
- **eMBMS Gateway**, which is the entity located between BM-SC and all eNB. Its principal function is to deliver MBMS user data packets to eNBs by means of IP Multicast. When an MBMS session arrives, it allocates IP mulicast address to which the eNB should join to receive MBMS data and maintains the IP Multicast group. Furthermore, eMBMS Gateway is responsible for MBMS session announcement and it also performs MBMS Session control Signaling (Session Start/Stop) toward E-UTRAN.
- **MCE (Multi-Cell Coordination Entity)**, which is the logical entity whose function is admission control and allocation of the radio resource use for MBSFN operation. The MCE is expected to be part of e-UTRAN and can be integrated as a part of a network element. MCE may be either part of eNB (a "distributed MCE architecture") or be a stand-alone MCE ("centralized MCE architecture").

Interfaces M1, M2 and M3 (see Fig. 1) connect the aforementioned new entities with each other. M1 interface is a user plane interface that connects e-MBMS-GW and eNB. M2 interface is a control plane interface between MCE and eNBs whereas M3 interface connects MME and MCE. M3-Application Part allows for MBMS Session Control Signaling on E-RAB level.

MBMS defines three MBMS-specific radio bearers, i.e. **MICH** (MBMS indicator channel) which is used to notify terminals about the imminent start of an MBMS transmission session, **MCCH** (MBMS control channel), which carries control information about all ongoing MBMS transmission sessions and **MTCH** (MBMS traffic channel), which carries the actual data of an MBMS transmission session. Since

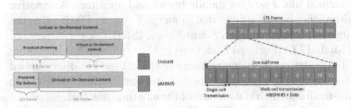

Fig. 2. Flexible spectrum allocation between unicast-broadcast services

HTTP cannot be used on unidirectional links, a unidirectional protocol was introduced for delivery of files or file segments, the so-called **FLUTE** protocol (File Delivery over Unidirectional Transport - RFC 3926). FLUTE protocol is carried over the User Datagram Protocol (UDP) and IP multicast toward end-user devices.

3 GPP Rel. 11 enhancements of eMBMS specify a mechanism that allows end-user devices to retrieve missing parts of a file after a broadcast session is over. This is due to the fact that it is impossible to use hybrid automatic repeat request (HARQ) on lower layers of unidirectional bearers. This mechanism is called file repair. Adaptive HTTP streaming, that is adopted by the Moving Pictures Experts Group as the baseline of MPEFG-DASH is specified in 3GP-DASH [15] as a compatible to MPEG-DASH profile. LTE resources reserved for eMBMS only when needed and there is no impact on LTE unicast capacity at other times (see Fig. 2).

2.2 Network Architecture and eMBMS Service Areas

A synchronous LTE network allows broadcast over a *Single Frequency Network* (MBSFN). An MBSFN area is defined as the set of cells participating in the same SFN transmission. Should different services be transmitted in the same cell but have different coverage areas, the cell transmits them in different time intervals where a different set of cells participate to the same SFN transmission (the cell/eNB may belong to up to eight MBSFN areas). The maximum extension of MBSFN areas is determined by the size of areas where cells are synchronized, which is called synchronization area. MBSFN transmissions not only include the data of the MBMS services but also periodic MBMS control and scheduling information messages indicating which services are currently transmitted and in which time intervals they may be transmitted. Efficient signal combining at user terminals achieves high operating SNR. Key Use Cases that take advantage of such an architecture include streaming video for real-time situational awareness, Push-To-Talk, file delivery for non-real-time services such as all points bulletin (e.g. amber alert), software updates/upgrades etc. Scalable push file delivery and delivery of content services are described in [16,17].

3   Comparison with Digital Video Broadcasting (DVB-NGH)

At the time MBMS over UMTS terrestrial radio access network (UTRAN) was standardized, the industry was focused on the mobile TV use case. Digital video broadcasting for handheld (DVB-H) networks was being deployed in several countries, and MBMS seemed like a way for mobile broadband operators. Alternative technologies for multimedia broadcasting that emerged were DMB (Digital Multimedia Broadcasting) deployed mainly in South Korea, ISDB 1-seg in Japan and in South America and MediaFLO (Forward Link Only) [18] in the USA. Qualcomm had conducted MediaFLO technical trials internationally at the time but discontinued development in 2010 whereas Japan has moved forward with ISDB-Tmm (Terrestrial mobile multi-media), which is a variant of the existing standard.

Digital video broadcasting for handheld (DVB-H) networks is a standard which is compatible with digital terrestrial TV (DVB-T) and transmits through digital terrestrial channels taking advantage of the digital dividend. Nevertheless DVB-H has been a commercial failure and the service is no longer on-air. Finland was the last country to switch-off signals in March 2012. The DVB group made a "call for technologies" for a successor system (DVB-NGH – Next Generation Handheld) in November 2009 in order to update and replace the DVB-H standard for digital broadcasting to mobile devices. The schedule was for submissions to be closed in February 2010. The new ETSI standard [19] published in 2013 and rollout of the first DVB-NGH devices is planned for 2015. DVB-NGH defines the next generation transmission system for digital and hybrid broadcasting to handheld terminals, i.e. a combination of digital terrestrial and digital satellite transmissions. The standard is based on the DVB-T2 system and reuses or extends lot of novel concepts introduced in the DVB-T2 specification. It adopts SFN transmission.

4   Cache Acceleration in CDNs and Dimensioning Considerations for Several Use Cases

Web is considered as a set of servers and clients, i.e. web browsers and any other software that is used to make a request of a Web server. Requests and responces of the hypertext transfer protocol (HTTP) before the establishment of a new connection determine the delays in retrieving some resource. It is estimated that the amount of time that is required can be approximated by two round-trip times plus the time to transmit the response and any additional DNS resolution delays. Caching is performed in various locations throughout the Web including at the two end locations. Proxy caches are intermediary caches between the client machine and the origin server. They generate new requests on behalf of the users if they cannot satisfy the requests themselves. Next generation wireless networks use edge located cache as well as core located cache in order to reduce traffic between the gateway and the internet and make mobile network response faster. This is refered to as hierarchical/distributed caching. *Internet Cache Protocol* (ICP) is an example of a popular mechanism used for coordinating caches in many different locations. The ICP protocol is described in RFC 2186 and its application to hierarchical web caching in RFC 2187. *Hypertext Caching Protocol* (TCP) (RFC 2756) is designed as a successor to ICP. *Object hit rate* and *byte hit rate* are used as measures in dimensioning cache size by network operators. Response time for mobile applications is determined by server processing time, delays in the network (bandwidth availability, round-trip time-RTT and tower connect time) as well as delays in the client device (parse time, resource fetches, layout and render processing and JavaScript delays). Less bandwidth, which is subject to frequent changes, is dedicated to downloading in mobile networks compared to the bandwidth which is dedicated to downloading in fixed networks. Furthermore processing speeds of mobile devices are up to ten times slower as compared to desktops. The path to faster mobile delivery therefore lies in reducing the number of round-trips as much as possible, reducing content payload size as much as possible, defer as much as possible and parse as little as possible.

One needs to classify and model traffic patterns at the edge of the mobile network in order to dimension cache. Chain ON/OFF models are adopted to this end. The design and development of a chain ON/OFF model relies on an accurate description of traffic entities from link level to application level. The model is generally used, when it is necessary to capture the scaling behaviors of network traffic. A chain ON/OFF model uses two states, namely ON & OFF for each state of a Markov model [9]. ON and OFF periods are usually heavy tailed processes. ON times correspond to resource transmissions while the OFF times correspond to intervals of browser inactivity. Furthermore, OFF times are classified either as *Active* (which account for client processing delays like document parsing and resource rendering) or as *Inactive* (i.e. user think time). The Weibull [20] and the Pareto [21] distributions are used to model ON and OFF times. Data generated by a specific type of application/service are discribed by a distinctive chain ON/OFF model. A queue that stores content at the edge of the mobile network is shared by different chain ON/OFF sources that are assumed to be statistically independent. The overall traffic through the edge aggregation link is modeled as a mixture of N chain ON/OFF sources belonging to M different groups, i.e. $\left\{ \frac{n_1}{N} Proc_1, \frac{n_2}{N} Proc_2 ... \frac{n_M}{N} Proc_M \right\}$. The chain ON/OFF sources which belong to a specific group are characterized by the transition probabilities between states as well as the mean number of packets/cells generated during the ON states and the relationship between ON and OFF periods. The probability of accessing one of the resources stored in a proxy cache resembles accessing a word in a written text of a natural language and can be calculated through the Zipf distribution [22,23]. Specifically, the number of references to cached item i, NR_i, is given as follows,

$$NR_i \sim \frac{\alpha}{rank(i)^Z}, \tag{1}$$

where $rank(i)$ denotes the position of cached item i after the sorting of the cache population on the basis of references (i.e., the most popular item is ranked first, the second most popular is ranked second and so forth), α is a constant value while Z is the Zipf parameter, which assumes a value close to 1. Should one choose α so as to normalize the sum of all references to a unit probability, he gets,

$$P_{hit} = \sum_{i=1}^{cache\ size} NR_i = \sum_{i=1}^{cache\ size} \frac{a}{rank(i)^z}, \tag{2}$$

LiveCity use cases include medical tele-monitoring for patients' treatment, city experiences (like interactivity in museums and cultural institutes, educational activities etc), municipal services and a school channel. One may use the chain ON/OFF model, which is illustrated in Fig. 3 as process of *Type* A, in order to describe such use cases as interactive guided tours in museums and other cultural institutes and the chain ON/OFF mo-del illustrated as process of *Type* B in order to describe tele-monitoring sessions and several municipal services. Streaming video for real-time situational awareness, *Push-To-Talk* municipal services and file delivery for non-real-time services such as all-points bulletin (e.g. amber alert), software updates /upgrades etc as well as a regional school channel

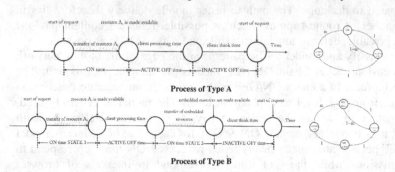

Process of Type A

Process of Type B

Fig. 3. Chain ON/OFF models used in the numerical simulations

may be implemented as broadcast services in the context of eMBMS deployment. Accessing a set of WWW resources (like HTML pages with embedded images) is modelled as a process of *Type* B in [24,25].

5   Numerical Simulations Using ON/OFF Chain Models

Not all content is cacheable. Personal data such as digital medical images and examinations as well as patients' records in the context of a telemedicine session are in general not cacheable. This holds true for personal data that may be transmitted in the context of a distance learning broadcasting. Overall 19.80-32.20% of unique URLs are uncacheable as estimated in [26]. This accounts for 21.50-28.32% of the total requests and 10.48-14.81% of the total bytes. HTML and JavaScript are dynamically generated and, thus, less cacheable. Analysis of real data in [26] indicates that, finally (after a certain buffer size) the slope of *byte_hit_rate*, which is denoted as $\lambda_{service}(buffer\_size)$, decreases as the size of the buffer increases. This leads to the following straightforward proposition regarding the optimal buffer size,

Proposition
The maximum value of the total (*byte_hit_rate* × *users* × *requests*) for a constraint size of a buffer used for caching requests for a mixture of *M* services is given as the sum of the buffer sizes allocated to each service for which the following condition holds,

$$\frac{\partial\,(byte\_hit\_rate \times users \times requests)}{\partial\,buffer\_size}\Big|_{per\,service}=\lambda_{service}\,(buffer\_size)=\lambda_{opt} \qquad (3)$$

for all of the M services.

A mixture of two services modeled as $\{0.67\,Process\,Type\,A, 0.33\,Process\,Type\,B\}$ is simulated. Three distinctive reference distributions are associated with *Process Type* B, one with state ON1 and two with state ON2. It is assumed that the sizes of the embedded objects are equal for each distribution. The selection of parameters of chain ON/OFF models is based upon actual measurements for web traffic caching like those presented in [26] and results for modelling data emanating from medical instruments in clinical environment ([27]). A 10 Gbps link is assumed to connect cached content at the gateway to the backbone network. The gateway is assumed to provide services for 500 concurrent users. Active ON times for non-cached objects are modeled according to Pareto distribution [21] with mean equal to *Object_size/User_bandwidth*. The delay for all cached objects is assumed to equal 200 ms. *Active* OFF or client processing delay is modeled according to the Weibull distribution [20]. It ranges from 1 msec to 1 sec. *Inactive* OFF times follow the Pareto distribution. All parameters are summarized in Table 1. Simulation results are illustrated in Fig. 4. Both hit rate and byte hit rates are given. Optimal buffer size is estimated as $s_1(\lambda')+s_2(\lambda')$. Total *byte hit requests* (*byte hit rate* × *users* × *requests per user*) are added for the same slope. The optimal slope decreases for higher budgets of buffer storage. Hit rates are up to 30%. Changing the mixture of the two services results in a different optimal value for the buffer size.

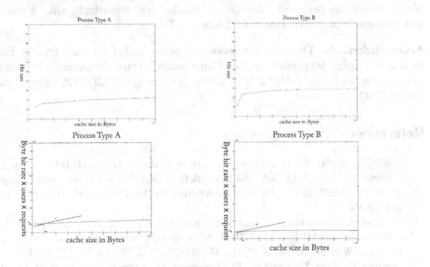

Fig. 4. Hit *rate* (upper row) & *byte hit rate* (lower row) per service (left for service modeled as *Type* A, right for service modeled as *Type* B). Buffer sizes indicate optimal allocation.

Table 1. Parameters of the chain ON/OFF models according to Fig. 3

	Process of type A	*Process of type B*		
Transitional probability ON state	0.25	0.6		
ON (Pareto)	shape=1.05 and scale=0.2	ON1:shape=4.5 -scale=0.2 (object 1e5) ON2:shape=1.5 -scale=0.2 (object 1e6) ON2:shape=1.05-scale=0.2 (object 1e7)		
ACTIVE_OFF (Weibull)	shape=3.0 and scale=0.045	shape=3.0 scale=0.045		
INACTIVE_OFF (Pareto)	shape=1.5 and scale=2.0	shape=1.5 scale=2.0		
Reference distribution(s) (**Zipf parameters**)	z=0.85 bytes per object 1e7	z=0.8 object bytes 1e5	z=0.9 object bytes 1e6	z=0.95 object bytes 1e7

6   Conclusion

LTE eMBMS takes advantage of single frequency network features and allows for flexible content delivery via unicast, multicast and broadband services. Live tests have been conducted by several companies. There is an increasing interest in launching commercial deployments and offering localized services through enhanced infrastructure. Service delivery and user experience may be boosted by web caching and acceleration by storing frequently accessed content at the edge of the network. A dynamic dimensioning scheme for optimal cache allocation is herein proposed. It requires traffic modeling of the services offered to the users. It assumes that the frequently accessed content is stored at the eMBMS gateway, which receives unicast streams and delivers MBMS user data packets to eNBs by means of IP multicast. Future work will emphasize upon dynamic buffer dimensioning according to the proposed approach taking into consideration such novel cache-accelerated techniques as request multiplexing and multi-resolution chunking. The proposed optimality condition may be applied in conjunction with such novel approaches as well. Simulation results are consistent with actual traffic measurements that are reported in the literature [27].

Acknowledgments. This research work has been funded by the *LiveCity* European Research Project supported by the Commission of the European Communities – *Information Society and Media Directorate General* (FP7-ICT-PSP, Grant Agreement No. 297291).

References

1. 3GPP TS 36.300: Evolved Universal Terrestrial Radio Access (E-UTRA) and Evolved Universal Terrestrial Radio Access Network (E-UTRAN); Overall description; Stage 2
2. 3GPP TR 25.905: Improvement of the Multimedia Broadcast Multicast Service (MBMS) in UTRAN
3. Alexiou, A., Asimakis, K., Bouras, C., Kokkinos, V., Papazois, A., Tseliou, G.: Cost optimization of MBSFN and PTM transmissions for reliable multicasting in LTE networks. Wireless Networks 18(3), 277–293 (2011), doi:10.1007/s11276-011-0399-7
4. Hartung, F., Horn, U., Huschke, J., Kampmann, M., Lohmar, T., Lundevall, M.: Delivery of Broadcast Services in 3G Networks. IEEE Trans. on Broadcasting 53(1), 188–199 (2007)
5. Hartung, F., Horn, U., Huschke, J., Kampmann, M., Lohmar, T.: MBMS – IP Multicast/Broadcast in 3G Networks. Int. J. Digital Multimedia Broadcasting, 1–25 (2009)

6. Nguyen, N.-D., Knopp, R., Nikaein, N., Bonnet, C.: Implementation and Validation of Multimedia Broadcast Multicast Service for LTE/LTE-Advanced in OpenAirInterface Platform. In: IEEE 38th Conference on Local Computer Networks Workshops (LCN Workshops), October 21-24, pp. 70–76. IEEE (2013), doi:10.1109/LCNW.2013.6758500, Print ISBN: 978-1-4799-0539-3

7. Wang, X., Wang, Y., Zhang, Y.: A Novel Transmission Policy for Reliable eMBMS Download Delivery. In: Wireless Communications and Networking Conference (WCNC), April 18-21, pp. 1–6 (2010)

8. Cohen, E., Kaplan, H.: Prefetching the Means for Document Transfer: A New Approach for Reducing Web Latency. In: 19th Annual Joint Conference of the IEEE Computer and Communications Societies, INFOCOM 2000. IEEE Proceedings, vol. 2, pp. 854–863 (2000)

9. Adas, A.: Traffic Models in Broadband Networks. IEEE Communications Magazine, 82–89 (July 1997)

10. Abrahamsson, H.: Traffic measurement and analysis. Swedish Institute of Computer Science (1999)

11. Williamsson, C.: Internet traffic measurement. IEEE Internet Computing 5(6), 70–74 (2001)

12. Celeda, P.: High-speed network traffic acquisition for agent systems. In: Proc. IEEE/WIC/ACM International Conference on High Speed Network Traffic Acquisition for Agent Systems, Intelligent Agent Technology, November 2-5, pp. 477–480 (2007)

13. Pezaros, D.: Network Traffic Measurement for the next Generation Internet. Computing Department Lancaster University (2005)

14. Lecompte, D., Gabin, F.: Evolved Multimedia Broadcast/Multicast Service (eMBMS) in LTE-Advanced: Overview and Rel.-11 Enhancements. IEEE Communications Magazine, 68–74 (November 2012)

15. 3GPP TS 26.247 Release 10: Transparent end-to-end packet-switched streaming Service (PSS); Progressive download and dynamic adaptive Streaming over HTTP (3GP-DASH), http://www.3gpp.org/ftp/Specs/html-info/26247.htm

16. Lohmar, T., Slssingar, M., Puustinen, S., Kenehan, V.: Delivering content with LTE Broadcast. Ericsson Review, 2–7 February (2013)

17. Lohmar, T., Ibanez, J.-A., Blockstrand, M., Zanin, A.: Scalable Push File Delivery with MBMS. Ericsson Review 1, 12–16 (2009)

18. ETSI TS 102 589: Forward Link Only Air Interface; Specification for Terrestrial Mobile; Multimedia Multicast, V1.1.1 (2009-02)

19. ETSI EN 303 105

20. http://en.wikipedia.org/wiki/Weibull_distribution

21. http://en.wikipedia.org/wiki/Pareto_distribution

22. Glassman, S.: A Caching Relay for the World Wide Web. Computer Networks and ISDN Systems 27(2) (1994)

23. Cunha, C., et al.: Characteristics of WWW Client-based Traces. Technical report BU-CS-95-010, Computer Science Department, Boston University (July 1995)

24. Hadjiefthymiades, S., Merakos, L.: Using Proxy Cache Relocation to Accelerate Web Browsing in Wireless/Mobile Communications. In: Proc. 10th International Conference on World Wide Web, pp. 26–35. ACM, New York (2001), doi:10.1145/371920.371927, ISBN:1-58113-348-0

25. Barford, P., Crovella, M.: Generating Representative Web Workloads for Network and Server Performance Evaluation. In: Proceedings of ACM Sigmetrics, Madison, Wisconsin, pp. 151–160 (July 1998)

26. Sunghwan, Ihm: Understanding and Improving Modern Web Traffic Caching. Ph. D. Thesis, Department of Computer Science, Princeton University (September 2011)

27. Ahmad, A., Riedl, A., Naramore, W.J., Chou, N.-Y.: Scenario-Based Traffic Modeling for Data Emanating from Medical Instruments in Clinical Environment. World Congress on Computer Science and Information Engineering, 529–533, ISBN 978-0-7695-3507-4

Video-to-Video E-Health Applications Supporting Medical Use Cases for Remote Patients

Panagiotis Diamantopoulos[1], Eleni Patouni[1], Nikolaos Bompetsis[1],
Nancy Alonistioti[1], João Gonçalves[2], Luís Cordeiro[2],
Ioannis P. Chochliouros[3], and George Lyberopoulos[4]

[1] Department of Informatics and Telecommunications
National and Kapodistrian University of Athens, Greece
{panos_10d,elenip,nbompetsis,nancy}@di.uoa.gr
[2] OneSource, Portugal
{joagonca,cordeiro}@onesource.pt
[3] Research Programs Section,
Fixed Hellenic Telecommunications Organization (OTE) S.A.
ichochliouros@oteresearch.gr
[4] COSMOTE S.A.
glimperop@cosmote.gr

Abstract. Information and Communication Technologies (ICT) relevant to health and healthcare systems can rise their effectiveness, expand quality of life and "reveal" innovation and novelty in modern health-related markets, thus implicating options for further growth. The broadly used term "e-Health" regularly implicates the use of ICT in health products, services to improve health of citizens, efficiency and productivity in healthcare delivery, and the economic and social value of health, *in general*. In this end, the LiveCity Project aims at empowering the citizens of a city to interact with each other in a more productive, efficient and socially useful way, by using high quality video-to-video (v2v) over the Internet. In this work, the LiveCity platform for e-Health is presented, along with two use cases for remote patients telemonitoring and emergency cases.

Keywords: e-Health, emergency, telemonitoring, video-to-video (v2v) communication.

1   Introduction

Over the past few years the vast advances on both wired and wireless technology has changed the face and aim of many service providers, including social ones. The uses of Internet and Video-to-Video (v2v) applications have provided users across the globe with the ability to overcome significant obstacles on their communication, including geographical and language barriers. Furthermore, the integration of Future Internet (FI) application in sectors of social welfare has become a subject of extensive research from many organizations. One of these sectors, which can be significantly enhanced by the use of FI technology, is the health care [1].

L. Iliadis et al. (Eds.): AIAI 2014 Workshops, IFIP AICT 437, pp. 24–29, 2014.
© IFIP International Federation for Information Processing 2014

Despite the advancements on medicine and health care systems, the rates of treatment failures remain significantly high for some cases where their treatment could be relatively easy. This is attributed mainly due to the inability of the patient to be physically present to the hospital and/or his difficulty to conform to the treatment ([2], [3]). Also, serious medical incidents are often failed to be addressed successfully due to delays between the incident and the provided treatment.

There are several systems that have managed to integrate successfully ICT applications to the health-care domain ([4], [5]). However, none of the aforementioned systems tackle the problem of high quality video calls across multiple devices. Existing video platforms that can be used for realizing v2v services do not provide the facility to integrate additional features. Furthermore, many of these platforms fail to guarantee the necessary data protection or QoS such applications require [6]. The LiveCity project aims on creating a set of plugins that integrate high quality video streaming that is used for the communication between patients and health care providers, while ensuring both security and high Quality of Service (QoS) ([7], [8]).

The rest of the paper is organized as follows: Section 2 provides an overview of the network infrastructure used for the realization of two different use cases. Section 3 and Section 4 present these use cases along with their respective applications. Finally conclusion remarks and directions for future research are drawn in Section 5.

2   LiveCity E-Health Platform

The LiveCity Network allows the communication of different plugins and provides the infrastructure for achieving high quality end-to-end video streaming. Furthermore, since this network is designed to accommodate use cases addressing serious medical conditions, as well as patient monitoring, the network incorporates a series of encryption and security mechanisms. These mechanisms implement state-of-the-art secure communication channels and protocols, thus ensuring data protection and preventing the extraction of sensitive medical information.

To guarantee the high quality bidirectional video streaming between two individual plugins, the network has integrated a QoS mechanism, called as "Virtual Path Slice" (VPS) engine. The VPS engine is a network resource reservation mechanism. It can be accessed through a specialized interface installed on every end-user plugin and allows the on-demand bandwidth reservation for providing constant bandwidth during video calls.

Figure 1 provides an overview of the underlying infrastructure of the related LiveCity network.

- The Client Plugin provides a simple Graphical User Interface (GUI) that exposes a series of commands that a user can give to the v2v Module. Also this plugin is responsible for presenting both local and remote video feed to the user.

- The v2v Module is the software component that is integrated to each Client Plugin and works as an intermediate layer between the end-user device and the v2v server. It is responsible for handling the communication between each plugins and the server, as well as, for initiating and maintaining the video stream during an active video session. Furthermore, the v2v Module is responsible for interacting with the VPS engine and request specific bandwidth reservation for a video call.

- The v2v server is the main message dispatcher between the individual plugins connected to the LiveCity network as well as the means of authenticating different connection requests. Finally, the v2v server integrates the aforementioned encryption mechanisms ensuring the protection of data.

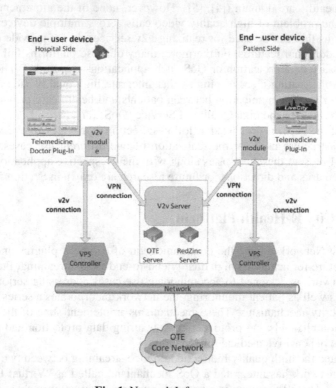

Fig. 1. Network Infrastructure

3   Video Solution for Patient Tele-Monitoring

3.1   Use Case Description

This use case allows the direct video-to-video connection between different healthcare providers and patients. Through the use of LiveCity Telemedicine plugin, a doctor can communicate and interact with patients, on remote or inaccessible locations, providing useful advice to persons in need. Furthermore, the use of public Internet as a main design axis renders the application easy-to-use, without the need of specialized hardware or software.

This use case was materialized for the monitoring of glaucoma patient in coordination of the Attikon Hospital in Athens, Greece and the National and Kapodistrian University of Athens.

3.2   Application

This LiveCity Telemedicine plugin provides two different end user applications, one addressing the need of the doctor and the other the needs of the patients.

Both applications have a simple, user oriented graphical interface, which exposes all the core functionality provided by the v2v module. By pressing the respective buttons, a user can connect or disconnect from the LiveCity network, provide his opinion and feedback regarding the application and reconfigure it.

The plugins also allow the modification of the characteristics of the transmitted video through a settings interface. To accommodate inexperienced user, four standardized profiles have been implemented that allow different quality of the transmitted video ranging from 360p to 1080p (Full high definition-HD). When a video call is active between two clients, each one can modify its respective settings of the received image in order to obtain better video and audio quality.

Despite their common functionality, the *Doctor Telemedicine plugin* differentiates from the one targeting patients, since it provide the means for establishing a direct video-to-video communication with connected patients. Furthermore, it can "adjust" the remote video that is transmitted from the patient by determining higher or lower video profile. Finally, this plugin allows the doctor to retrieve a snapshot of the patient as complement information to his examination.

On the other hand, the *Patient Telemedicine plugin* aims on providing a notification system that allows the immediate notification of the doctor, when a patient requires his assistance via the application or a SMS (Short Message Service). Finally, this plugin does not provide the aforementioned functionality for changing the remote video received from the doctor or for retrieving a snapshot of the video feed.

4   Video Solutions for Emergency Use Cases

4.1   Use Case Description

This use case aims at accommodating patients in emergency medical situation, such as heart attacks or strokes, during the "golden hour" (that is the critical time period when a patient is transferred, e.g. via an ambulance, to the nearest emergency center of a hospital for immediate medical care). Through the use of video-to-video application, emergency personnel in ambulances can communicate directly with qualified personnel to the hospital and ask for proper advice. With this kind of communication the time to provide important treatment is significantly reduced, thus increasing the chances of patient's successful recovery.

4.2   Application

This use case was materialized through the use of specialized equipment for the emergency personnel, while the hospital side used commodity laptops and simple xDSL[1] lines.

---

[1] The term "xDSL" stands for any type of Digital Subscriber Line access technology.

This specialized equipment is a wearable backpack computer that consists of off-the-self items and provides all the necessary components for achieving high definition video transmission thought wireless networks. It is comprised by:

- A microcomputer which acts as the main component for transmitting the video feed to the hospital as well as for receiving the audio feed from the hospital.
- A modem that allows the connection of the backpack computer to any wireless network.
- A simple 4 button interface that allows the immediate issue of commands to the computer.
- The necessary hardware (HD camera, headset) for receiving and transmitting the best possible quality of video and audio.

The software used on this use case was designed as a command line interface mainly for energy efficiency reasons. It receives input through the buttons and provides audio feedback to the user. Upon activating the backpack, it connects to the hospital side allowing the immediate communication between the two ends. Finally, a set of additional commands can be issued from the user that allows him to notify the hospital side, requesting immediate attention.

5   Conclusions

Fueled by advances in mobile and wireless communications, the last decade has seen an unprecedented proliferation in the service delivery and devices interconnection across the globe. This is particularly important towards the vision of "Smart Cities", where physical infrastructure is complemented by the availability of intellectual and social capital, increasing both urban competitiveness and quality of life. However, before such a paradigm shift can be realized, significant challenges need to be resolved to support remote communications with high QoS / QoE (Quality of Experience) "anywhere and anytime", thus serving remote patients.

Such gaps are investigated by the EU-funded LiveCity Project ("*Live Video-to-Video Supporting Interactive city Infrastructures*"). In this paper, we presented the LiveCity platform and two respective e-health use-cases. The first use case builds upon the use of v2v for remote telemonitoring of glaucoma patients while the second one analyzes the benefits of v2v for emergency cases. In the next steps, we plan to analyze the results of these two ongoing pilots.

Acknowledgments. The present article has been structured in the context of the LiveCity Project and has been supported by the Commission of the European Communities - DG CONNECT (FP7-ICT-PSP, Grant Agreement No.297291).

References

1. European Commission: Communication on "e-health Action Plan 2012-2010 – Innovative Healthcare for the 21s Century" [COM(2012) 736 final, 06.12.2012], Brussels, Belgium. European Commission (2012)

2. Pawar, P., Jones, V., van Beijnum, B.J.F., Hermens, H.: A Framework for the Comparison of Mobile Patient Monitoring Systems. J. Biomed. Inform. 45, 544–556 (2012)
3. Doukas, C., et al.: Digital Cities of the Future: Extending@ Home Assistive Technologies for the Elderly and the Disabled. Telematics and Informatics 28(3), 176–190 (2011)
4. Vargiu, E., Fernández, J.M., Miralles, F.: Context-Aware Based Quality of Life Telemonitoring. In: Lai, C., Giuliani, A., Semeraro, G. (eds.) Distributed Systems and Applications of Information Filtering and Retrieval. SCI, vol. 515, pp. 1–23. Springer, Heidelberg (2014)
5. Corchado, J., Bajo, J., Tapia, D., Abraham, A.: Using Heterogeneous Wireless Sensor Networks in a Telemonitoring System for Healthcare. IEEE Trans. Inf. Technol. Biomed. 14(2), 234–240 (2010)
6. International Telecommunication Union- Telecommunication Standardization Sector (ITU-T): ITU-T Technology Watch Report on E-Health Standards and Interoperability (April 2012), http://www.itu.int/en/ITUT/techwatch/Pages/ehealth-standards.aspx
7. Chochliouros, I.P., Stephanakis, I.M., Spiliopoulou, A.S., Sfakianakis, E., Ladid, L.: Developing Innovative Live Video-to-Video Communications for Smarter European Cities. In: Iliadis, L., et al. (eds.) Artificial Intelligence Applications and Innovations, Part II. IFIPAICT, vol. 382, pp. 279–289. Springer, Heidelberg (2012)
8. Patouni, E.: CIP/LiveCity - Live Video-to-Video Supporting Interactive City Infrastructure. In: The Workshop on Network Virtualization, Future Internet Assembly (FIA) 2014, Athens, Greece (2014)

The Potential of Telemedicine

Bibiana Metelmann[1,2,*], Camilla Metelmann[2,*], Konrad Meissner[2], Michael Wendt[2],
Joao Goncalves[3], Peadar Gilligan[4], Ahjoku Amadi-Obi[4], Donal Morris[5],
Eleni Patouni[6], and Martin von der Heyden[2]

[1] Klinik und Poliklinik für Anästhesiologie und Intensivmedizin,
Universitätsmedizin Greifswald, Fleischmannstraße 42-44, 17475 Greifswald, Germany
bibiana.metelmann@uni-greifswald.de
[2] Klinik und Poliklinik für Anästhesiologie und Intensivmedizin,
Universitätsmedizin Greifswald, Greifswald, Germany
[3] One Source, Consultoria Informatica Lda., Coimbra, Portugal
[4] The Royal College of Surgeons in Ireland, Dublin, Ireland
[5] RedZinc Services Ltd., Dublin, Ireland
[6] National and Kapodistrian University of Athens, Athens, Greece

Abstract. Telemedicine as a communication technology to overcome geographical distances can increase the quality of medicine. The prerequisite for telemedicine is that two persons or groups of persons are connected with each other. Often this connection is built beforehand during a face-to-face-meeting when both partners apportion the communication devices. If an emergency patient is not already part of a telemedicine project, the connection has to be newly created and the device to build the connection has to be brought to the patient. In the EU-funded LiveCity-Project the hypothesis was evaluated, that in emergency situations a telemedicine connection between a patient and a remote medical doctor can be accomplished by a device brought to the patient by paramedics. It was to be established if communication with a head mounted video-camera coupled with a LTE internet connection was feasible for this purpose.

Keywords: Telemedicine, emergency medicine, head mounted video-camera, simulation, LiveCity.

1   Introduction

Telemedicine represents communication and information technologies that, when together, can help diagnose and treat diseases of patients over a geographical distance [1, 2]. The word "telemedicine", which is a hybrid word containing the Greek "tele" and Latin "medicus", was first used by Kenneth Thomas Bird in the 1970s and is often translated as "medicine at a distance"[3, 4]. Albeit there are more than 100

---

* Authors Contributed equally to this work.

L. Iliadis et al. (Eds.): AIAI 2014 Workshops, IFIP AICT 437, pp. 30–37, 2014.
© IFIP International Federation for Information Processing 2014

different peer-reviewed definitions of the word "telemedicine", there is no single agreed definition.[5] Telemedicine is an important future topic, which is for example highlighted on a national scale in the coalition agreement of the new German Government[6] and on a global scope by the World Health Organization, which established a Global Observatory for eHealth[2].

There are two dimensions of communication in telemedicine. One is a horizontal communication between two or more medical experts (e.g. medical doctors in different hospitals) and the other is a vertical communication between a medical expert and a patient (e.g. medical doctor and patient at home). The vertical dimension is also between two healthcare-providers with different levels of medical expertise (e.g. emergency doctor and paramedic). Based on that, there is the possibility to have a medical supervisor for both medical staff and medical non-experts. There are several situations in which such help from a distance might be useful, e.g. a nurse in a family practice helping a patient at home with his daily treatment of chronic diseases. One goal of telemedicine is to reduce hospital admissions and to provide care to a patient in his familiar surroundings.[7] Another goal is to treat urgent worsening of chronic diseases or emergencies.[8, 9]

The German medical emergency system consists of two partners: the paramedics and the emergency doctor[10]. While the paramedics respond to every emergency-call, the emergency doctor is only alerted in certain situations. Criteria to alert an emergency doctor are severe damage of the body or danger to life.[10] The evaluation of severity of every emergency case is done based on the information established during the emergency call and according to determined regulations.[11] The paramedics and the emergency doctor are then alerted independently.

In 49,5% of all emergency cases an emergency doctor is alerted, too. [12] In the last decade there was an increased need for an emergency doctor detectable. [12] In 99,1% the paramedics and emergency doctors meet via the rendezvous system. [12] This means that the paramedics and the emergency doctor reach the emergency site with different vehicles independently to start working together at the emergency site[12, 13]. The concept of the rendezvous system was established, because there are more paramedics than emergency doctors. Therefore it is possible to maintain additional ambulance stations with paramedics.[12]. In average the paramedics in Germany arrive after 8.7 minutes while the emergency doctor reaches the emergency site after 12.3 minutes[12]. Thus in most cases in the German medical emergency system, the paramedics reach the patient earlier than the emergency doctor.

The prerequisite for telemedicine is that two persons or groups of persons are connected with each other by the means of a communication technology. Often this connection is built beforehand during a face-to-face-meeting when both partners apportion the communication devices. When using telemedicine in emergency medicine, there is no possibility to meet beforehand and establish a connection. If an emergency patient is not already part of a telemedicine project, the connection has to be newly created and the devices to build the connection have to be brought to the patient.

2   Research Context

In the EU-funded LiveCity-Project this hypothesis was evaluated, where in emergency situations a telemedicine connection between the patient and a remote medical doctor can be accomplished by the means of a device brought to the patient by paramedics. LiveCity, short term for "Live Video-to-Video Supporting Interactive City Infrastructure", studies how live video connections via LTE can improve the quality of life of European Citizens in various situations.

The Greifswald subgroup studied the use of this LTE video connection in simulated medical emergency situations. As shown above in most cases the paramedics reach the patient earlier than the emergency doctor does. Therefore there is a certain period of time, in which the paramedics are without guidance of an emergency doctor. In addition there are other situations in which the paramedics at the emergency site are without an emergency doctor although one is required. As explained above the emergency dispatcher team makes an initial assessment of the situation based on the alert-call. Depending on this initial assessment, the decision is made, whether to alert an emergency doctor or not. There are some cases, in which the need for an emergency doctor is not apparent from the beginning, so that the emergency doctor isn't alerted.

The LiveCity-Project examines these situations – paramedics being without the help of an emergency doctor at the emergency site. If the paramedics want to consult a medical doctor, they could do that via telemedicine.

There is for instance the possibility to connect via a live-video so that the medical doctor can see the patient and his vital signs as done in the LiveCity-Project. Based on that, he is able to help with diagnostics and treatment, therefore minimizing the time it takes to start the right treatment. Starting the right treatment as early as possible is one of the main goals of emergency medicine, e.g. when treating myocardial infarction, stroke or trauma [14-16].

In 2002, the European Resuscitation Council (ERC) established during a conference held in Florence, Italy, the term "First Hour Quintet"[17-19]. This concept states five life threatening diseases, in which fast medical treatment is crucial for medical outcome concerning morbidity and mortality. Examples are stroke, myocardial infarction and traffic accidents, which are among the leading causes of death in Europe[20, 21]. To improve the therapy of stroke the American Heart Association and American Stroke Association gave recommendations to implement the use of telemedicine in emergency treatment of stroke in 2009.[22]

3   Methodology

In the LiveCity-Project the telemedicine-connection between the patient experiencing an emergency and the emergency doctor stationed at a hospital was to be established with a video-camera with a LTE Internet connection. In this project the remote emergency doctor is called "Tele Emergency Doctor". The video-camera, which was especially developed for this purpose, was brought to the patient by the paramedics. One paramedic wears the camera with a headband, enabling the Tele Emergency Doctor to see the same things the paramedic is seeing. Therefore the paramedic was able to show to the Tele Emergency Doctor the patient, his vital signs and all other important information while having both hands free to work. Moreover the Tele Emergency Doctor was thus able to help the paramedic with diagnosing and treating the emergency. For ethical reasons the study was not done with patients but was simulated in realistic and praxis-oriented scenarios in a simulation center.

The study was designed as a randomized, two-armed crossover study trial. 2 paramedics treated together as a team 5 "patients" without the Tele Emergency Doctor (control cases) and 5 "patients" with help of the Tele Emergency Doctor (experimental cases).

The paramedics approached 10 different scenarios, which were especially developed for this study. Those 10 scenarios were matched pairs belonging to one of the five categories: stroke, myocardial infarction, traffic accidents, pregnancy and rare diseases. While the first three categories were chosen, because they are part of the First Hour Quintet (see above), the last two were created due to the fact that they are a special challenge for the paramedics and emergency doctors and require exceptional expertise. Every scenario was independently evaluated by 3 paramedics and 3 medical doctors regarding practicability.

To minimize influences of confounders, there was a special randomization regarding the chronological order of scenarios. Each scenario was equally tested as a control case and experimental case. Every first group started with the control cases and every second group with the experimental cases.

For all scenarios the mannequin called Laerdal Resusci Anne was used. In every scenario it was initially placed either on a sofa or on the floor, depending on the emergency simulated, but the paramedics were able to move it around as needed.

During the study the paramedics and emergency doctors were asked to fill out specially developed questionnaires. The questions covered different topics, including medical, technical, practical and psychological aspects.

4   Results

10 emergency doctors and 21 paramedics aged between 18 and 50 years were recruited. More than 350 questionnaires were completed.

The paramedics were asked, whether they would prefer in their daily work to have more help with practical, manual skills (e.g. endotracheal intubation) or with finding the right diagnosis and establishing the right therapy. More than two-thirds answered, they encounter more often situations, in which they need theoretical rather than practical support (Fig. 1).

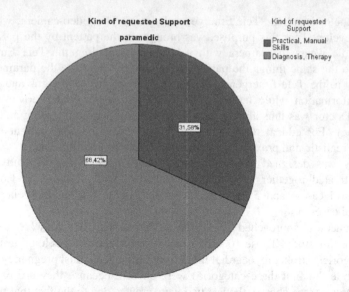

Fig. 1. Kind of requested Support

Furthermore roughly two-thirds stated that they would call the Tele Emergency Doctor in cases, where they normally wouldn't call an emergency doctor (Fig.2).

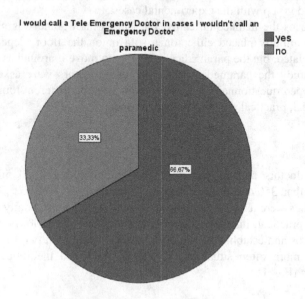

Fig. 2.   I would call a Tele Emergency Doctor in cases I wouldn't call an Emergency Doctor

After finishing all the simulations, the paramedics were asked to rate the assertion "I considered the Tele Emergency Doctor as helpful." 71% answered "agree", 29% "partly agree" and no one "partly disagree" or "disagree" (Fig. 3).

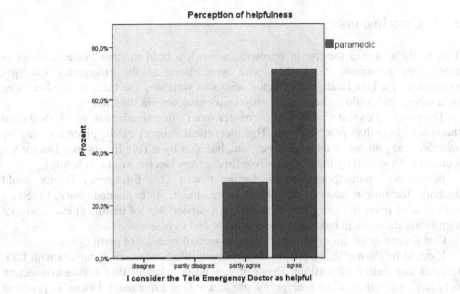

Fig. 3. Perception of helpfulness

As shown by Fig. 4 paramedics and emergency doctor think the Tele Emergency Doctor could lead to a faster start of treatment.

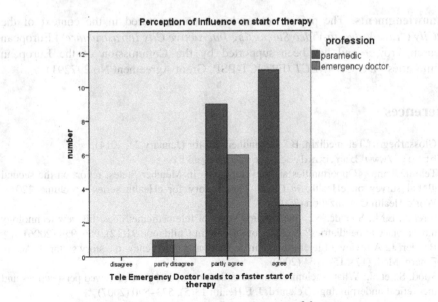

Fig. 4. Perception of influence on start of therapy

A communication between the paramedics and the Tele Emergency Doctor via the head-mounted camera was realizable and audio/picture quality was good enough so that the Tele Emergency Doctor could help with the diagnostics and treatment.

5   Conclusions

This study analyzed the use of telemedicine with a head-mounted video camera in emergency medicine. Paramedics, who were alone at the emergency site, got assistance of a Tele Emergency Doctor, who was watching the video in another room on a laptop and could communicate with the paramedics via the camera.

The paramedics stated, that they encounter more often situation in which they need theoretical help than practical help. This theoretical support can be given not only by an emergency doctor at the emergency site, but also by a Tele Emergency Doctor via a camera. As shown by Figure 1-3 a Tele Emergency Doctor would be helpful.

Furthermore participants got the feeling that a Tele Emergency Doctor could decrease the time it takes to start the right treatment. As explained above, time is a very crucial point in emergency medicine. An earlier start of treatment can lead to a significant decrease in morbidity and mortality and save life.

One limitation of this study is the relatively small number of participants.

It could be shown, that a communication with this headmounted camera with LTE Internet connection is feasible. Hence it can be concluded, that a videoconnection between paramedics at the emergency site and a Tele Emergency Doctor is practical and that it could potentially be accomplished with help of LiveCity technology.

The authors of this study would further like to conclude, that a Tele Emergency Doctor might be a very good addition to the existing German medical emergency system. Further studies are on the way.

Acknowledgments. The present article has been structured in the context of the *LiveCity* ("*Live Video-to-Video Supporting Interactive City Infrastructure*") European Research Project and has been supported by the Commission of the European Communities - *DG CONNECT* (FP7-ICT-PSP, Grant Agreement No.297291).

References

1. Glossarbegriff Telemedizin, B.f. Gesundheit, Editor (January 24, 2014), http://www.bmg.bund.de/glossarbegriffe
2. Telemedicine – Opportunities and developments in Member States: report on the second global survey on eHealth, in Global Observatory for eHealth series - Volume 22011, World Health Oragnization (2009)
3. Strehle, E.M., Shabde, N.: One hundred years of telemedicine: does this new technology have a place in paediatrics? Archives of Disease in Childhood 91(12), 956–959 (2006)
4. Benger, J.: A review of telemedicine in accident and emergency: the story so far. J. Accid. Emerg. Med. (17), 157–164 (2000)
5. Sood, S., et al.: What is telemedicine? A collection of 104 peer-reviewed perspectives and theoretical underpinnings. Telemed. J. E Health 13(5), 573–590 (2007)

6. Deutschlands Zukunft gestalten; Koalitionsvertrag zwischen CDU, CSU und SPD; 18. Legislaturperiode, C.-L. CDU Deutschlands, SPD, Editor Dezember 2013: Union Betriebs-GmbH, pp. 99–100

7. Koch, B., et al.: Regional Health Care (RHC). Notfall + Rettungsmedizin 12(6), 461–466 (2009)

8. Shah, M.N., et al.: High-intensity telemedicine-enhanced acute care for older adults: an innovative healthcare delivery model. J. Am. Geriatr. Soc. 61(11), 2000–2007 (2000)

9. Saleh, S., et al.: Re-admissions to hospital and patient satisfaction among patients with chronic obstructive pulmonary disease after telemedicine video consultation - a retrospective pilot study. Multidiscip. Respir. Med. 9(1), 6 (2014)

10. Harding, U., et al.: Schlaganfall immer mit Notarzt? Medizinische Klinik - Intensivmedizin und Notfallmedizin 108(5), 408–411 (2013)

11. Schilling, M., et al.: Zuweisungskonzept bei akutem Schlaganfall. Der. Nervenarzt. 83(6), 759–765 (2012)

12. Reinhard Schmiedel, H.B., Leistungen des Rettungsdienstes 2008/09, B.f. Straßenwesen, Editor 2011: Dr. Schmiedel GmbH, Bonn

13. Wahlperiode, D.B., Bericht über Maßnahmen auf dem Gebiet der Unfallverhütung im Straßenverkehr 2008 und 2009 (Unfallverhütungsbericht Straßenverkehr 2008/2009), B.u.S. Bundesministerium für Verkehr, Editor 2010

14. Members, A.T.F., et al.: ESC Guidelines for the management of acute myocardial infarction in patients presenting with ST-segment elevation: The Task Force on the management of ST-segment elevation acute myocardial infarction of the European Society of Cardiology (ESC). European Heart Journal (2012)

15. Lyden, P.: Thrombolytic Therapy for Acute Stroke — Not a Moment to Lose. New England Journal of Medicine 359(13), 1393–1395 (2008)

16. Sharma, A., Jindal, P.: Principles of diagnosis and management of traumatic pneumothorax. J. Emerg. Trauma Shock 1(1), 34–41 (2008)

17. Krafft, T., et al.: European Emergency Data Project EMS Data-based Health Surveillance System Project Report

18. Nilsen, J.E., Improving quality of care in the Emergency Medical Communication Centres (EMCC) in Konferanse for medisinsk nødmeldetjeneste 7. - 8.nov. 20122012: Sola, Norway

19. Lackner, C.K., et al.: Von der Rettungskette zum akutmedizinischen Netzwerk. Notfall + Rettungsmedizin 12(1), 25–31 (2009)

20. Fischer, M.: Factors influencing outcome after prehospital emergencies - a european perspective. In: Euroanesthesia 2007, München, Deutschland (2007)

21. Fischer, M.: Emergency medical services systems and out-of-hospital cardiac arrest Cardiac Arrest. Cambridge University Press (2007)

22. Schwamm, L.H., et al.: A review of the evidence for the use of telemedicine within stroke systems of care: a scientific statement from the American Heart Association/American Stroke Association. Stroke 40(7), 2616–2634 (2009)

Software-Defined Networking: Guidelines for Experimentation and Validation in Large-Scale Real World Scenarios

Joao Goncalves[1], David Palma[1], Luis Cordeiro[1], Sachin Sharma[2], Didier Colle[2], Adam Carter[3], and Paulo Simoes[4]

[1] OneSource, Consultoria Informatica, Lda.
{john,palma,cordeiro}@onesource.pt
[2] Department of Information Technology (INTEC), Ghent University, iMinds
{sachin.sharma,didier.colle}@intec.ugent.be
[3] Edinburgh Parallel Computing Centre (EPCC), University of Edinburgh
adam.carter@ed.ac.uk
[4] CISUC, Department of Informatics Engineering, University of Coimbra
psimoes@dei.uc.pt

Abstract. This article thoroughly details large-scale real world experiments using Software-Defined Networking in the testbed setup. More precisely, it provides a description of the foundation technology behind these experiments, which in turn is focused around OpenFlow and on the OFELIA testbed. In this testbed preliminary experiments were performed in order to tune up settings and procedures, analysing the encountered problems and their respective solutions. A methodology consisting of five large-scale experiments is proposed in order to properly validate and improve the evaluation techniques used in OpenFlow scenarios.

Keywords: OFELIA, OpenFlow, Software-Defined Networking.

1 Introduction

The research towards the Future Internet raises many challenges that require demanding infrastructures and setups to allow their assessment and clarification. However, testing new protocols and services in existing network facilities is not typically considered as an option. Consequently many sectors are not willing to experiment with new protocols and applications that might affect their services.

Software-Defined Networking (SDN) has been acknowledged as suitable answer to provide necessary functionalities and flexibility. This is achieved by decoupling the networking decisions from the hardware enforcing them, new routing protocols or on-demand routing decisions can be easily introduced in production networks, mitigating the effects of typical changes or updates in existing networks. Moreover, the definition of an open protocol such as OpenFlow breaks the barriers of hardware dependant routing mechanisms, enabling a larger dissemination of well-structured networks and services.

L. Iliadis et al. (Eds.): AIAI 2014 Workshops, IFIP AICT 437, pp. 38–47, 2014.

In this work a detailed compendium of the preliminary experiments performed on The OpenFlow in Europe: Linking Infrastructure and Applications (OFELIA) testbed [1] is presented, that ultimately led to the definition of the five large experiments, which are set out to expose the properties and limitations of a SDN employing the OpenFlow standard [2]. More specifically, this OpenFlow network is coupled with a Quality of Service (QoS) system towards an emulated city of one million inhabitants.

In Section 2 of this document, all the tools and mechanisms used throughout the experiments are explained in detail. Followed by the definition of the relevant Key Performance Indicators (KPI) and of the proposed experimentation process which are described in Section 3, with particular attention given to the preliminary tests and the problems identified while executing them. In Section 4 an analysis of the expected results is given and finally in Section 5 the main conclusions are drawn and future steps considered.

2   Experiment Tools and Mechanisms

The OFELIA project consists of a collaborative creation of an experimental research facility, involving several testbed islands. This large-scale facility is based on OpenFlow, allowing its users and researchers a unique opportunity to dynamically configure, control and experiment the whole network infrastructure using different technologies without limitations.

The architecture followed in these experiments, sets out a goal for future OpenFlow-based networks, enabling them with the possibility to dynamically configure paths with guaranteed traffic characteristics. The separation between data and control planes followed by the SDN paradigm fits this goal, allowing additional business intelligence to be included on top of the control plane, which in turn enforces the necessary decisions on the data plane.

2.1   OFELIA Testbed Islands

The motivation for this work also resulted from previous experiments performed resorting to OFELIA. On these particular experiments, six different interconnected testbeds were used, each one with very specific characteristics. The defined scenarios were exploited so that an extensive use of these islands could provide meaningful results and insights for future large-scale experiments [1].

In particular the following OFELIA islands were and will be used in the performed tests:

– **i2Cat (Barcelona):** Providing L2 switches and optical equipment, enabling power aware OpenFlow experiments;
– **ETHZ (Zurich):** Integration within the perimeter of the campus network and also provides the possibility of an operational research track in parallel with an experimental track; available there is the provision of connections to OneLab and GENI;

- **CNIT-Cantania (Cantania):** Based on NetFPGA and Open vSwitch (OvS) technologies while focusing on Information Centric Networking;
- **Create-Net (Trento):** Citywide distributed island based on Layer 2 switches and NetFPGAs, with opt-in users via heterogeneous access technologies;
- **TUB (Berlin):** A few OpenFlow switches are integrated in the campus network;
- **iMinds (Ghent):** Central hub of the federated facility that allows for large scale simulation with at least 100 interconnected servers, with a total of 10 NetFPGA 1Gb cards.

Each of these pan European islands possess their own characteristics but are seen as a whole within these experiments, building large-scale scenarios for the performed evaluations, using the OFELIA provided facilities.

2.2   Main Components

A whole set of different components needed to be defined for further use in the desired large-scale experiments. The detailing of each component appears in the following sections.

Open vSwitch. OvS [3] is a production quality, multilayer virtual switch. It is designed to enable massive network automation through programmatic extension, while still supporting standard management interfaces and protocols. OvS is targeted at multi-server virtualization deployments, a landscape for which the current stack is not well suited. These environments are often characterized by highly dynamic end-points, the maintenance of logical abstractions, and – sometimes – integration with or offloading to special purpose switching hardware.

Floodlight. OpenFlow is an open standard managed by Open Networking Foundation. It specifies a protocol through which a remote controller can modify the behaviour of networking devices through a well-defined "forwarding instruction set"; and the Floodlight Open SDN Controller [4] is an enterprise-class, Apache-licensed, Java-based OpenFlow Controller. Floodlight is designed to work with the growing number of switches, routers, virtual switches, and access points that support the OpenFlow standard.

Queue Installer. With the intention of providing an interface inside the Floodlight controller to ease the process of queue creation within an OF enabled switch, in particular the OvS, a queue installer was created. This entity presents itself as an extension to Floodlight, a module per se, rather than as a standalone application; following this approach allows this module to be easily embedded in every Floodlight installation and also to enable Floodlight to handle all the event processing, avoiding the creation of additional overheads on the controller communication.

The required communication for queue configuration, between an OpenFlow Controller and its corresponding switches, is not foreseen in the current implementation of the OvS used, therefore we have designed an architecture capable of generating the appropriate queue configuration messages, following the Open vSwitch Database Management Protocol (OVSDB) standard [5]. It should be noted also that the Open Networking Foundation has recently created a working group for the Configuration and Management of queues and other hardware related configurations.

RouteFlow. RouteFlow [6] is an open source project to provide virtualized IP routing services over OpenFlow enabled hardware. Mimicking the connectivity created by a physical infrastructure and running IP routing engines, RouteFlow is composed by an OpenFlow Controller application, an independent RouteFlow Server, and a virtual network environment.

The main issue of RouteFlow for large-scale experiments is that an administrator needs to devote a long time in manual configurations: (1) creating a virtual environment, (2) creating mapping between a virtual environment and the physical infrastructure, and (3) writing routing configuration files for each machine of the virtual environment. Therefore, for large-scale experimentation, we implemented a framework to configure RouteFlow automatically. In our framework [7], we use an additional module that gathers configuration information by sending probe messages in the physical infrastructure. The configuration information is then sent to RouteFlow using configuration messages. Using these messages, RouteFlow configures itself.

QoS Platform. The QoS platform is service-based, meaning that everything being done within the platform is about setting up services to provide for clients. This will be achieved in the following manner:

1. Setting up basic data connectivity for every equipment – *Routers*;
2. Setting up basic service models – *Trunk, Interconnected Trunk, Node*;
3. Adding equipment to establish the service;
4. Setting up service specific configurations for the aforementioned equipment;
5. Publishing the information to the appropriate path table, or to an interconnected partner;
6. Setting up session services for service instantiation;
7. Publishing session services to the appropriate clients

The aforementioned clients are mostly applications that use a provided Application Programming Interface (API) to request service operations, like triggering, stopping or modifying existing services.

2.3 Supporting Tools

The following tools were mainly designed and created to be of service to the experimentation process, predominately aiding in the generation and collection of data.

Pulse Generator. The Pulse Generator is a standalone application that was defined and implemented in order to realistically emulate a large number of users. It generates both data plane and control plane traffic and uses a customizable profile to allow for varying traffic characteristics.

The Pulse Generator reads a traffic profile and schedules each connection profile described therein to take place at some point in the future. A connection profile is a set of parameters that describe a repetitive connection to the QoS Platform (in the control plane) and optionally also a data-traffic generator (in the data plane). The period of this repetitive connection can be as low as tens of milliseconds. The connection profile also indicates the duration of the repetitive connection (or if only a single connection is required), the lifetime of each service and if data-traffic is required. As there are two parts to a service event on the control plane (invocation/trigger and stop/termination), the pulse generator handles both.

Measurement System. A proper data collection and processing mechanism must to be required in order to present concrete and detailed results from pre-established metrics within the experiments. Although mainly focused on Java technology, existing profilers are not optimized for performance using the Java Virtual Machine, therefore creating an unwanted overhead if used to assess the capability of a given module.

Striving to provide deterministic and accurate methods for measuring data within the software being used, a tailored library was designed and developed for distribution and integration with any software that may be used in the testbed. The decision to produce a library in lieu of a standalone application, for the most part, comes from the requirement of measuring actions that take place inside the core of the used software for each scenario, in order correctly understand the impact of each component and its performance.

Proprietary formats were used to store the information so as the development time could be reduced to a minimum while providing some degree of modularity to the whole system, facilitating integration into newer or different applications.

3   Experimentation

Within the paper, the focus has been on the assessment of large-scale OpenFlow networks and in particular take into account a QoS platform in a dynamic control environment, where a high event arrival-rate is expected as we intend to emulate 1 million inhabitants. Figure 1 depicts the defined large-scale topology for reaching the experiment's goals.

Whether developing a new platform, or constructing and assessing a new set of components, a validation process must take place. This will be guaranteed by defining relevant KPIs and different experiments that targeted at several distinct goals.

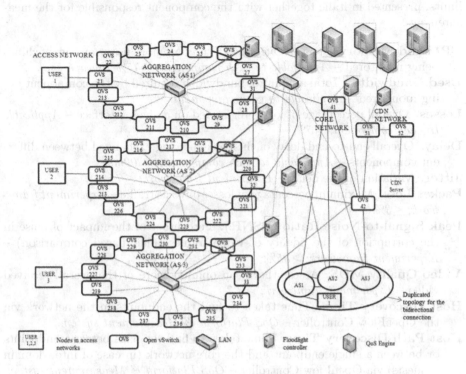

Fig. 1. Large-scale topology

3.1 Key Performance Indicators

The selection of appropriate Key Performance Indicators is crucial for a suitable evaluation of the defined scenarios. Both qualitative and quantitative KPIs may be considered in the selection process as long as they are relevant for the assessment being conducted. The overall set of KPIs should also consider all of the aspects of the solution under consideration. Moreover, the identified KPIs should not be redundant and their measurement must be possible either through objective or subjective (like when assessing Quality of Experience) techniques [8].

In addition to the selection of an appropriate set of KPIs, it is also important to understand what are the expected outcomes of the performed experiments and what constitutes a successful experiment. By assessing the performance results obtained for the selected indicator, the experiment will be considered successful if they are within the pre-defined lower and upper limits. These KPIs reflect not only the competence of each constituent module but also the performance of both voice and video traffic.

These and other network related parameters are further detailed in the following bullet points that present the metrics to be considered and their expected values. One considers a parameter to be successful whenever it is within these limits, presented in italic together with the component responsible for the measurement.

CPU and Memory Usage. Overall CPU and Memory usage on the machines being monitored – *Applicable to each component; ≤75% (on average).*

Used Bandwidth. Total amount of bandwidth required by the component being monitored – *Applicable to each component.*

Losses. Overall percentage of losses registered on a given interface – *Applicable to each component; ≤5%.*

Delay. Overall end-to-end delay of the packets being exchanged between different components – *Applicable to each component; ≤300ms.*

Jitter. Variation of the delay – *Experiment framework; ≤75ms.*

Packet Loss. Maximum percentage of losses of voice traffic – *Experiment framework; ≤3%.*

Peak Signal-to-Noise Ratio (PSNR). Percentage of the impact of noise in the corruption of the fidelity of video frames (pixel-based comparison) – *Experiment framework; ≥35%.*

Video Quality MOS. Assess the mean opinion score of the overall received video – *Experiment framework; ≥3.7.*

Host Discovery Time. Time taken to find the endpoints in the network via the OpenFlow Controller – *QoS Platform & Measurement system.*

Host Path Discovery Time. Time taken to find the path between endpoints or between a single endpoint and the core network (in case of inter-domain requests) via OpenFlow Controller – *QoS Platform & Measurement system.*

Queue Installation Time. Time taken to install a queue in a switch via the OpenFlow Controller – *QoS Platform & Measurement system.*

Queue Removal Time. Time taken to remove a queue in a switch via the OpenFlow Controller – *QoS Platform & Measurement system.*

Flow Installation Time. Time taken to install a flow in a switch via the OpenFlow Controller – *QoS Platform & Measurement system.*

Flow Removal Time. Time taken to remove a flow in a switch via the OpenFlow Controller – *QoS Platform & Measurement system.*

Route Creation Time. Time taken by RouteFlow to add a new route in a switch – *Route Flow; It depends on the defined hello interval.*

Route Deletion Time. Time taken by RouteFlow to delete a route in a switch – *RouteFlow; It depends on the defined router dead interval.*

By having a very specific set of KPIs, a more detailed study of the whole platform can be accomplished, enabling an easier and faster improvement over each batch of trials towards the platform optimization.

3.2   Preliminary Tests

Before setting into automated testing, a proper platform validation was required, so a scenario was set up in order to assess the inner workings of the selected tools and how they would cope in these scenarios.

A set of simple tests at low scale will be undertaken. The purpose of this experiment is to assess the functionality of all components and their interoperability, therefore validating the existing platform for the upcoming experiments. These experiments would be executed manually, always taking into consideration the feedback received from each individual run before proceeding with the next set of trials. For the first experiment, to validate the designed topology and verify if the packets where travelling through the expected paths, single triggers on the network were manually issued.

The results will be obtained via a script that parses the logs to create a list of average response times based on the 85%, 90%, 95% and 100% best results. These results have to be considered very preliminary, as their main intent is to further validate the platform. Having said, the first run will have a duration of 10 seconds, with triggers spaced by one second. Provided the platform copes well with the previous validation tests, these will be extended to a total duration of one hour, with triggers spaced by 200 milliseconds.

3.3   Main Experiments

Following the preliminary tests, a larger set of experiments to understand the performance of the components employed was designed. In total five experiments were designed, with the initial four aimed at stress testing the platform and verifying its resiliency. In the fifth and final experiment, the previous scenarios will be extended as to encompass as many islands of the OFELIA testbed as possible.

1st Experiment. The main purpose is to improve over the preliminary experiments, solving any problems that might have risen from an experimental operation point of view. This experiment can be perceived as an extension to the integrated experimental platform validation, where a whole assessment of the system is performed in order to prepare it for the next five, fully featured, experiments. It will aim to confirm that the OpenFlow connectivity and bandwidth guarantee can be achieved, and that each of the 50 OvS can be set-up and brought down again.

2nd Experiment. This experiment sets out to stress test the signalling in the context of high signalling load. Single and multiple domain scenarios will be stress tested. Variations regarding connection duration will be introduced. The maximum performance will be obtained under various network conditions. The experiment will be repeated following any improvements identified and implemented during the testing.

3rd Experiment. Here the resilience of the unidirectional communication will be stress-tested to support right of way for World Wide Web HD video service in the presence of background traffic (i.e., assessing if the system maintains a sustainable bandwidth level for the invoking application). The stress test will be conducted in the context of various network conditions of background traffic load, focusing on unidirectional traffic from a Content Delivery Network (CDN) to an end user, and later on using instead bi-directional video-to-video traffic from one end-user to another.

4th Experiment. This experiment will assess how the system behaves under several different failure scenarios. A framework will be designed so that the controller will detect data plane failures, with data plane traffic (high-priority and best-effort) being rerouted on failure free paths. In this experiment, three failure recovery scenarios will be evaluated. In the first scenario, enough bandwidth will be assumed so that neither high-priority nor best-effort traffic will be affected after all the traffic has been re-routed to a failure-free path. In the second scenario, the available data path bandwidth will be restricted so that all priority traffic can be rerouted as soon as all flows are recovered. However, some of the best-effort traffic flows will experience packet loss in order to meet the requirements of high-priority traffic. In the third scenario, the capacity is squeezed further so high-priority traffic will also be starved for bandwidth after all traffic is redirected to a failure-free path.

5th Experiment. This experiment is based on expanding the previous learning experience of the experiments described above, in order to expand the acquired knowledge to a deployment based on a set of federated distributed island testbeds. It will take a version of one of the small-scale experiments designed, and implement it on the other available islands from the OFELIA project. The experiment will try to assess the behaviour of the control plane in a multi Autonomous System (AS) scenario where each AS is located in a different OFELIA island. It will be addressed the question of how operational issues faced in a heterogeneous environment impact the components in the software stack.

3.4   Methodology

Through the controlled generation of pulses – where a pulse is everything between the trigger creation, call duration and trigger removal – with the Pulse Generator, their number in a given interval will be gradually increased between runs. The value will be increased until a breaking point is reached, i.e., where the platform stops responding to the trigger creation requests due to the amount of currently instantiated triggers.

   These pulses will be distributed within a two-hour interval according to a Poisson distribution, providing a more accurate representation of a real world scenario. Upon identifying the breaking point, a series of repetitions will be performed; later assessing the validity of the results obtained by the Measurements framework and the Pulse Generator logs.

4   Results

As previously mentioned preliminary results were obtained from performing a small scale evaluation on top of the used OpenFlow testbed. From these results we established a methodology that enables a thorough and rigorous method of assessment for each of the developed and used components. These components and structure definition will allow a thorough assessment of realistic future networks resorting to OpenFlow.

Despite the defined approach, the actual performance evaluation and results collection for the large-scale testbed are scheduled but not presented in this work. The results obtained in upcoming iterations will further validate the proposed methodology which currently stands as the main contribution and outcome from this work, which includes not only the definition of 5 realistic large-scale scenarios but also the necessary tools and components for the desired evaluation.

5   Conclusion

In order to establish a methodology and performance analysis of a real large-scale Software-Defined Networking testbed a set of preliminary tests was performed. During these preliminary tests, some issues arose and were promptly identified and circumvented. They provided insights and directions towards the presented final scenarios, which allow emulating future smart cities with millions of inhabitants.

In addition to the defined scenarios, a set of necessary components was also identified and implemented as explained in the paper, being now ready to be used in order to undertake the totality of the experiments described above. The utmost purpose will be to perform these tests at the higher possible performance levels, identifying bottlenecks on the current setup by analysing the results meticulously, while also validating the proposed solution and contributing for the development of more robust and innovative solutions for the assessment of the Future Internet.

References

1. OpenFlow in Europe: Linking Infrastructure and Applications,
   http://www.fp7-ofelia.eu
2. McKeown, N., Anderson, T., Balakrishnan, H., Parulkar, G., Peterson, L., Rexford, J., Shenker, S., Turner, J.: Openflow: Enabling innovation in campus networks. SIGCOMM Comput. Commun. Rev. 38, 69–74 (2008)
3. Open vSwitch: An Open Virtual Switch, http://openvswitch.org
4. Floodlight OpenFlow Controller,
   http://www.projectfloodlight.org/floodlight/
5. Open Networking Foundation, Configuration and Management working group,
   https://www.opennetworking.org/working-groups/configuration-management
6. Nascimento, M.R., Rothenberg, C.E., Salvador, M.R., Corrêa, C.N.A., de Lucena, S.C., Magalhães, M.F.: Virtual routers as a service: The routeflow approach leveraging software-defined networks. In: Proceedings of the 6th International Conference on Future Internet Technologies, CFI 2011, pp. 34–37. ACM, New York (2011)
7. Sharma, S., Staessens, D., Colle, D., Pickavet, M., Demeester, P.: Automatic configuration of routing control platforms in openflow networks. SIGCOMM 43, 491–492 (2013)
8. Schatz, R., Hossfeld, T., Janowski, L., Egger, S.: Datatraffic monitoring and analysis, pp. 219–263. Springer, Heidelberg (2013)

Remote Video-to-Video Eye Telemonitoring Use Case for Glaucoma Patients

Dimitrios Alonistiotis[1], Evgenia Kontou[1], Nikolaos Karachalios[1], Eleni Patouni[2],
Panagiotis Diamantopoulos[2], Nikolaos Bompetsis[2], Nancy Alonistioti[2]
and Ioannis P. Chochliouros[3]

[1] 2nd Department of Ophthalmology, University of Athens,
"Attikon" Hospital,
1, Rimini Street, Haidari, GR-124 62, Athens, Greece
dalonis@yahoo.gr
[2] Department of Informatics and Telecommunications
National and Kapodistrian University of Athens,
6, Panepistimiopolis, Ilissia, GR-15784, Athens, Greece
{elenip,panos_10d,nbompetsis,nancy}@di.uoa.gr
[3] Research Programs Section, Fixed
Hellenic Telecommunications Organization (OTE) S.A.,
99, Kifissias Avenue, GR-151 24, Athens, Greece
ichochliouros@oteresearch.gr

Abstract. Glaucoma is the second leading cause of blindness globally and the second most common cause of avoidable visual impairment. It also holds a record in noncompliance to therapy from the patients in up to 50% of the subjects treated with anti-glaucoma eye drops. LiveCity e-Health is a European research program, which aims to provide better treatment and follow up of glaucoma patients at their home, through telemonitoring with high definition video-to-video (v2v) communication from the University Hospital. Secondly, it aims to reduce the cost of health and improve the city environment by decreasing the number of visits to the Hospital. For this purpose, a software application has been developed; the latter is easy to use for elderly people at home, and capable of keeping the medical history and digital records of every patient in the Glaucoma Department. In addition, a specific web camera with snapshot ability of high quality photo of the eye has utilised. Two patients have been initially enrolled in the study and the preliminary results are so presented.

Keywords: e-Health, glaucoma, telemonitoring, video-to-video (v2v) communication.

1   Introduction

According to the World Health Organization (WHO) and all the available surveys, glaucoma is the second leading cause of blindness globally as well as in most regions of the world, following only cataract, and the third in the developed world, following

L. Iliadis et al. (Eds.): AIAI 2014 Workshops, IFIP AICT 437, pp. 48–55, 2014.
© IFIP International Federation for Information Processing 2014

Age-related Macular Degeneration and Diabetic Retinopathy ([1]-[3]). It is also considered to be the second most common cause of avoidable visual impairment [4].

Glaucoma is an optic neuropathy with certain features including an intraocular pressure that is too high for the continued health of the eye, which leads to ganglion cell death. Ophthalmology so far has focused in the management of glaucoma through the lowering of the intraocular pressure and the preservation of the remaining visual ability [5]. This is achieved primarily by eye drop instillation daily and only in few selected cases by surgery or laser. Normal intraocular pressure values vary from 10 to 21 mmHg, and most glaucoma patients tend to have more than 24 before diagnosis [6]. Regaining some of the impaired vision due to glaucoma is impossible.

At the same time, glaucoma treatment heavily reduces the quality of life of the patient. The European Glaucoma Society has insisted that glaucoma therapy must abide by a good quality of life for the patient, according to the stage of the disease, the patient's age, general health and even life expectancy and also his abilities and needs [7]. The main reason that Glaucoma influences the patient's attitude towards his therapy is that it stays an asymptomatic disease until the end-stage of the glaucomatic damage (when preservation of vision is impossible by then) and can only be detected through preventive examination of the intraocular pressure and the optic nerve. Secondly, chronic use of eye drops causes topical side effects like redness of the eye, discomfort, itching, conjunctivitis, blurry vision, even hyperpigmentation of the iris and enlargement of the eye lashes in the most common anti-glaucoma drug category. Even systemic side effects from long lasting anti-glaucoma therapy, including depression, respiratory and cardiac side effects and sexual impotency have been noticed. So, in most cases, a healthy otherwise and asymptomatic individual who goes for a check-up to the Ophthalmologist and is suddenly considered and treated as glaucomatic ends up with many different symptoms not from the disease itself but from the medications with an aim to maintain – and not restore- his vision. Last but not least, the cost of therapy for the entire lifetime becomes an economic burden for the citizen.

For all the above reasons Glaucoma holds a record in noncompliance to therapy from the patients. Several surveys have shown that glaucomatic patients do not take their medications correctly in a percentage up to 50% [8]. Besides that, because of the nature of the disease, it is estimated that there is also a 50% of undiagnosed glaucomatic patients in the general population and that almost half of the people under anti glaucoma therapy may not need it after all. Moreover, because many glaucoma patients need more than one medication (eye drop) and in different daily frequency or time of instillation (once or twice by day), are geriatric patients or with specific difficulties with applying medication, (e.g. due to tremor), noncompliance to therapy increases according to the number of eye drops per day, number of medications, age, general mental condition ([9]-[10]).

Pharmacy companies have tried to improve this by producing more effective drugs, reducing the daily need for eye drops, introducing drugs with less side effects and preservative free single use blisters. Nevertheless, glaucoma patients need close follow up for their obedience to therapy and even training for the correct instillation of eye drops [11].

The rest of the paper is organised as follows: Section 2 analyses the LiveCity Project research in the area of e-health, focusing on the description of the telemedicine platform and software application for remote patients' telemonitoring. The telemonitoring use case for glaucoma patients is introduced in section 3. Finally conclusion remarks and directions for future research are drawn in section 4.

2   The LiveCity e-Health Research for Smart Cities

The LiveCity Project [12] is based on live high-definition interactive video communication easily available on user equipment. LiveCity e-Health for glaucoma is a part of this European program, which aims to improve the quality of life of chronic patients, reduce their need for transportation to the Hospital and facilitate communication with the doctor through a high definition video-to-video telemonitoring from the patient's home [13]. More benefits from this include reduction of morbidity, limiting of therapy's side effects, reduction of healthcare cost because of increased compliance to therapy, less car traffic in the cities, reduced greenhouse emissions, less fossil fuel consumption, less air pollution.

For the glaucoma patients in particular, the study involves the Hellenic Telecommunications Organization S.A. (OTE), the Informatics & Telecommunications Department of the National and Kapodistrian University of Athens and the Glaucoma Department of the 2nd Ophthalmological University Clinic of Athens in Attikon Hospital. The essential aim is to communicate with glaucomatic patients at their home, from the Hospital, both in regular appointments and in case of emergency. Regular appointments include monitoring of their compliance to therapy, eye drop instillation reminding, eye drop instillation training and display [14], common examination with a high definition video-to-video communication and even the use of camera snapshots. At the same time, the patient can call the Glaucoma Department in case of emergency for a 24/7 follow-up. This may include cases of a common but very impressive subconjuctival haemorrhage which needs no transportation to the hospital and can be easily diagnosed from the camera, any acute symptoms from the eyes or side effects from therapy, even postsurgical examination. Authorized doctors have also the ability to send communication request to a specific patient, if needed.

2.1   LiveCity Telemedicine Platform

The LiveCity Telemedicine platform supports high definition video streaming between an eye doctor and a patient over public Internet. In this end, two different end-user applications, the Doctor Plugin and the Patient Plugin, have been designed and developed to address the needs of the doctor and the patient respectively. The plugin allows the doctor to establish high definition quality video-to-video communication with a chosen patient. The core functionality of the application is provided by a simple graphical user interface. Specifically, the following functionality is supported:

- Connect and Disconnect: The doctor can register the client software in the LiveCity network and obtain access to the LiveCity services by pressing the button Connect and filling-in his certified credentials. By pressing the button Disconnect the doctor unregisters his client from the LiveCity network.
- Call: This button allows the doctor to establish video-to-video communication with patients that are currently connected to the LiveCity network.
- Notifications: The notifications area is used to inform the doctor that a patient requires his attention. When a patient issues a call request the application shows the caller's name in the notifications area informing the doctor.
- Feedback: This button provides the doctor the ability to report to the developers issues arising during the application use. The doctor can issue a report for a bug or unexpected application behavior providing a short description of the problem.
- Adv. Settings: Provides a set of additional configurations that extend the overall flexibility of the application. Through these configurations the doctor can change the network parameters of the application, determine new video and audio devices and even define the video resolution.

During an active video session the doctor can manually adjust the settings of his own video resolution. Through the Adv. Settings the doctor can choose between four individual video profiles: platinum (1080p video streaming), gold (720p video streaming), silver (480p video streaming) and bronze (360p video streaming).

In addition, the doctor can adjust the volume, contrast, brightness and saturation of the remote video. This functionality is available by pressing the button Controls. Moreover the doctor may save a snapshot of the received video by pressing the button Screenshot. The image is stored in the LiveCity database and can be used later by the doctors to assist the patient examination. The doctor may change the resolution of the streamed video using the button Remote Video. Finally, the doctor can terminate the call session by pressing the button Hang up.

The LiveCity Telemedicine Patient Plugin is a simple and lightweight application used by the patients and it mainly targets the inexperienced users. It provides exactly the same high definition video-to-video communication profiles and the same core functionality of the Hospital side plugin. It should be noted however, that the Patient Plugin does not provide the same call functionality. By pressing the call button the patient application sends a notification to the doctor application. Thus, the doctors can initiate a call, so that they have the full control of the video communication establishment.

The patient's access to the v2v service is realized through a fixed and low-cost xDSL connection. In addition, at the hospital side, a high quality and bandwidth vDSL line has been established with the network provider for ensuring high quality video delivery to the medical personnel. The upload bandwidth utilization of the patient and, *as a consequence,* the download utilization of the doctor, is nearly linear at 1.3 Mbps. On the other hand, the upload utilization of the doctor is gradually increasing up to 3 Mbps, due to the additional traffic generated by the doctor movement during examination.

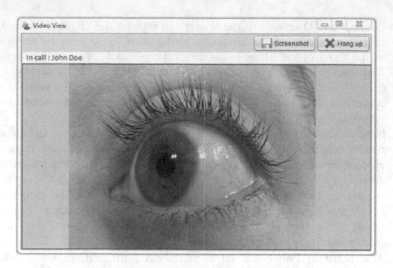

Fig. 1. Eye monitoring camera snapshot

Patients are equipped with common laptops supplied with the software developed for the LiveCity e-Health Project and with individual web cam. Camera snapshots have reached a quality of analysis better than the human eye and sometimes equal to that of an examination slit-lamp and can even detect a contact lens wearer (as shown on Fig.1).

3   Telemonitoring Use Case for Glaucoma Patients

The telemedicine plugin has been developed from the Informatics Department of the University of Athens and used from the Attikon Hospital in practise including: patient's medical history, baseline slit-lamp examination, intraocular pressure measurements, list of all available anti-glaucoma drugs, individual file for each patient with his previous appointment findings, comments and eye snapshots.

The technical part has absorbed most of the preparation time for this use case, allowing for only a short period for the patients' telemonitoring up to now. Training of the patients for the use of laptops, the web cam and the software is necessary, despite the fact that both the hardware and the software are user friendly. This is due to the geriatric patients [15] and their limited experience with the use of computers and Internet. Regular appointments are weekly and more often in the first month. The doctor-patient communication scenario for the eye telemonitoring is shown in Table 1.

For quality control we have decided to include in the study patients with borderline control of their glaucoma, with the purpose to improve the results of conservative therapy via increased compliance. Also, advanced stage patients are included in the study group with the aim to postpone surgical treatment through closer follow-up. Numerical measurements like initial and intermediate intraocular pressure values, number of drugs and the number of eye drops used for each eye are monitored throughout follow-up, in order to reach medical conclusions. Also, record is kept for the number of regular appointments, emergency calls and the number of visits that have been avoided because of video-to-video telemonitoring, for the final evaluation ([16], [17]). Subjective satisfaction report of each patient and doctor for the program and the quality of the video is asked. An unavoidable two (2) second delay in sound and video image has been noticed during the v2v communication with the patients.

Table 1. Scenario of Doctor-Patient Communication in the LiveCity Project

Questions of Regular Appointments	Questions of Emergency Call	Additional Questions for the First Call
General condition of the treated eye.	What is the cause of the call?	Ease of use of the hardware/software.
Remark any troublesome symptoms after starting treatment	When did the disturbing symptoms appear?	Level of satisfaction with the new type of communication.
Difficulty in applying eye drops?	What is the duration of the symptoms?	
Demonstrate eye to the camera.	Is there any decrease in vision?	
Application of eye drops demo.	Feeling any pain?	
Follow instructions for implementing drip-hours?	Demonstrate eye to camera.	
Confirm details for next appointment.	Possible questions from the patient.	
Possible questions from patient.	Confirm details for next appointment.	

Because of the advanced technical infrastructure needed for the program and the cost for each patient included in this, so far only 2 patients (females, 70 and 81 years old respectively) have been enrolled and 2 more have been recruited and waiting for the connectivity establishment. A control group for comparison, with similar demographic characteristics but without access to LiveCity, has been formed. Because of the small number of participants so far, no comparative conclusions can be reached. Both the enrolled patients are scheduled for cataract extraction in the next 6 months and postoperative examination with video-to-video communication will be assessed (One patient is treated for glaucoma since 10 years with one medication - the second patient is newly diagnosed and under triple therapy).

4   Conclusion

Telemonitoring of glaucoma patients with high definition video-to-video communication is a novel approach. Various imaging examinations exist in everyday practice in Ophthalmology and the bases of the ophthalmological diagnosis. Glaucoma patients besides being suitable for tele-examination with high definition video and imaging have also a record in noncompliance to therapy. Telemonitoring is expected to improve training, reminding, correct instillation of eye drops and therefore compliance to therapy. Existing video platforms candidates for realizing v2v services (e.g. Skype, Oovoo) do not offer any facility to integrate additional features to support such e-health use cases. The LiveCity Project develops a user-friendly software plugin that can be easily used in wide scale for community connection with the reference Hospital in smart cities [18]. A larger number of patients will be enrolled in the project in the next months for deriving more accurate statistical results.

Acknowledgments. The present article has been structured in the context of the *LiveCity* ("*Live Video-to-Video Supporting Interactive City Infrastructure*") European Research Project and has been supported by the Commission of the European Communities - *DG CONNECT* (FP7-ICT-PSP, Grant Agreement No.297291).

References

1. Resnikoff, S., Pascolini, D., Etya'ale, D., Kocur, I., Pararajasegaram, R., Pokharel, G.P., Mariotti, S.P.: Global Data on Visual Impairment in the Year 2002. Bulletin of the World Health Organization 82, 844–851 (2004)
2. World Health Organization (WHO): Prevention of Blindness and Deafness. Global Initiative for the Elimination of Avoidable Blindness. Geneva, WHO (2000). (WHO document WHO/PBL/97.61 Rev2)
3. World Health Organization: The World Health Report 2003: Shaping the Future. Geneva, WHO (2003)
4. Quigley, H.A., Broman, A.T.: The Number of People with Glaucoma Worldwide in 2010 and 2020. British Journal of Ophthalmology 90, 262–267 (2006)
5. Leske, M.C., Heijl, A., Hyman, L., Bengtsson, B., Komaroff, E.: Factors for Progression and Glaucoma Treatment The Early Manifest Glaucoma Trial. Current Opinion in Ophthalmology Journal 15, 102–106 (2004)
6. Heijl, A., Leske, M.C., Bengtsson, B., Hyman, L., Hussein, M.: Early Manifest Glaucoma Trial Group. Reduction of Intraocular Pressure and Factors for Progression and Glaucoma Progression: Results from the Early Manifest Glaucoma Trial. Archives of Ophthalmology 120, 1268–1279 (2002)
7. Kholdebarin, R., Campbell, R.J., Jin, Y.-P., Buys, Y.M.: Multicenter Study of Compliance and Drop Administration in Glaucoma. Canadian Journal of Ophthalmology 43(4), 454–461 (2008)
8. European Glaucoma Society: Terminology and Guidelines for Glaucoma, 3rd Edition. Savona, Editrice Dogma S.r.l (2008)
9. Sleath, B., Robin, A.L., Covert, D., Byrd, J.E., Tudor, G., Svarstad, B.: Patient-reported Behavior and Problems in Using Glaucoma Medications. Ophthalmology Journal 113, 431–436 (2006)
10. Bloch, S., Rosenthal, A.R., Friedman, L., Caldarolla, P.: Patient Compliance in Glaucoma. British Journal of Ophthalmology 61, 531–534 (1977)

11. Busche, S., Gramer, E.: Improved Eyedrop Administration and Compliance in Glaucoma Patients. A Clinical Study. Klinische Monatsblatter fur Augenheilkunde Journal 211, 257–262 (1997)
12. FP7-ICT-PSP LiveCity project, http://www.livecity-psp.eu/
13. Patouni, E.: CIP/LiveCity - Live Video-to-Video Supporting Interactive City Infrastructure. In: The Workshop on Network Virtualization, Future Internet Assembly 2014 (FIA 2014), Athens, Greece (2014)
14. Patel, S.C., Spaeth, G.L.: Compliance in Patients prescribed Eyedrops for Glaucoma. Ophthalmic Surgery 26, 233–236 (1995)
15. Gurwitz, J.H., Glynn, R.J., Monane, M., et al.: Treatment for Glaucoma: Adherence by the Elderly. American Journal of Public Health 83, 711–716 (1993)
16. Tsai, J.C.: Medication Adherence in Glaucoma: Approaches for Optimizing Patient Compliance. Current Opinion in Ophthalmology Journal 17, 190–195 (2006)
17. Zimmerman, T.J., Zalta, A.H.: Facilitating Patient Compliance in Glaucoma Therapy. Survey of Ophthalmology Journal 28, 252–258 (1983)
18. Merentitis, A., Gazis, V., Patouni, E., et al.: Virtualization as a Driver for the Evolution of the Internet of Things: Remaining Challenges and Opportunities Towards Smart Cities. Accepted in the International Journal on Advances in Internet Technology 7(1&2) (2014)

ICT in the Future Classrooms and Teaching: Preparing the Knowledge Workers of the 21st Century

Dimitrios Charalambidis

ICT Coordinator, "Saint Joseph" Hellenic-French School,
2, Thessalonikis str., Pefki 151 21, Athens, Greece
dimitrios.charalambidis@stjoseph.gr

Abstract. A knowledge worker is a person that adds value to an organization by processing existing information to create new information and knowledge that could be used to define and solve problems. The future classroom is the cradle (for students) and the action field (for teachers) of a knowledge worker. It incarnates the vision for learning and teaching with the use of Information, Communications and Technology tools (ICT) in order to provide to the students significant 21st century skills such as personal and social responsibility, critical thinking, digital competence, as well as collaboration and communication abilities. From the teacher's point of view it demands understanding and creative use of ICT tools, organizational and administrative competences, use of up-to-date teaching scenarios, intuitive assessment methods and most of all a "lead and teach by example" attitude. Building the future classroom is all about delivering competent and effective people to the society, people that will be the key stakeholders in a world that seeks innovation and creativity in order to keep going socially, scientifically, economically, etc. In this paper we will provide insights about the reasons why a future classroom teacher should afford 21st century skills to his students and continuous professional development to himself.

Keywords: ICT in education, knowledge workers, 21st century skills, 1 to 1 computing, future classroom scenarios, innovation in teaching and learning, professional development.

1   Introduction

Why is so important for a classroom teacher to afford 21st century skills to his students? To help us get a better feel for the changes coming to schooling and learning, let's take a few minutes and do a thought test, submitting four questions to ourselves.

First, imagine that we have a child whom we love and care about. Moreover, this child is about to start kindergarten this year. Question #1: *What will the world be like twenty years from now when our child finishes school?* We can easily answer to that question if we try to remember what life was like twenty years ago and all the

L. Iliadis et al. (Eds.): AIAI 2014 Workshops, IFIP AICT 437, pp. 56–62, 2014.

changes we have seen taking place until today. Question #2: *What skills will our child need to be successful in a world twenty years from now?* Inexorably, this question generates most of the 21st century skills covered in this article, including hard and soft skills as well as knowledge, values and behaviors [1].

Secondly, let us change our point of view and imagine that we run a small business, an R&D department or a university research laboratory. Furthermore, this business, R&D department or university lab is about to compete other innovative companies, R&D departments or labs locally and abroad. Question #3: *What skills would we require from our staff in order to achieve innovative goals?* This question also generates most of the times 21st century skills required from competitive business or academic environments.

Finally, let us look over our answers to the first three questions and consider how most students currently spend their time each and every day in school. Question #4: *What would teaching and learning be like if it were designed and implemented around our answers to the first three questions?*

2 Schooling: A Time for High-Performance Cultures

The final stage of the upper secondary Hellenic educational system called *Lykeio* is primarily focused on teaching and learning what will be tested in the big government exams (*Panellinies*). In this case the spirit of innovation is dead, considering that the main goal is to get good examination marks that will eventually lead to enrollment in a high educational institution. Ironically, students and their parents think that *Lykeio* is unable to fulfil this preparatory role in a satisfying way. As a result, they seek resource to private tutorial schools (*frontistiria*). *Frontistiria*, a purely Greek phenomenon, undergo intensive preparations for the country-wide exams (*Panellinies*) [2].

This innovation-hostile schooling culture deprives the students as well as the teaching staff of all the benefits that innovation can bring to an unchanging, and thus obsolescent and "human-eating", system. Why is this educational system obsolescent? Because, it delivers highly educated underskilled professionals. Therefore, it produces long-term unemployed or underpaid people. Besides, long-term unemployment leads to an erosion of skills and knowledge. Why is this educational system "human-eating"? Because, it produces greatly unhappy and desperate people.

An underdeveloped school is striving to succeed by standard methods and it is unable to make or sustain improvements. Additionally, a traditional school is effective by standard methods and seeks improvement within old-style pedagogical and managerial methods. On the other hand, an innovative school goes beyond standard methods and measures of success. It constantly evaluates how effectively it is preparing students with core academic and 21st century skills. Twenty-first century schools are first and foremost effective organizations by design. They are schools where innovation and continuous improvement are core values. They have strong leaders who create learning environments, which motivate and attract the best educators. They also use metrics and analytics, because they know what gets measured gets done efficiently. Overall, innovative schools build high-performance cultures. However, as mentioned before, what mostly characterizes these schools is a determination to prepare students for 21st century professions. The jobs of the future will be complex and will require innovative and creative thinking. Jobs that are not, will be done by computers.

3   21st Century Skills: The World within and Beyond the Classroom

In order to thrive in a digital world, students will need digital age proficiencies. It is important for every educational system to fulfill its mission in society, namely the preparation of students for the world beyond the classroom. Therefore, schools must comprehend and embrace 21st century skills within the context of rigorous academic standards. The concept of 21st century skills does not have a precise definition, but it is intended to convey the idea that changes in technology and culture are leading to changing demands in the workplace, and so the skills that are required in current and future workplace are different from those required in the past [3].

3.1   Digital Literacy

In the early 1900s, a person who had acquired simple reading, writing, and calculating skills was considered literate. Today, all students need to learn to read critically, write persuasively, think and reason logically, and solve complex problems in mathematics and science. Digital cameras, graphics packages and streaming video are the means to communicate ideas in a visual and effective way. Students need good visualization skills to be able to decipher, interpret, detect patterns, and communicate using imagery. Digital literacy includes accessing information efficiently and effectively, evaluating information critically and competently, using information accurately and creatively [4].

3.2   Inventive Thinking

The interconnectedness of today's world brings with it unprecedented complexity. Interaction in such an environment requires individuals to be able to identify and react to changing conditions independently. Students should become self-directed learners who are able to analyze new conditions as they arise. Professionals should identify new skills that will be required to deal with these conditions and independently chart a course that responds to these changes. They should be able to take into account contingencies, anticipating changes, and understanding interdependencies within systems [4].

3.3   Global Awareness

Global awareness and international collaboration during schooling period results in more rounded individuals, encouraging students to see things from different perspectives and helping them to make informed decisions, acquiring transferable skills that will be useful to them and will remain with them for life. According to the UKs Association of Graduate Recruiters [5] companies cannot find enough applicants with the requisite skills to operate in an international market place, indicating that greater efforts by schools in fostering global awareness and international collaboration are needed to best prepare students for life and work in the 21st century [6].

3.4   Creativity and Risk-Taking

Innovation is impossible to achieve without taking a necessary amount of risk. Inevitably, every success sees failures along the way. A teacher should act and teach as an effective innovation leader, encouraging creativity and risk taking, while also practicing a tolerance for failure. Obsolete schooling systems punish failure with low grades. Instead, innovative schooling methods consider the fear of failure as an "innovation killer". Accordingly, failure and risk-talking is being seen and recognized as a learning experience [7]. Creativity and risk-taking leads to a sense of initiative and entrepreneurship. The student acquires the ability to turn ideas into action as well as the ability to plan and manage projects in order to achieve objectives [8].

3.5   Teamwork

Information technology plays a key role in the ease with which individuals and groups collaborate. Email, faxes, voice mail, audio and video conferencing, chat rooms, shared documents, and virtual workspaces can provide timely, iterative collaborations. In today's wired, networked society it is imperative that students understand how to communicate using technology. This includes asynchronous and synchronous communication such as person-to-person email interactions, group interactions in virtual learning spaces (such as Learning Management Systems or LMS), chat rooms, interactive videoconferencing, etc. Such interactions require knowledge of etiquette often unique to that particular environment. Other dimensions introduced through global communication include scheduling over time zones, cultural diversity and language issues [4].

3.6   Learning to Learn

Learning to learn is the ability to pursue and persist in learning, to organise one's own learning, including through effective management of time and information, both individually and in groups. This competence includes awareness of one's learning process and needs, identifying available opportunities, and the ability to overcome obstacles in order to learn successfully. This competence means gaining, processing and assimilating new knowledge and skills as well as seeking and making use of guidance [8].

4   Preparing the Knowledge Workers of the Knowledge Age

The adoption of the 21st century skills in schooling will eventually create a new work force known as *knowledge workers*. Peter Drucker, the management guru, is credited with popularizing the term *knowledge worker* as long ago as 1968 [9]. Back then he argued, *"Today the center is the knowledge worker, the man or woman who applies to productive work ideas, concepts, and information rather than manual skill or brawn...Where the farmer was the backbone of any economy a century or two ago...knowledge is now the main cost, the main investment, and the main product of the advanced economy and the livelihood of the largest group in the population"*.

The knowledge workers of an organization produce knowledge capital, meaning experience, information, knowledge, learning, and skills. Of all the factors of production, knowledge capital creates the longest lasting competitive advantage of an innovative organization. A knowledge worker, as a producer and bearer of the knowledge capital and also as part of an unseparated whole, becomes the most value asset of his team, his company, his society, his country and therefore of the globalized world. One could argue that a knowledge worker simply signifies a more modern conception of a good job where workers are viewed as more than what they produce, but ultimately it is more than that [10].

Our digital world is currently formed in the Digital Age but moves quickly towards change and gradually is being transformed into the Knowledge Age. Knowledge Age alters the form of entrepreneurship. In Knowledge Age, knowledge and ideas will become the main source of economic growth (more important than land, labor, money, or other tangible resources). As well as this (and this is very important for education), knowledge's meaning is changing. Knowledge is no longer being thought of as 'stuff' that is developed (and stored) in the minds of experts, represented in books, and classified into disciplines. Instead, it is now thought of as being like a form of energy, as a system of networks and flows, something that does things, or makes things happen. In Knowledge Age, knowledge is defined and valued not for what it is, but for what it can do. It is produced, not by individual experts, but by a collective intelligence, meaning groups of people with complementary expertise who collaborate for specific purposes. These changes will eventually alter and define education systems and inexorably will prepare students to become the knowledge workers of tomorrow [11].

5   Building the Future Classroom of the Knowledge Age

Technologically speaking, the school classrooms will become the cradle of the knowledge workers, thus the student and his teachers. From the teachers' point of view it demands understanding and creative use of ICT tools, organizational and administrative competences, use of up-to-date teaching scenarios, intuitive assessment methods, project, change and conflict management skills and most of all a "lead and teach by example" attitude. Building the future classroom is all about delivering competent and effective people to the society, people that will be the key stakeholders in a world that seeks innovation and creativity in order to keep going socially, scientifically, economically, etc. From the students' point of view all that is needed is courage, willingness and perseverance. Finally, from the parents' point of view an open-minded and proactive outlook that will permit them to spot and choose the proper schooling environment for their children.

In order to provide to the students significant 21st century skills, a pleiade of ICT tools is needed such as an efficient infrastructure (internet access, multi-touch LCD interactive boards, tablets, email accounts, collaboration and e-learning platforms, video recording and editing equipment, live video-to-video supporting interactive infrastructures like those proposed and implemented by LiveCity EU program, etc.) as well as the proper educational content. The book-based paradigm, which has dominated the organization of schooling for two centuries, can't really become much more efficient [12] and therefore new approaches such as internet research, interactive e-books and teacher-student content co-creation are needed.

6   Conclusion

How well we educate our children, whether or not they learn the skills now needed to participate and thrive in global economy, will determine the future health, wealth, and welfare of everyone. This is the good scenario, answering the *question #2* of the introduction: *What skills will our child need to be successful in a world twenty years from now?* The bad scenario would be to live in conditions of an endless economic depression where unemployment changes from a microeconomic problem into a structural one. In this case the unemployed workers lack the skills needed for the jobs, or they are not living in the part of the country or the world where the jobs are available. This is how our country would simply become in the long run; a no man's / no worker's land. Also, the recent global recession has given us a painful glimpse of what life could be like if we don't succeed in providing our children a 21st century education. Though the causes of the global recession are not directly related to education, the results provide us with an important lesson.

References

1. Trilling, B., Fadel, C.: 21st Century Skills: Learning for Life in Our Times, Wiley Desktop Editions, San Francisco (2009)
2. Vretakou, V., Rousseas, P.: Vocational education in Greece. Cedefop Panorama series, vol. 59, p. 13. Office for Official Publications of the European Communities, Luxemburg (2003)
3. Kyllonen, P.C.: Measurement of 21st Century Skills within the Common Core State Standards, In: Invitational Research Symposium on Technology Enhanced Assessments, p. 4. K-12. Center at ETS (2012)
4. The North Central Regional Educational Laboratory, http://enGauge.ncrel.org
5. Association of Graduate Recruiters, http://www.agr.org.uk
6. Why students need a global awareness and understanding of other cultures. The Guardian Teacher Network, http://www.theguardian.com/teacher-network

7. Innovation is Creativity x Risk Taking,
   http://www.innovationexcellence.com
8. Key Competences for Lifelong Learning. A European Reference Framework. Official Journal of the European Union,
   http://ec.europa.eu/dgs/education_culture/publ/educ-training_en.html
9. Drucker, P.F.: The age of discontinuity: Guidelines to our changing society. Transaction Publishers, London (1968)
10. Brinkley, I., Fauth, R., Mahdon, M., Theodoropoulou, S.: Knowledge Workers and Knowledge Work. A Knowledge Economy Programme Report. The Work Foundation, London (2009)
11. The Knowledge Age. New Zealand Council for Educational Research,
   http://www.shiftingthinking.org/?page_id=58
12. Lloyd, M.: Schooling at the Speed of Thought: A blueprint for making schooling more effective. Spiderwize, London (2010)

Promoting ICT Skills through Online Services: Case Study of Video Use for Adult Education in Municipalities

Andreea Molnar[1], Vishanth Weerakkody[2], and Ahlam Almuwil[2]

[1] Arizona State University
andreea.molnar@asue.edu
[2] Brunel University
{vishanth.weerakkody,ahlam.almuwil}@brunel.ac.uk

Abstract. The usage of video communication in public services can alleviate some of the existing problems these services face in their adoption and also help in increasing the level of outreach to citizens. In this article we focus on the usage of video-to-video (V2V) communication in municipality services for ICT education of senior citizens. For this purpose, our study focuses on determining criteria that are important for both the teachers and citizens in using such an innovative service. This paper reports on the findings from focus group consultations conducted with teachers and students, in this case senior citizens, of a municipality in Spain who was piloting the use of V2V services for adult education. The research identified several criteria relevant for consideration when deploying V2V as part of a local government service portfolio, of which, trust, quality of service and convenience emerged as common for all stakeholders involved.

Keywords: public services, adult education, video-to-video, smart cities.

1 Introduction

With the increased digitization of society, the usage of information and communication technology (ICT) in public services has seen a rapid increase during the last few years. Although governments have continued to push digital applications to the front line of public services, not all public services are feasible to be offered online or when offered, their adoption rate has not always been as high as expected [1, 2]. It is likely that some services have not been used due to the lack of awareness and interest on the part of citizens or those citizens were trying to use some of these online services but failed to fully complete them and ended up using other means [3]. For some other services, deep-rooted issues were found such as accessibility, lack of inclusion of the elderly and the difficulty to provide the same rich interaction as they find in face-to-face services. This is particularly true when dealing with complex services such as health and education and social or domestic services such as social security, housing or employment [4, 5]. This research focuses on education for senior citizens.

The poor adoption of eGovernment has led to public services being available both online and also mirrored via other channels [3]. However, it is not always possible to

L. Iliadis et al. (Eds.): AIAI 2014 Workshops, IFIP AICT 437, pp. 63–72, 2014.

ensure that human resources are constantly available or that it is possible to reach people that may need a certain service, but are not located closely to the centres that offer these services. This is particularly true in the current environment, where local governments are facing growing deficits, loss in revenues and increasing costs [6] while at the same time dealing with citizens whose usage of ICT in the private sector has led to creating new expectations [3], of high quality from services available in the public sector [7].

A potential solution to these problems is the usage of high definition (HD) video technology in the provisioning of public services [8]. This can imitate the richness found in face-to-face communication while preserving the efficiency, speed and outreach of e-government services [8]. Moreover, the involvement of citizens from the beginning of a service design and diffusion and finding out what their needs are could increase the likelihood for a service to be adopted [9]. This research seeks to explore criteria for the adoption and the usage of HD V2V communication in public services, focusing on continuous education for senior citizens. To do so, we performed focus groups with the municipality staff, the senior citizens using these services and volunteers that were helping for these services to be delivered, in Spain. This empirical research was performed in the context of the LiveCity project which is funded by the European Commission's ICP programme (EU FP7 CIP grant agreement No. 297291). The remaining sections of the paper are structured as follows. The next section presents the background of technology usage in public services and focuses on V2V communication in particular. The next section briefly introduces the proposed service and how V2V will be ensured. The following section presents the study performed. We continue with conclusions and the discussion, and the paper ends with future work.

2   Background

The increasing availability of the Internet and the development of ICT have made electronic services an attractive solution for delivering public services. Several benefits of using ICT in public services are cited in the literature as factors driving adoption: increasing efficiency and decreasing the cost of providing services [10], availability anytime and anywhere [11], being able to provide more personalised solutions that put citizens in the center [12], and providing more transparent and account-able services [13]. It is considered that an e-government service is successful when that service is highly adopted.

The usage of HD V2V has been seen as a solution for providing personalised services and as a means to offer help for citizens that have problems using current e-government services [8]. Several preliminary studies have been performed on determining criteria for the usage of V2V in education [14, 15] but to the best of our knowledge they have not yet focused on local e-government services such as continuous education for senior citizens. In addition, although prior research has identified some common criteria for the usage of HD video, it is necessary to define these criteria based on the context in which the service is used [16] as prior research has shown failed attempts to measure new contexts when using general scales [17].

Education increases the well-being of an individual and strengthens their independence [18] and for senior citizens, it makes them live more productively and less isolated [19]. When lifelong education is offered to senior citizens, it is mostly focused on the usage of information technology [20, 21]. To do so, these communities need to have some basic ICT skills. Being able to access a program that teaches basic ICT skills is significantly influenced by the access to training by senior citizens [20]. It has been shown that having computer skills can positively affect the life of citizens and improve their social relationships [22]. However the resources allocated to teach senior citizens basic information technology skills are usually limited [20]. Therefore it is necessary to find new ways to give access for more citizens to these types of services. To the best of our knowledge the study covered in this paper is the first study that attempts to use V2V technology to teach senior citizens computer skills.

3   Research Context

This research is done in the context of the LiveCity project. It seeks to provide HD video over public infrastructure [8]. LiveCity is user community driven and is aimed at improving core public services such as education, public administration, health and city experience in different European cities. The way LiveCity aims to ensure HD video delivery has already been discussed in our previous studies [8] and it is out of the scope of this article to examine this again. This paper focuses on education for senior citizens. The usage of V2V for these services will happen in Valladolid, Spain.

The usage of V2V for providing education for senior citizens would allow one teacher to connect to several citizens at any given time. For example, V2V will allow teaching two classes at the same time, one face-to-face and another through a video connection or having senior citizens connecting from their home. This would allow more senior citizens to benefit from lifelong learning and eliminate the need for having several teachers or facilitators, which from a municipality perspective will be one of the biggest challenges when delivering such a programme. This could also allow people that do not live in the proximity of a municipality center where these courses take place, to connect from their home. The challenge in this instance is whether the senior citizens will accept this type of technology and if they will be able to use it. This is a significant challenge as the citizens will need to make use of the ICT skills they are learning through V2V in order to receive the training. Therefore, the acceptability and the way one would design the interface to make it suitable for learners are among the challenges one faces when offering such remote training.

The service focused in this paper was originally designed by the Municipality of Valladolid as a means of helping adult citizens enhance their ICT skills. At present, volunteers teach the classes that take place in one of the venues offered by the municipality using face-to-face communication. The motivations for considering video to facilitate this service has been influenced by the fact that these classes are typically overbooked and currently there are not enough volunteers to teach them on a face-to-face basis. Due to the current economic climate there are no plans for the municipality to increase the number of classes or to hire teachers to compensate for the demand, therefore V2V communication is seen as an appropriate solution to increase the service offered without additional cost to the municipality. Moreover, the elderly population is increasing and the usage of ICT can allow them to live a more independent, productive and less isolated life [19]. Senior citizens, particularly those whose families live away from home, are often motivated to take these classes as they use ICT skills gained to communicate with their families.

4   Study

The aim of our study was to determine which criteria would be important for the adoption of V2V services, if they were offered to citizens as part of their municipality services. To achieve this aim, we focused on consulting the stakeholders using a qualitative approach. The motivation for adopting a qualitative, rather than a quantitative (such a survey) is because it allowed the researchers to get close to reality as possible and explore criteria that need to be considered from a user perspective for such a new type of service to be successful.

4.1   Methodology

Exploratory research was chosen due to the scarcity of empirical work related to the usage of V2V in the public sector for this particular service. Case study research is useful when dealing with broad and complex phenomenon, and the existing body of knowledge is insufficient to permit posing casual questions [23]. Due to the lack of understanding and the complexity involved in the area of research under investigation, focus groups were used as the primary data collection method. It was not necessary to design a structured agenda to ask questions in a specific order [23], but literature reviews provided the themes to be explored during the empirical work. Focus groups are useful in providing the necessary focus needed to probe a research domain that is exploratory at present [23]. As people have different views and perspective over the same things focus groups are a suitable method to understand user views [24]. Moreover, focus groups, when used in an e-government context, were found to be helpful in finding citizen's needs and expectations for new services [9]. Two focus groups were organised and recorded as this allowed for an easier analysis of the information. A thematic analysis process [25] was used for the analysis, where emerging issues were linked to the criteria identified through literature and documenting any new issues and assigning labels to these.

The participants were given an information sheet containing information about the project, what is expected from them and their rights to withdraw from the study anytime without any prior notice or explanation and/or not to respond to the questions they do not feel content in responding. Participants demographic data were collected prior to the interview through a questionnaire, however the participants did not have to disclose their demographic data if they did not want to. The focus group was facilitated by a moderator that introduced the scope of the focus group and guided the interview.

The participants in the study were recruited by the local partners in the LiveCity project. The focus groups were held in Spanish and the results translated to English with the help of the Municipality of Valladolid.

4.2   Participants

We held focus groups with the teachers and the students all senior citizens who are part of the ICT education pilot. The teachers were males and over 56 years of age. A focus group was held with the students who took part in one of the classes. A total of fifteen students participated in the focus groups, 40% were males and 60% females. They were all over 56 years of age (53% belonged to the 56-65 age group and 47% were over 65 years). Their education level varied from those who have finished primary education (47%), to those who had postgraduate university education (7%). Most of them reported having used an online (e-government) municipality service for less than one year (94%) and the rest reported using it between two to five years. The teachers were volunteers and the students (senior citizens) lived in the municipality and its suburbs. All students in the ICT class were invited to participate in the focus group, no incentives were offered and all the students participated in the study.

4.3   Results

Considering the importance of involving relevant stakeholders in the implementation of ICT in public services [26], in this study we aim to involve the volunteer teachers and the senior students. In this respect it is important to find out what criteria is necessary for these services to be successfully adopted. We followed a qualitative approach by performing focus groups with the teachers and students. We used both an inductive and deductive approach when analysing the data. When criteria emerged it was linked with the previously found criteria from the literature. If any new criteria were found, they were added to the existing list.

Teachers. A focus group was held with teachers involved in the programme. The teachers were senior citizens who knew how to use technology, some of them being retired teachers. They mentioned as their concern problems with the technology and the fact that the citizens/students are typically beginners in using a computer and therefore distrust them, but they do not see teaching these courses as difficult or that the students have a problem in getting involved. They were keen to ensure that the V2V technology used allowed them to interact with individual students as and when they had a question or needed individual attention. The teachers highlighted the need for the V2V system to incorporate features that will allow individual students to raise their hand or alert the teacher when they needed to speak in a manner that would not be intrusive to the other participants in the class, but would allow them to listen and benefit from any questions raised by others. In addition, quality of the video and voice transmission was identified by the teachers as one of the most important criteria that they were looking for in a remote, online education programme. The teachers viewed the proposed system as different to other online and distance education programmes as it allows real time two way V2V making the programme as close to face-to-face as possible. Table 1 summarizes the main criteria identified by the teachers.

Table 1. Criteria for V2V based Education Services as Identified by Teachers

Criteria	Proposed Description
Trust	Students confidence in the system, believing that the system is reliable
Access and Availability	The system must be available and easily accessible by the citizens/students
Quality of Service	The system must allow high quality live video and voice
Two way interaction	The system has to facilitate seamless two way communication between teacher and student
Ease of use and presence	The system should be as close to a real class room experience for the teachers and students

Senior Citizens/Students. The focus group took place with the students of one class involved in learning computer use as part of their municipality service. A total of 15 citizens participated in the service. The students were beginners and have little or no experience in using V2V services. Trust is mentioned as one of the biggest obstacles for them in using online services as well as the fear of "doing something wrong", "breaking something". They mentioned that having the V2V connection would allow more students to be enrolled and allow other students who are on the waiting list to participate in adult education. Overall, all the students were supportive of the idea, "I like the idea". They stated that they would not mind being either in the class with the teacher or accessing the course online, as long as the teacher is interacting with them and is available "to answer questions". Some of the students suggested that video based learning will be convenient for them as they presently have to travel long distances to be physically present at the adult education class. They also thought that video based learning will be better than having to be part of a large class with others where senior citizens can easily be distracted due to the social interactions that

Table 2. Summary of Criteria for V2V services based on Citizens/Students' Views

Criteria	Proposed Description
Trust	Confidence in the system, believing that the system is reliable and secure
Confidence	Students belief that they could use the system
Number of courses	Number of courses available through the V2V service should exceed those currently offered using classroom based teaching
Individual feel of the system	The system should allow individual interaction irrespective of the number of students that are taught through the V2V service
Convenience	The video based learning should be convenient for those finding it difficult to travel to a physical class room due to distance
Ease of use	The system should be easy to use and should feature live video of both the teacher and his/her computer/keyboard when needed
Cost	Should be cost effective for citizens when comparing internet usage against travel expenses incurred to attend a physical class

occur in the classroom. In terms of the 'look and feel' of video based learning, the students suggested that the system should have the option where they could switch or split the screen when needed between seeing the live video of their teacher and the specific actions that need to be carried out involving the keyboard to complete an ICT function. In general, the students' needs and expectations mirrored those of the teachers as they were keen to see a system that was as close to a real classroom experience as it can be. Table 2 outlines the main criteria discussed during the focus groups.

5   Discussion and Conclusion

5.1   Summary

This paper presents a possible usage of video in public services, focusing on lifelong learning education for senior citizens. Focus groups were performed with volunteer teachers and senior citizens taking part in an education program offered by the municipality. The results show that all participants are positive about the introduction of V2V services in their municipality. The usage of V2V was seen as a way to allow more citizens to participate in education activities. Fig. 1 summarises the criteria important both from the teachers and students point of view. Trust in technology is common to both groups, whereas the rest of the criteria are group specific.

Fig. 1. Citizens generated Criteria for using V2V Communication for Adult Education

5.2   Discussion

Although from a supplier's perspective, the usage of ICT in public services is seen as a tool for automating and streamlining service delivery [27], in the example studied in this paper ICT is seen as a means to reach out and provide access to services to as many citizens as possible. This was highlighted in the service offered to the senior citizens in the Valladolid municipality. In the context of their adult education programme, the municipality of Valladolid is expected to teach more people at the same time, hence increasing the teacher's (and the municipality's) productivity. As the teachers are all volunteers no direct cost saving are seen for the municipality, but the education programme offered by the volunteers is seen as a value adding service to the municipality in terms of public perception. Citizens see the usage of V2V communication in a very positive light.

Considering the dual role of public services, to increase efficiency and be useful to citizens [9], this article has been focused on the later by aiming to better understand the needs of citizens when using a new service and determine criteria to be taken into account during the implementation and deployment of such services. By consulting the stakeholders (users and voluntary teachers) during the design and delivery of V2V services, we have tried to align the needs of the municipality with citizens' needs and expectations. It is hoped that this approach will result in better acceptance of online services in general [9] and V2V in particular, as a mechanism for adult education as well as other wider municipality services.

5.3 Theoretical and Practical Implications

This study adds to the state of the art by determining criteria that have to be considered for the adoption of V2V communication in the public sector as a mechanism for online service delivery. In particular, it adds to the literature of public services and adult education. To the best of our knowledge this is the first study to explore criteria necessary from the citizens' perspective when using V2V for learning ICT as part of a lifelong learning program for senior citizens.

The results presented in this research can be used by policy makers and managers and municipalities providing such services to determine what criteria are seen as relevant from the citizens' perspective in order to introduce new services or to improve the already existing ones. These criteria could also be used as a means of evaluating existing services.

5.4 Limitations

The results of this study are based on potential criteria as determined from focus groups with teachers and students. None of them have used the V2V system before or were involved in a similar project. Although the teachers were ICT savvy and used similar V2V services (e.g. Skype), the students did not. This may have led to difficulties in expressing what they expect from such a system [28]. The students may have also overestimated their ability to use the system without having a teacher in the class given their low level of ICT skills. The study is performed in a single location in one country and therefore cultural differences may lead to different results if the study was to be performed in another location.

6 Future Work

The results of this study have been taken into account when the service was implemented and deployed. Currently a small pilot is taking place in Spain. The study performed here was used to inform the design of the services and the evaluation process.

The data about citizens and the municipality experience with the service is currently being collected and will be reported. Based on these results, improvements to the two services will be made and other criteria that may be found as a result of the participants' interaction with the system will be taken on board to continuously improve the systems. We aim for the end result of this research to be a complete taxonomy of criteria that are relevant for V2V services in the public sector. The taxonomy will be evaluated in the last stages of this research when this service will be offered on a larger scale as part of the Smart City infrastructure implemented by Valladolid in Spain.

Acknowledgements. The authors acknowledge the contributions made to this article by the LiveCity consortium of partners who are funded by the European Commission, especially to National & Kapodistrian University of Athens for helping in organising the study presented in the paper.

References

1. Cordella, A., Contini, F.: Socio Technical Regimes and e-Government Deployment: The Case of Italian Judiciary. In: European Conference on Information Systems, Paper 27 (2012)
2. Norris, D.F., Reddick, C.G.: Local e - Government in the United States: Transformation or Incremental Change? Public Administration Review 73(1), 165–175 (2013)
3. Andersen, K.N., Medaglia, R., Henriksen, H.Z.: Frequency and Costs of Communication with Citizens in Local Government. In: Andersen, K.N., Francesconi, E., Grönlund, Å., van Engers, T.M. (eds.) EGOVIS 2011. LNCS, vol. 6866, pp. 15–25. Springer, Heidelberg (2011)
4. Andreassen, H.K., Bujnowska-Fedak, M.M., Chronaki, C.E., Dumitru, R.C., Pudule, I., Santana, S., Wynn, R.: European Citizens' use of e-Health Services: A Study of Seven Countries. BMC Public Health 7(53) (2007)
5. Santana, S., Lausen, B., Bujnowska-Fedak, M., Chronaki, C., Kummervold, P.E., Rasmussen, J., Sorensen, T.: Online Communication between Doctors and Patients in Europe: Status and Perspectives. Journal of Medical Internet Research 12(2), e20 (2010)
6. Rosenbloom, J.: Government Entrepreneurs: Incentivizing Sustainable Businesses as part of Local eEconomic Development Strategies. In: Salkin, P., Hirokawa, K. (eds.) Greening Local Government, pp. 11–22 (2011)
7. Hazlett, S.A., Hill, F.: E-Government: The Realities of Using IT to Transform the Public Sector. Managing Service Quality 13(6), 445–452 (2003)
8. Weerakkody, V., Molnar, A., Irani, Z., El-Haddadeh, R.: Complementing e-Government Services Through the Use of Video: The LiveCity Project. In: E-Government Conference, pp. 124–131 (2013)
9. Axelsson, K., Melin, U.: Talking to, not about, Citizens–Experiences of Focus Groups in Public e-Service Development. In: Wimmer, M.A., Scholl, J., Grönlund, Å. (eds.) EGOV. LNCS, vol. 4656, pp. 179–190. Springer, Heidelberg (2007)
10. Affisco, J.F., Soliman, K.S.: E-Government: A Strategic Operations Management Framework for Service Delivery. Business Process Management Journal 12(1), 13–21 (2006)
11. Bekkers, V., Homburg, V.: The Myths of e-Government: Looking Beyond the Assumptions of a new and better Government. The Information Society 23(5), 373–382 (2007)

12. Dunleavy, P., Margetts, H., Bastow, S., Tinkler, J.: New Public Management is Dead - Long Live Digital Era Governance. Journal of Public Administration Research and Theory 16(3), 467–494 (2006)
13. Dwivedi, Y.K., Weerakkody, V., Janssen, M.: Moving towards Maturity: Challenges to Successful e-Government Implementation and Diffusion. ACM SIGMIS Database 42(4), 11–22 (2012)
14. Molnar, A., El-Haddadeh, R., Hackney, R.: Facilitating the Adoption of Public Services Using High Definition Video: The Case of Primary Education. In: American Conference on Information Systems, vol. 10 (2013)
15. Weerakkody, V., Molnar, A., El-Haddadeh, R.: Indicators for Measuring the Success of Video Usage in Public Services: The Case of Education. In: American Conference on Information Systems (2014)
16. Dabholkar, P.A., Thorpe, D.I., Rentz, J.O.: A Measure of Service Quality for Retail stores: Scale Development and Validation. Journal of the Academy of Marketing Science 24(1), 3–16 (1995)
17. Dagger, T.S., Sweeney, J.C., Johnson, L.W.: A Hierarchical model of Health Service Quality Scale Development and Investigation of an Integrated Model. Journal of Service Research 10(2), 123–142 (2007)
18. Bronswijk, J.V., Bouma, H., Fozard, J.L.: Technology for Quality of Life: An Enriched Taxonomy. Gerontechnology 2(2), 169–172 (2002)
19. van Dyk, T., Gelderblom, H., Renaud, K., van Biljon, J.: Mobile Phones for the Elderly: a design framework. In: International Development Informatics Association Conference, pp. 85–102 (2013)
20. Zumarova, M., Cerna, M.: Senior Citizens as a Specific non-Traditional Group of Students in a Lifelong Education. In: Global Engineering Education Conference, pp. 771–776 (2011)
21. Dianti, M., Parra, C., Casati, F., De Angelli, A.: What's Up: Fostering Intergenerational Social Interactions. In: Designing for Inter/Generational Communities, pp. 21–27 (2012)
22. Zumarova, M.: The role of ICT in the Lives of Senior Citizens. Educational Technologies, 77–82 (2010)
23. Yin, R.K.: Case Study Research: Design and Methods. Incorporated, London, UK (2003)
24. Kitzinger, J., Barbour, R.S.: Introduction: the Challenge and Promise of Focus Groups. Developing Focus Group Research:Politics, Theory and Practice, 1–20 (1999)
25. Boyatzis, R.E.: Transforming Qualitative Information: Thematic Analysis and Code Development. SAGE Publications, Thousand Oaks (1998)
26. Goel, S., Dwivedi, R., Sherry, A.: Role of Key Stakeholders in Successful e-Governance Programs: Conceptual Framework. In: American Conference on Information Systems (2012)
27. Grant, G., Chau, D.: Developing a Generic Framework for e-Government. Journal of Global Information Management 13(1), 1–30 (2005)
28. Flak, L.S., Moe, C.E., Sæbø, Ø.: On the evolution of e-government: The user imperative. In: Traunmüller, R. (ed.) EGOV 2003. LNCS, vol. 2739, pp. 139–142. Springer, Heidelberg (2003)

Modification of Colors in Images for Enhancing the Visual Perception of Protanopes

Polyxeni Sgouroglou[1] and Christos-Nikolaos Anagnostopoulos[2]

[1] School of Science & Technology, Hellenic Open University, Patras, Greece
std041740@ac.eap.gr
[2] Cultural Technology & Communication Dpt., University of the Aegean, Mytilene, Lesvos, Greece, 81 100
canag@ct.aegean.gr

Abstract. The purpose of this paper is to design and to implement proper color modification methods for enhancing visual perception of digital images for people with protanopia (one of the types of Color Vision Deficiency). This paper proposes one simulation method and two daltonization methods for the modification of digital colored images, implementing the additional step of quantization for a faster modification of colors which are not correctly perceived. In order to avoid color confusion a color checking module is implemented and optimal daltonization parameters are selected. The ultimate objective of this paper is to minimize the total processing time to such an extent that it is possible to apply the proposed methods to real-time sequences (online video). The entire application is tested in a set of artificial and real images, while the performance of the proposed algorithms is evaluated in an appropriately selected set of colored images.

Keywords: Color Modification, Color Quantization, Real-Time Processing.

1 Introduction

Color Vision Deficiency (CVD) is quite common since about 8 - 12% of the male population and 0.5% of the female European population show in a certain degree, some type of CVD. Even though there is no known medical method which corrects such deficiencies in human vision, these people are not considered to suffer from a serious malfunction. Nevertheless there are certain cases in everyday life where color confusion may become annoying (e.g. reading a website) or even critical (e.g. road signs). Consequently there is intense research interest around the problematic color perception, resulting in the development of various techniques and algorithms for the digital simulation of such malfunctions. In addition several enhancement/modification methods (daltonization techniques) for the color information in digital images have been proposed in order to reduce in certain cases color confusion.

Many researchers have conducted research for appropriately modeling the visually impaired vision, presenting algorithms to simulate what individuals with vision deficiency see. The problem of color adaptation according to user's perception is also

L. Iliadis et al. (Eds.): AIAI 2014 Workshops, IFIP AICT 437, pp. 73–84, 2014.

addressed in [1,2]. In [1] paper, among other issues, the problem of tailoring visual content within the MPEG-21 Digital Item Adaptation (DIA) framework to meet users' visual perception characteristics was addressed. The image retrieval aspect for people with CVD was discussed in [3]. Even a physiologically motivated human color visual system model which represents visual information with one brightness component and two chromatic components was proposed for testing the color perception of people suffering from CVD [4].

One of the most common techniques is presented in [6], in which the daltonization parameters are selected with a trial-and-error process. In [5], which was based on the work of [6], a method was proposed for the automatic modification of the daltonization parameters in such a way that no conflict among colors that were daltonized and colors that remained intact is ensured.

Although many algorithms have been proposed for the color modification of images in order to become more visible to people with CVD, relatively few steps have been made in the field of video and specifically in the field of real time processing, due to the high computational time required for the color modification of a single image/frame. For this purpose a color quantization technique is implemented by this paper.

In this paper the proposed methods are subject to the modification of the methods presented in [6] and [5] and their parallel combination with a color quantization technique. In fact the additional step of color quantization is implemented in [5] by this paper. Other methods where a color quantization technique has been applied is [7] although no RGB similarity checking routine is implemented and [8] although still relatively slow for use in real-time sequences due to the complexity of the selected clustering Fuzzy-C-means algorithm.

In the proposed methods, as presented in [9], a color reduction technique is initially applied to the image and a smaller set of colors/clusters is created. This smaller set of colors is then divided into two main groups: Colors that a person with protanopia can perceive right and don't need modification and colors that a person with protanopia cannot perceive right and do need modification. The colors of the second set are modified using specific initialization daltonization parameters and are placed in a new set. Then to avoid confusion an rgb similarity checking module is applied, so that there are no similar colors between the set of colors created after the modification and the set of colors that remained intact. If a similarity is found then the daltonization parameters are iteratively modified until similarity is eliminated.

The proposed algorithms were implemented in the programming environment MATLAB. As mentioned, the ultimate objective of this paper is to minimize the total processing time under the perspective of future application in real-time sequences. For this reason as appropriate quantization method the technique used in [10] was selected, as simpler and faster to use and with relatively reliable and comparable to k-means results. The daltonization methods described in this paper are focused only on a specific type of color vision deficiency, namely protanopia.

2   Proposed Method

The original adaptive algorithm contains three steps [5]. Since the original adaptive algorithm is quite time consuming, in order to resolve this problem, the methods of this paper are proposed.

2.1   Simulation

Initially the number of the colors in the image is reduced using the color quantization method proposed in [10]. The resulting palette is an Nx3 matrix 'map', where the number of rows is equal to the number of quantized colors. Similarly to [5,6], the proposed methods follow the transformation matrices from RGB to LMS and vice versa. The improvement which is introduced in this paper is that in the methods proposed in this paper the simulation is applied only in the matrix 'map', and not in every pixel of the image [6]. After performing the simulation for each row in the matrix 'map' (i.e. N times total), the matrix 'map_sim' is produced, which is the color palette of the quantized color image as perceived by a person with protanopia.

For each row of matrix 'map', namely for every color of the color palette, the following steps are repeated:

Transformation of RGB Values to LMS Values. The transformation from RGB to LMS is achieved by a multiplication with matrix A_1 (1):

$$A_1 = \begin{bmatrix} 17.8824 & 43.5161 & 4.1193 \\ 3.4557 & 27.1554 & 3.8671 \\ 0.02996 & 0.18431 & 1.4670 \end{bmatrix} \tag{1}$$

The linear transformation from RGB to LMS is actually the multiplication described in (2):

$$[L \quad M \quad S]^T = A_1 \cdot [R \quad G \quad B]^T \tag{2}$$

Finding the Modified L'M'S' Values. Then a linear transformation [11] is applied, in order to reduce the normal color domain (LMS) to the protanope color domain $(L_pM_pS_p)$. The values that correspond to the loss of color information in the protanope vision are subtracted from the LMS values.

Reverse Transformation of L'M'S' Values to R'G'B' Values. The conversion from $L_pM_pS_p$ to $R_pG_pB_p$ is again achieved through a multiplication proposed in [11]. The new modified $R_pG_pB_p$ values demonstrate the way a protanope perceives each color represented by the corresponding row of the color palette.

After the execution of the described simulation sub-steps for every row of matrix 'map' (i.e. N times), the result is a matrix which represents the color palette of the color-quantized image, as it is perceived by a protanope.

2.2   Daltonization

It consists of the following sub-steps:

Calculation of the Palette's Error Matrix. The error matrix is the absolute value of the subtraction between the color palette of the quantized image and the color palette that emerged after the simulation (map - map_sim). Per row we have the following equations (3), (4), (5).

$$E_R = \left| R - R_p \right| \tag{3}$$

$$E_G = \left| G - G_p \right| \tag{4}$$

$$E_B = \left| B - B_p \right| \tag{5}$$

After performing the subtractions for all the rows of the palette (N times total), we get the table 'error_map'.

$$error\_map = \begin{bmatrix} \left| R_1 - R_1' \right| & \left| G_1 - G_1' \right| & \left| B_1 - B_1' \right| \\ \vdots & \vdots & \vdots \\ \left| R_N - R_N' \right| & \left| G_N - G_N' \right| & \left| B_N - B_N' \right| \end{bmatrix} \tag{6}$$

Creating Sets 'S1' and 'S2'. The first set is called 'S$_1$' and contains all the colors of the quantized image that do not need modification and are correctly perceived by a person with protanopia. The second set is called 'S$_2$' and contains the colors of the quantized image that require modification. The separation is based on (6) and (7).

$$rightPerceived\left(RGB, R_p G_p B_p \right) = \begin{cases} true \ if \ \left(E_R < c, \ E_G < c, \ E_B < c \right) \\ false \ otherwise \end{cases} \tag{7}$$

There is a subtraction of the components of a color before and after the simulation (RGB and $R_p G_p B_p$ respectively). If the subtraction of the three components is less than 20.48, then the RGB color is correctly perceived by a person with protanopia, not modified and placed in the set 'S$_1$'. Conversely, if the subtraction of even one of the three components is bigger than the value 20.48, then the RGB color is not properly perceived and placed in the set 'S$_2$' for daltonization. Basically we check if for all three values of each row of the error matrix (error_map) applies <0.08. Simultaneously the matrix 'error_S$_2$' is created. If a row of the color palette must be placed in 'S$_2$', then the corresponding row of the error matrix is added into the error matrix of 'S$_2$' (error_S$_2$).

Daltonization of 'S2'. Key to the daltonization procedure is not the error matrix of the whole color palette as in [6] and [5], but the error matrix only of those rows that need modification, namely the rows that belong to 'S$_2$' (error_S$_2$). Actually the values of the error matrix of 'S$_2$' are added to the initial values of the color palette of the quantized image, so that the values that indicate a high error level are redistributed to the blue side of the spectrum. This process is achieved with the help of the parameter matrix 'M' of [6]. In the proposed methods the parameter matrix 'M' is always initiated according to the values presented in [5], as seen in (8).

$$M = \begin{bmatrix} m_1 & m_2 & m_3 \\ m_4 & m_5 & m_6 \\ m_7 & m_8 & m_9 \end{bmatrix} = \begin{bmatrix} -1 & 0 & 0 \\ 1 & 1 & 0 \\ 1 & 0 & 1 \end{bmatrix} \tag{8}$$

Initially a multiplication of matrix 'M' with every row of 'error_S$_2$' is executed.

$$\begin{bmatrix} E_R' & E_G' & E_B' \end{bmatrix}^T = M \cdot \begin{bmatrix} E_R & E_G & E_B \end{bmatrix}^T \tag{9}$$

The modified values E_R', E_G', E_B' which arise after the multiplication in (9), are added to the values of the three components of the elements of 'S$_2$', resulting the final daltonized values R_d, G_d, B_d (10).

$$\begin{bmatrix} R_d & G_d & B_d \end{bmatrix} = \begin{bmatrix} R & G & B \end{bmatrix} + \begin{bmatrix} E_R' & E_G' & E_B' \end{bmatrix} \tag{10}$$

The colors R_d, G_d, B_d that result after the modification of the elements of 'S$_2$' are placed in a new set named 'S$_3$'.

2.3   RGB Similarity Checking Module

The RGB similarity checking module consists of the following sub-steps:

Simulation of 'S3'. For all the colors in 'S$_3$' a simulation is performed and the new colors which result from the simulation are placed in the set 'S$_{3\_sim}$'.

RGB Similarity Checking Module. Then the RGB similarity checking module is applied. It must be ensured that among new daltonized colors there is no color perceived by a person with protanopia, similar to a color that from the beginning remained intact. In other words, we must ensure that there is no similarity between the elements of the sets 'S$_1$' and 'S$_{3\_sim}$'. This requirement is technically expressed by equations (11), (12).

$$similar(RGB, R'G'B') = \begin{cases} true \ if \ (|R-R'|<d, |G-G'|<d, \ |B-B'|<d) \\ false \ otherwise \end{cases} \tag{11}$$

$$confusion(S_1, S_{3\_sim}) = \begin{cases} true \ if \ (\exists RGB \in S_1, R'G'B' \in S_{3\_sim} : similar(RGB, R'G'B')) \\ false \ otherwise \end{cases} \tag{12}$$

As 'd' we denote an appropriately chosen threshold, which in this paper has the same value as in [5]. As similar are considered all the combinations for which applies (r±10, g ±10, b±10). This value in MATLAB corresponds to 0.04 (in the range 0-1). We check the subtraction between the components of each color of 'S$_1$' and 'S$_{3\_sim}$' (RGB and R′G′B′ respectively). If the subtraction of the three components is below the value 0.04 then the two colors are perceived as identical. Conversely, if the subtraction of even one of the three components is bigger than the value 0.04 then the two colors are perceived as distinct. If even one color in 'S$_1$' is found similar to a color in 'S$_{3\_sim}$' then the procedure of daltonization is repeated with modified parameters of the matrix 'M'. At each subsequent iteration the parameter 'm$_4$' is reduced by amount 's', while the parameter 'm$_7$' increases by the same amount. In this paper we used the same value as in [5], namely s = 0.05. Thereby the redistribution of the red color in the blue channel increases, while the same amount of green is transferred to the blue channel.

Repetition if Necessary. For the repetition of the daltonization routine when the RGB similarity check locates a pair of similar colors between 'S$_1$' and 'S$_{3\_sim}$' two methods were developed in [9] and are presented in this paper.

Method 1: Daltonization by modification of entire 'S2' and 'error_S2'. In the first proposed method the entire modified matrix 'M' and the matrix 'error_S$_2$' are subjected to the daltonization routine. This method results into more rapid changes of the image, as each time that a pair of similar colors is found, the entire content of 'S$_2$' is being modified.

Method 2: Daltonization by modification of 'S2' and 'error_S2' per row. In the second proposed method the entire modified matrix 'M' and only one row of matrix 'error_S$_2$' are subjected to the daltonization routine. This row is the one that appears similar to some element of 'S$_1$'. The results of this method show milder coloration results in the final image.

Termination Condition. The iterative RGB similarity checking routine ends when there is no similar color found between 'S$_1$' and 'S$_{3\_sim}$' or when the parameter m$_4$ of matrix 'M' is ≤ 0.05.

Image Reconstruction. Before termination the algorithm proceeds to the creation of the final color palette 'map$_1$' and reconstructs the image by replacing in the original color palette map all the colors that belong to the set 'S$_2$' with the respective colors in 'S$_3$'. Finally, the color palette 'map$_1$' is subjected to a final simulation in order to obtain the final daltonized image, as it will be perceived by a person with protanopia.

3   Experiments

At first the algorithm was tested on the artificial image "test.png" for a 4-cluster quantization without dithering. It has to be noted that in this specific image the results of the algorithm remain the same even for a 256-cluster quantization. When the original image includes fewer colors than those specified from the user through the quantization method, the produced color palette will also include less colors and the final image will include all the colors of the original image. Furthermore, in this specific image both Method 1 and Method 2 produce the same results, since the error matrix contains only one row. The results are presented in Figure 1.

Subsequently, the simulation routine is executed for figure 1b and figure 1c is created, while a simulated color palette of figure 1b with the name 'map_sim' is produced (13). As shown in figure 1c, left '1' is almost invisible for a protanope, as the respective color for the background (color A) is now perceived as (69,69,205), very close to the values of the color of the left '1' (color B).

$$map\_sim = \begin{bmatrix} 69 & 69 & 205 \\ 193 & 193 & 255 \\ 73 & 73 & 203 \\ 255 & 255 & 255 \end{bmatrix} \tag{13}$$

$$error\_map = \begin{bmatrix} 141 & 18 & 1 \\ 0 & 0 & 0 \\ 0 & 0 & 0 \\ 0 & 0 & 0 \end{bmatrix} \tag{14}$$

At this point we calculate the error of the color palette 'map' by finding the absolute value of the difference between the palette 'map_sim' and the original palette 'map' of the quantized image. The result is stored in the matrix 'error_map'.

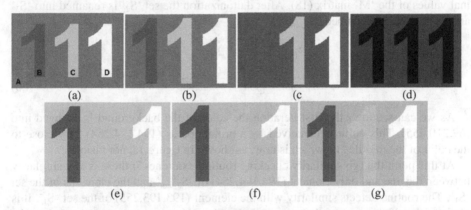

Fig. 1. (a) Original image 'test.jpg', (b) quantization of (a) to 4 colors, (c) protanope simulation of (b): left '1' is hardly visible, (d) error_map: the area of the image which presents high error value and needs daltonization appears red, (e) protanope simulation of (b) after the 1st iteration of the proposed algorithm: middle '1' is not visible, (f) protanope simulation of (b) after the 2nd iteration of the proposed algorithm: middle '1' remains imperceptible for protanopia, (g) protanope simulation of (b) after the 3rd iteration of the proposed algorithm: middle '1' becomes visible

Having defined the error palette, the algorithm continues with the classification of colors to those which are correctly perceived by a protanope and do not need daltonization (S_1) and to those which are not correctly perceived and do need daltonization (S_2). It is obvious in (14) that only the R component of the first row in the 'error_map' matrix shows a high value of error and therefore daltonization is necessary. In this way the colors of palette 'map' are divided into the sets 'S_1' and 'S_2', while the error matrix 'error_S_2'of the set 'S_2' is produced.

$$S_1 = \begin{bmatrix} 193 & 193 & 255 \\ 73 & 73 & 203 \\ 255 & 255 & 255 \end{bmatrix} \tag{15}$$

$$S_2 = \begin{bmatrix} 210 & 51 & 204 \end{bmatrix} \tag{16}$$

$$error\_S_2 = \begin{bmatrix} 141 & 18 & 1 \end{bmatrix} \tag{17}$$

$$M = \begin{bmatrix} m_1 & m_2 & m_3 \\ m_4 & m_5 & m_6 \\ m_7 & m_8 & m_9 \end{bmatrix} = \begin{bmatrix} -1 & 0 & 0 \\ 1 & 1 & 0 \\ 1 & 0 & 1 \end{bmatrix} \tag{18}$$

In fact the set 'S$_2$' contains the color A which is the color of the background, since that color is confused with the color B of the left '1' in the case of a protanope.

Thereafter the first iteration of the proposed daltonization algorithm is executed. The 'error_S$_2$' matrix (17) is subjected to the daltonization routine as well as the original values of the 'M' matrix (18). After daltonization the set 'S$_2$' is renamed into 'S$_3$' and its simulation is named 'S$_{3\_sim}$'.

$$S_3 = \begin{bmatrix} 69 & 210 & 255 \end{bmatrix} \tag{19}$$

$$S_3\_sim = \begin{bmatrix} 194 & 194 & 254 \end{bmatrix} \tag{20}$$

As we can see after the first iteration the color of the background is changed into (69,210,255). This color is perceived by a protanope as (194,194,254) very close to the color of the middle '1', which is now, as shown in figure 1e, not visible.

At this point the rgb similarity checking routine examines if there is any similarity between the one and only element (194,194,254) of 'S$_{3\_sim}$' and the elements of the set 'S$_1$'. The routine detects similarity with the element (193,193,255) of the set 'S$_1$'. It is obvious that if the step of the rgb similarity checking routine was omitted then the final image would be that of figure's 1e and the middle '1' would be invisible by a protanope. The parameters of the 'M' matrix are modified, as shown in (21). Thereafter comes the second iteration of the algorithm, by subjecting to the daltonization routine whether the whole 'error_S$_2$' matrix (Method 1) or only that row of the matrix which shows the similarity (Method 2). To the daltonization routine we also subject the new matrix 'M' with the modified parameters (21).

$$M_2 = \begin{bmatrix} m_1 & m_2 & m_3 \\ m_4 & m_5 & m_6 \\ m_7 & m_8 & m_9 \end{bmatrix} = \begin{bmatrix} -1 & 0 & 0 \\ 0.95 & 1 & 0 \\ 1.05 & 0 & 1 \end{bmatrix} \tag{21}$$

The results of the second iteration are described in the following:

$$S_3 = \begin{bmatrix} 69 & 203 & 255 \end{bmatrix} \tag{22}$$

$$S_3 \_ sim = \begin{bmatrix} 188 & 188 & 254 \end{bmatrix} \tag{23}$$

After the second iteration the color of the background is transformed into (69,203,255). This color is perceived by a protanope as (188,188,254). However, once again there is no significant difference compared to the color C of the middle '1' which, as shown in figure 1f, remains indistinguishable.

Once more the rgb similarity checking routine is activated and examines whether the color (188,188,254) of the new matrix 'S_{3_sim}' is noticeably different from the elements of the set 'S_1'. Once more a similarity with the element (193,193,255) of the set 'S_1' is detected. We should remind that according to the rgb similarity checking routine as similar we consider all the color combinations for which stands $(r\pm10, g\pm10, b\pm10)$. The parameters of matrix 'M' are once more adapted as shown in equation (24) which results to the third iteration of the algorithm.

$$M_3 = \begin{bmatrix} m_1 & m_2 & m_3 \\ m_4 & m_5 & m_6 \\ m_7 & m_8 & m_9 \end{bmatrix} = \begin{bmatrix} -1 & 0 & 0 \\ 0.9 & 1 & 0 \\ 1.1 & 0 & 1 \end{bmatrix} \tag{24}$$

The results of the third iteration of the algorithm are:

$$S_3 = \begin{bmatrix} 69 & 196 & 255 \end{bmatrix} \tag{25}$$

$$S_3 \_ sim = \begin{bmatrix} 182 & 182 & 254 \end{bmatrix} \tag{26}$$

As we see, after the third iteration of the proposed algorithm, the color of the background (A) is changed into the color (69,196,255) which a protanope perceives as (182,182,254).

Once more the rgb similarity checking routine examines if the shade (182,182,254) of 'S_{3_sim}' matrix is considerably different from the elements of the set 'S_1'. This time fortunately no similarities are found and the algorithm continues with the reconstruction of the color palette 'map' by replacing the elements of the set 'S_2' with the elements of the set 'S_3'. In this way the palette 'map_1' of the final image is produced (27) which is perceived by a protanope as shown in (28).

$$map_1 = \begin{bmatrix} 69 & 196 & 255 \\ 193 & 193 & 255 \\ 73 & 73 & 203 \\ 255 & 255 & 255 \end{bmatrix} \tag{27}$$

$$Final \_ sim = \begin{bmatrix} 182 & 182 & 254 \\ 193 & 193 & 255 \\ 73 & 73 & 203 \\ 255 & 255 & 255 \end{bmatrix} \tag{28}$$

Figure 2 demonstrates the result of applying both methods in Van Gogh's "Field of Poppies". It is evident that Method 1 has more a distinguishable effect from Method 2 and therefore it is considered more adequate to model the color difference for the specific image.

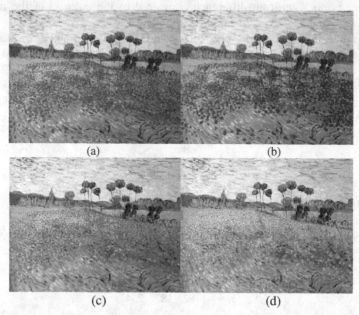

(a)   (b)

(c)   (d)

Fig. 2. For quantization to 256 colors: (a) Initial image, (b) Protanope perception, (c) Method 1: Protanope perception, (d) Method 2: Protanope perception

4   Experimental Results

The implementation of the proposed algorithm resulted in a very good computational time, which was drastically reduced compared to that of [5]. Indicatively, in an Intel Core Duo, 2 GHz, 2 GHz, each iteration of the proposed algorithm's Method 2 in MATLAB when applied in a 24 bit 400x300 image with quantization to 256 colors, lasts about 0.04 seconds.

Since the computational time show up to 15×10^{-3} variance in each execution of the algorithm and in order to determine more clearly if it is suitable for real-time processing in image sequences with the PAL/SECAM standard (25 frames/sec), a function that executes the algorithm 25 times iteratively was created. The results show that the application of Method 2 of the proposed algorithm with quantization to 256 colors, can achieve real-time processing in a live broadcasted video, as for the processing of 25 frames they remain under the threshold of 1 second.

From these results we also find that Method 2 of the proposed algorithm is slightly faster than Method 1 and this is especially evident in quantization with more than 128 colors. This is expected as in Method 1 many matrix multiplications are performed at each iteration of the rgb similarity checking module, in contrast to Method 2 where in each iteration only one multiplication is performed. Moreover, Method 1 after each iteration requires simulation of the entire 'S$_3$' matrix, since the entire 'S$_2$' changes, while in Method 1 simulation needs to be performed only at one row of 'S$_3$' matrix, namely the one that corresponds to that row in 'S$_2$' which was modified.

Consequently, the minimization of the overall processing time in order to establish real time processing is successfully achieved.

The proposed algorithm could be modified in the future so after the initial daltonization of the image, the iterative daltonization of the RGB similarity checking module would only be performed if there is similarity and adjacency of relatively large areas among similar colors of 'S$_1$' and 'S$_{3\_sim}$'.

In conclusion, the proposed algorithm presents several desirable features such as: (i) It is fast and deterministic (i.e. it always produces the same result for a given input image), (ii) It satisfies a global consistency property (i.e. all pixels of the same color in the original image will be mapped to the same recoloration in the final image), (iii) It preserves the original image luminance and natural appearence (i.e. false colors are avoided) and (iv) It can be efficiently implemented on modern GPUs.

5   Conclusions

In this paper an improved daltonization algorithm for protanopia is proposed. An intelligent iteration technique is suggested for the selection of the adaptation parameters, incorporating a color quantization technique. Computational time is drastically reduced. Consequently, the minimization of the overall processing time in order to establish real time processing in live broadcasted/streaming videos is successfully achieved.

References

1. Nam, J., Ro, Y.M., Huh, Y., Kim, M.: Visual Content Adaptation According to User Perception Characteristics. IEEE Trans. Multimedia 7(3), 2205, 435–445
2. Yang, S.: Roy. M.: Visual contents adaptation for color vision deficiency, Image Processing. In: Proc. Of Int. Conf. on Image Proc., vol. 1, pp. I - 453-6 (2003)
3. Kovalev, V.: Towards Image Retrieval for Eight Percent of Color-Blind Men. In: Proc. of 17th Int. Conf. on Pattern Recognition (ICPR 2004), vol. 2, pp. 943–946 (2004)
4. Martin, C.E., Keller, J.G., Rogers, S.K., Kabrisky, M.: Color Blindness and a Color Human Visual System Model. IEEE Trans. on Systems, Man, and Cybernetics—Part A: Systems and Humans 30(4), 494–500 (2000)
5. Anagnostopoulos, C., Anagnostopoulos, I., Tsekouras, G., Kalloniatis, C.: Intelligent modification for the daltonization process of digitized paintings. In: Proc. of 5th International Conference on Vision Systems, ICVS 2007, Bielefeld, Germany, March 21-24 (2007), accessible on line:
   http://biecoll.ub.uni-bielefeld.de/volltexte/2007/52/
6. Fidaner, O., Poliang, L., Ozguven, N.: Analysis of color blindness,
   http://scien.stanford.edu/pages/labsite/2005/psych221/
   projects/05/ofidaner/colorblindness_project.htm (last date of access: June 21, 2014)

7. Kuhn, G.R.: An Efficient Naturalness – Preserving Image – Recoloring Method for Dichromats. IEEE Trans. On Visualization and Computer Graphics 14(6), 1747–1754 (2008)
8. Doliotis, P., Tsekouras, G., Anagnostopoulos, C.A., Athitsos, V.: Intelligent Modification of Colors in Digitized Paintings for Enhancing the Visual Perception of Color-blind Viewers. In: Iliadis, L., Vlahavas, I., Bramer, M. (eds.) AIAI 2009. IFIP, vol. 296, pp. 293–301. Springer, Boston (2009)
9. Sgouroglou, P.: Modification of Colors in Colored Images for Enhancing the Visual Perception of Protanopes, Thesis, Hellenic Open University (2013)
10. Spencer, W.T.: Efficient Inverse Color Map Computation. In: Arvo, J. (ed.) Graphics Gems II. Academic Press, Boston (1991)
11. Vienot, F., Brettel, H., Mollon, J.: Digital Video Colourmaps for Checking the Legibility of Displays by Dichromats. Inc. Col. Res. Appl. 24(4), 243–252 (1999)

Extracting Knowledge from Collaboratively Annotated Ad Video Content

Manolis Maragoudakis[1], Katia Lida Kermanidis[2], and Spyros Vosinakis[3]

[1] Department of Information and Communication Systems Engineering,
University of the Aegean, 83200 Karlovasi, Samos, Greece
[2] Department of Informatics, Ionian University, 49100 Corfu, Greece
[3] Department of Product and Systems Design Engineering, University of the Aegean,
84100 Ermoupoli, Syros, Greece
kerman@ionio.gr, {mmarag,spyrosv}@aegean.gr

Abstract. Creative advertising support tools have relied so far on static knowledge represented in creativity templates and decision making systems that indirectly impose restrictions on the brainstorming process. PromONTotion is a system under development that aims at creating a support tool for advertisers for the creative process of designing a novel ad campaign that is based on user-driven, generic, automatically mined and thus dynamic semantic knowledge. Semantic terms and concepts are collaboratively provided via crowdsourcing. The present work describes data mining techniques that are applied to the collected annotations and some interesting initial results regarding ad content, ad genre, ad style and ad impact/popularity information.

1   Introduction

Creative advertising is the process of capturing a novel idea for an ad campaign and designing its implementation. As it constitutes one of the highest-budget enterprises today, several studies have been published regarding the impact of advertising (Amos et al., 2008; Aitken et al., 2008), as well as creativity in advertising (Hill and Johnson, 2004).

Several tools have been proposed in the literature for supporting the creative brainstorming process of ad design. A number of creativity support tools have been proposed, to help ad designers come up with novel ideas for setting up a new campaign. They usually focus on using creativity templates (Goldenberg et al., 1999), decision making systems, like ADCAD (Burke et al., 1990), linguistic wording schemata (Blasko and Mokwa, 1986), databases of predefined concepts and their associations, like IdeaFisher (Chen, 1999). Idea Expander (Wang et al., 2010) is a tool for supporting group brainstorming by showing pictures to a conversing group, the selection of which is triggered by the conversation context. Opas (2008) presents a detailed overview of several advertising support tools.

L. Iliadis et al. (Eds.): AIAI 2014 Workshops, IFIP AICT 437, pp. 85–95, 2014.
© IFIP International Federation for Information Processing 2014

Based on the belief that total freedom is not the most efficient way for enhancing the creative process, but constraining it with the use of a limited number of idea-forming patterns is, these tools force upon the advertiser a certain restricted way of thinking. Concepts, associations are static and predefined, and the user is guided through the problem entities to make connections between them. Required facts and rules are expert-dependent and usually quite hard to craft. Dictionaries and databases are static and non-expandable. According to Opas (2008), static, passive, expert-dependent knowledge models can hurt creativity.

PromONTotion is a creative advertising support tool that relies on generic, automatically acquired knowledge. The core of the tool is comprised of a semantic thesaurus, i.e. an ontology of terms and concepts that are related to the content, style, genre, artistic features, consumer impact and popularity of television ad videos. These terms and concepts are provided via crowdsourcing by players of a multi-player quiz/combat/action style computer game, House of Ads, developed especially for this purpose and described in detail in Kermanidis et al. (2013).

The present work describes in detail the mining experiments performed on an initial set of annotations. Mining techniques are applied to the provided data for the extraction of correlation information, of interdependencies between various features of the ad, of knowledge related to what increases the ad impact and popularity.

PromONTotion relies on very limited predefined knowledge, i.e. the ontology backbone, its hierarchical structure. The ad knowledge provided by the tool is data-driven and automatically derived making PromONTotion generic, dynamic, scalable, expandable, robust and therefore minimally restricting in the creative process and imposing minimal limitations to ideation or brainstorming.

This article is structured as follows: Section 2 presents a sketch of the ontological backbone used to identify key elements of the advertisement domain, while Section 3 describes the data mining approaches used upon collected data, in order to identify meaningful patterns and associations between objective ad parameters and subjective user feedback. Finally, the paper concludes in section 4.

2   Modeling the TV Ad Domain

The domain model in this work is represented as a hierarchical ontology of TV ad video concepts and terms. The ontological backbone structure is the only piece of knowledge that is manually crafted, by advertising domain experts. It contains terms and concepts that are related to the video content, the ad style and genre, its production values, its artistic features, as well as its impact and its impression print on consumers. The structure is scalable. Some representative product and services types are included currently, but the list is easily expandable. In detail, the ontology structure and categories are shown in Figure 1.

1. Artistic features
   (a) Sound
       (i) Music/Song recognisability
       (ii) Song Music type
   (b) Filming
       (i) Photography
       (ii) Style
2. Location
   (a) Indoors
   (b) Outdoors
3. Ad impact
   (a) Convincing power
   (b) Opinion
   (c) Improvement suggestions
4. Production
   (a) Producer
   (b) Director
   (c) Production quality
5. Participating elements
   (a) Main character
       (i) Recognisability
       (ii) Type
           (1) Human
               (a) Gender
               (b) Age
               (c) Occupation
           (2) Animal
           (3) Inanimate
   (b) Key participants

6. Message communication
   (a) Ad structure
   (b) Linguistic schemata
   (c) Indirect Critique on competition
   (d) Humoristic elements
   (e) Tag lines
   (f) Brand name
7. Product type
   (a) Product
       (i) Food
       (ii) Beverage
       (iii) Electric device
           (1) Device type
           (2) Energy class
       (iv) Electronic device
       (v) Store
       (vi) Vehicle
           (1) Type
           (2) Value
       (vii)   Household
   (b) Service
       (i) Telecommunications
       (ii) TV
       (iii) Banking
       (iv) Insurance
       (v) Healthcare
8. Target group
9. Product origin

Fig. 1. The Ontology backbone

Following the trend of serious games paved by von Ahn (2006) and game design for ontology populating (Siorpaes and Hepp, 2008), the *House of Ads* multiplayer browser-based action game (Kermanidis et al., 2013) was designed and developed for collaboratively providing ad content annotations, i.e. for ontology populating.

3   Mining through Advertisement Terms

An initial set of annotations of 74 videos were used for performing preliminary mining experiments. The data were provided by 21 players. For the preliminary experiments 20 different ad videos were selected and every player was assigned the annotation of 3 to 5 videos.

Exploratory as well as predictive data mining is applied to the collected data, for detecting patterns and interdependencies among ad features, as well as their association with the ad's impact on consumers/players and its popularity. All categories proposed by the domain experts are considered features (41 in number) that define the ad domain space, and every ad video constitutes a data instance. It needs to be noted that 13 out of the 41 features present a missing value percentage of more than 80%, due to the hierarchical structure of the ontology. The input of lower-level annotations depends on the corresponding value of the parent node in the hierarchy. If the specific parent value is not provided, values of the sub-nodes are missing.

3.1   Mining Associations

The Apriori algorithm (Liu et al., 1998) on the Weka[1] machine learning workbench was used for extracting association rules from the data. Given the small number of distinct videos involved in the preliminary experiments, rules concerning only the objective (content/style/genre) data regarding the ad, e.g.

```
Service=telecom=>GenderOfMainCharacter=male(Confidence=1)
TargetGroup=women=>Location=inside(Confidence=1)
```

have not been given attention. Although they are interesting. the small number of videos does not allow for their generalization. Rules related to the subjective data, however, reveal interesting findings. For example, they show an absolute correlation between the convincing ability of the ad and how the player liked the ad:

```
Opinion=LikeNo => Convincing=No (Confidence=1)
Opinion=LikeALittle => Convincing=So-So (Confidence=1)
Opinion=LikeALot => Convincing=ALot (Confidence=1)
```

A strong correlation between these two features was expected, but the absolute correlation constitutes an interesting finding. These rules emerged after removing attributes (e.g. ProductType) that led to trivial, uninteresting rules, like

```
Service=telecom => ProductType=service
```

3.2   Classification

Predictive experiments were run on the Weka platform for identifying the correlation of the various ad features with the ad's impact on the players. To this end the *Opinion* feature denoting how the player liked the ad (with values *No, ALittle, Considerably, ALot*), the *Convincing* feature, indicating how persuaded the player felt by the ad

---

[1] http://www.cs.waikato.ac.nz/ml/weka/

(with values *No, SoAndSo, ALot*), and the *Changes* feature, that lists the ad features that the player does not like and would change if he could (with values *Everything, Nothing, TheCharacters, TheLocation, TheContent, TheMusic, TheGenre, Other*), were used as class labels. The main goal of this task is to focus on exploratory data analysis and reveal interesting interdependencies. As the following experiments show, the data is too few, the percentage of missing values too high and the feature space too large to expect high accuracy predictive performance at this phase.

Due to their comprehensibility and their ability to cope with missing values using distribution-based imputation (Saar-Tsechansky and Provost, 2007), decision trees are the first algorithm employed for inducing a classification model. Support Vector Machines (Cortes and Vapnik, 1995) were also experimented with, as they have been proven to perform well on problems with few learning items in large feature spaces. The Sequential Minimal Optimization algorithm (Platt, 1998) is used for training the SVM classifier and a third degree polynomial kernel function is chosen. Tables 1, 2 and 3show the results for the three class attributes and the two classification algorithms. All experiments were run using 10-fold cross validation.

For the Changes attribute, the results for the remaining values are zero, or close to zero, due to their sparse occurrence in the data. In the last line of Table 3, the average precision and recall value over all class values is shown instead.

Table 1. Results for the *Opinion* class

Opinion Attribute	C4.5		SVMs	
	Precision	Recall	Precision	Recall
No	0.389	0.467	0.462	0.4
ALittle	0.524	0.55	0.464	0.65
Considerably	0.35	0.412	0.429	0.353
ALot	0.8	0.545	0.632	0.545

Table 2. Results for the *Convincing* class

Convincing Attribute	C4.5		SVMs	
	Precision	Recall	Precision	Recall
No	0.579	0.524	0.632	0.571
SoAndSo	0.718	0.7	0.727	0.8
ALot	0.5	0.615	0.636	0.538

Table 3. Results for the *Changes* class

Changes Attribute	C4.5		SVMs	
	Precision	Recall	Precision	Recall
Everything	0.357	0.455	0.3	0.273
Nothing	0.676	0.758	0.5	0.848
Content	0.364	0.364	0	0
Average (all classes)	0.449	0.5	0.268	0.419

Keeping in mind that the available learning items in these preliminary experiments are very few, the number of features is significant, the percentage of missing values is very high and the classification task is not binary, these results at this stage are quite promising for the experiments to be conducted when the annotation phase has progressed and learning items have accumulated.

The 'simple' SVMs classifier used is in most cases outperformed by C4.5. This is largely attributed to the missing values problem. It has been reported in the literature (Pelckmans et al., 2005) that SVMs require additive models (e.g. component wise SVMs and LS-SVMs) to cope adequately with missing values.

The attributes that are significant for predicting the convincing power of the ad, included in the decision tree model, include (except the opinion of the player) the fame of the main character (famous characters seem to be more convincing) and the genre of the ad (cartoon ads seem to imprint their message more efficiently on consumers compared to realistic genre ads). Regarding the prediction of the player's opinion of the ad, except for the convincing attribute, the location of the ad story (urban environment, home, country), as well as the artistic features (e.g. music recognition) play an important role. Finally, regarding the prediction of changes the players propose, the opinion is understandably the most important feature, with a negative correlation. The use of taglines and famous logos is also strongly correlated to the prediction of the content value of changes, possibly indicating that they are a communication tool consumers get easily tired of.

3.3   Bayes Simulation

For the purpose of our study, Bayesian Networks (BN) are used for modelling of a marketing process. More specifically, subjective advertisement attributes, such as User Opinion, Convincing Level and Changes Required are modelled in parallel with objective attributes, as thoroughly described in Section 2. Unlike classification analysis, BN modelling can support two semantic operations, namely diagnostic inference (i.e. from effects to causes) and causal inference (i.e. from causes to effects) (Friedman and Goldszmidt, 1996). As regards the former operation, we are mostly interested in providing the most probable value for the subjective attributes, given the values of remaining objective ones and benchmark its performance against the well-known classification algorithms. Concerning the latter operation, the aim is to model the dependency between input and output parameters, and based on the extracted modelling outcome, calculate the objective input parameters that guarantee an optimal subjective output (i.e. with a maximal probability). The aforementioned task can be accomplished by entering the intended output as evidence and then calculate the most probable instantiation of the remaining input parameters that caused the intended output.

3.3.1. Diagnostic Inference
In this setting, a BN was built from the initial dataset, using the K2 learning approach (Cooper and Herskovits, 1992). Upon, learning, the BN was used in order to generate

data instances. The generation process is as follows: a node without parents is selected as a starting point. According to its conditional probability table (CPT), a value is drawn using a roulette wheel strategy. This value is passed to the child node and again, using the CPT from the BN, a value for that child node is also selected. This process is progressing to all nodes in the BN until a sample containing values for all attributes is generated. The new, augmented dataset was evaluated using a variation of Naïve Bayes classifier which incorporates elements from Bayesian Networks, namely a variation of Tree-Augmented Naïve Bayes (TAN) algorithm (Friedman et al., 1997). In our approach, the class node is kept separately from the other, attribute nodes. A BN is learned and then the class node is placed as the parent of all other nodes in order to favor classification. For comparative analysis, tables 4, 5 and 6 show classification results of TAN against C4.5 and SVMs (parameter setup identical to the previous experiments) with the simulated dataset.

Table 4. Classification results with the simulated dataset for the *Opinion* class

Opinion Attribute	C4.5		SVMs		TAN	
	Precision	Recall	Precision	Recall	Precision	Recall
No	0.553	0.638	0.556	0.542	0.545	0.602
ALittle	0.452	0.207	0.328	0.321	0.423	0.342
Considerably	0.386	0.686	0.354	0.329	0.401	0.312
ALot	0.808	0.619	0.569	0.623	0.762	0.687

Table 5. Classification results with the simulated dataset for the *Convincing* class

Convincing Attribute	C4.5		SVMs		TAN	
	Precision	Recall	Precision	Recall	Precision	Recall
No	0.691	0.732	0.649	0.626	0.661	0.673
SoAndSo	0.797	0.723	0.689	0.711	0.691	0.715
ALot	0.665	0.784	0.6	0.577	0.701	0.873

Table 6. Classification results with the simulated dataset for the *Changes* class

Changes Attribute	C4.5		SVMs		TAN	
	Precision	Recall	Precision	Recall	Precision	Recall
Everything	0	0	0.153	0.182	0.134	0.152
Nothing	0.436	1	0.447	0.522	0.356	0.466
Content	0	0	0.172	0.181	0.231	0.199
Average (all classes)	0.19	0.436	0.259	0.292	0.232	0.243

The TAN classifier is considered superior than Naïve Bayes (Friedman et al., 1997), mainly because of the fact that input attributes are not considered as independent among each other, given the class. On the contrary, the input attributes in a TAN representation form a BN that could explain possible inter-relations found in data.

From the above results, it is clear that *Convincing*, when used as the class, could provide the most satisfactory outcomes, compared with the other two attributes,

Opinion and *Changes*. The TAN methodology performs slightly better than SVM and is very close to C4.5, although in some cases TAN seem to outperform both learners.

3.3.2. Causal Inference

The aim of this experiment is to model the dependency between input and output parameters, and based on the extracted modeling outcome, to calculate the input parameters that guarantee an optimal output with a maximal probability. The structure of the BN allows for such an operation by entering target values on selected nodes and then estimating the most probable instantiation of the remaining input parameters that provides the entered output.

The exact scenario is as follows: a domain expert is interested in observing the effect that *Opinion* and *Convincing* have to the other variables, given that *Changes* is known. The first scenario (optimistic) assumes that *Opinion=Alot* and *Convincing=Alot* given Changes=**nothing**, while the second scenario (pessimistic) considers the hypothesis that *Opinion=No* and *Convincing=No* given Changes=**everything.**

By measuring the progression of the values for each input variable given the scenario assumptions, one can draw significant conclusions on the effect of each parameter. For reasons of readability, only the variables that belong to the Markov Blanket of the nodes *Opinion* and *Convincing* are shown in Tables 7 and 8. The Markov Blanket of a node contains the set of parent, children and children's parent of that node. It is proven that any node that is not within the Markov Blanket, is not influencing or not being influenced by that node. Table 7 tabulates the most probable attribute instantiations for the input attributes, given evidence described in scenario 1, while Table 8 shows the same information given the evidence representing the pessimistic scenario.

Taking a closer look at the first scenario, one could observe that when a user has a positive opinion and states that an advertisement is convincing, the most probable product being advertised is a drink. Moreover, the main character is human, middle aged and not a famous actor or celebrity. Since the most probable product is a drink, the advertisement does not promote something novel and moreover, the location is indoor and therefore the photography is neither picturesque nor landscape. Finally, the advertisement is an isolated episode and there are no famous utterances used.

On the contrary, using the pessimistic scenario, in which a user holds a negative opinion and is not persuaded by the advertisement, one could observe that the product is not listed (note: the listed products were: electronic services, clothes, drink, electronic device, household, vehicle and store). An interesting remark is that despite what most would assume, the presence of famous persons, young in age and the inclusion of a humoristic scenario do not guarantee user satisfaction in terms of convincement and opinion. On the other hand, these instantiations seem to cause a negative impact. Finally, it is noteworthy that in a pessimistic scenario, the location is a workplace.

Table 7. The most probable attribute instantiations for the optimistic scenario. (i.e. *Opinion=Alot* and *Convincing=Alot* given Changes=**nothing**)

Attribute	Value
Product	Drink
Genre	Realistic
MainCharacter	Human
Fame	Everyday character
Age	MiddleAged
Occupation	Employee
Humor	noHumor
SongRecognizable	Yes
Novelty	Known product/service
Location	Indoor
Photography	Other
Structure	Isolated episode
FamousLines	No famous lines

Table 8. The most probable attribute instantiations for the pessimistic scenario. (i.e. *Opinion=No* and *Convincing=No* given Changes=**everything**)

Attribute	Value
Product	Other
Genre	Other
MainCharacter	Human
Fame	Famous person
Age	Young
Occupation	Other
Humor	Humoristic scenario
SongRecognizable	No
Novelty	New product/service
Location	Workplace
Photography	Notvalid
Structure	Isolated episode
FamousLines	No famous lines

4   Conclusion

In this paper we described the results of some preliminary mining experiments that were performed on an initial set of ad content annotations. More specifically, Association Discovery, Classification and Bayesian network analysis were incorporated in order to examine potentially meaningful patterns. The major obstacle that the limited amount of training data was confronted by training a Bayesian network and then applying a simulation step that generated a plethora. Experiments with the initial and the augmented dataset using various classifiers showed that classification accuracy may be improved using Bayesian modelling. Moreover, causal inference was also applied using the BN structure, simulating two different scenario settings, which revealed some useful insight on input attributes and their impact on subjective attributes such as User opinion and Convince level.

More interesting and confident findings will be revealed after processing an annotation set of significant size, when the annotation process of PromONTotion has progressed. The use of mining algorithms that are better suited to cope with data consisting of a significant percentage of missing values is also planned for future research.

Acknowledgements. This Project is funded by the National Strategic Reference Framework (NSRF) 2007-2013: ARCHIMEDES III – Enhancement of research groups in the Technological Education Institutes.

References

Aitken, R., Gray, B., Lawson, R.: Advertising Effectiveness from a Consumer Perspective. International Journal of Advertising 27(2), 279–297 (2008)

Amos, C., Holmes, G., Strutton, D.: Exploring the Relationship between Celebrity Endorser Effects and Advertising Effectiveness. International Journal of Advertising 27(2), 209–234 (2008)

Blasko, V., Mokwa, M.: Creativity in Advertising: A Janusian Perspective. Journal of Advertising 15(4), 43–50 (1986)

Burke, R., Rangaswamy, A., Wind, J., Eliashberg, J.: A Knowledge-based System for Advertising Design. Marketing Science 9(3), 212–229 (1990)

Chen, Z.: Computational Intelligence for Decision Support. CRC Press, Florida (1999)

Cooper, G.F., Herskovits, E.: A Bayesian Method for the Induction of Probabilistic Networks from Data. Machine Learning 9 (1992)

Cortes, C., Vapnik, V.: Support-vector Networks. Machine Learning 20(3), 273–297 (1995)

Friedman, N., Goldszmidt, M.: Building Classifiers and Bayesian Networks. In: Proceedings of the National Conference on Artificial Intelligence. AAAI Press, Menlo Park (1996)

Friedman, N., Geiger, D., Goldszmidt, M.: Bayesian Network Classifiers. Machine Learning 29, 131–163 (1997)

Goldenberg, J., Mazursky, D., Solomon, S.: The Fundamental Templates of Quality Ads. Marketing Science 18(3), 333–351 (1999)

Hill, R., Johnson, L.: Understanding Creative Service: A Qualitative Study of the Advertising Problem De-lineation, Communication and Response (APDCR) Process. International Journal of Advertising 23(3), 285–308 (2004)

Kermanidis, K., Maragoudakis, M., Vosinakis, S., Exadaktylos, N.: Designing a Support Tool for Creative Advertising by Mining Collaboratively Tagged Ad Video Content: The Architecture of PromONTotion. In: Papadopoulos, H., Andreou, A.S., Iliadis, L., Maglogiannis, I. (eds.) AIAI 2013. IFIP AICT, vol. 412, pp. 10–19. Springer, Heidelberg (2013)

Lee, M., Shen, H., Huang, J.Z., Marron, J.S.: Biclustering Via Sparse Singular Value Decomposition. Biometrics 66, 1087–1095 (2010)

Liu, B., Hsu, W., Ma, Y.: Integrating Classification and Association Rule Mining. In: Fourth International Conference on Knowledge Discovery and Data Mining, pp. 80–86 (1998)

MacCrimmon, K., Wagner, C.: Stimulating Ideas Through Creativity Software. Management Science 40(11), 1514–1532 (1994)

Opas, T.: An Investigation into the Development of a Creativity Support Tool for Advertising. PhD Thesis. Auckland University of Technology (2008)

Pelckmans, K., de Brabanter, J., Suykens, J.A.K., de Moor, B.: Handling Missing Values In Support Vector Machine Classification. Neural Networks 18(5-6), 684–692 (2005)

Platt, J.: Sequential Minimal Optimization: A Fast Algorithm for Training Support Vector Machines. Advance. In: Kernel Methods-Support Vector Learning (1998)

Saar-Tsechansky, M., Provost, F.: Handling Missing Values When Applying Classification Models. Journal of Machine Learning Research 8, 1625–1657 (2007)

Siorpaes, K., Hepp, M.: Games With A Purpose For The Semantic Web. In: IEEE Intelligent Systems, pp. 1541–1672 (2008)

Vapnik, V.: The Nature of Statistical Learning Theory. Springer (1995) ISBN 0-387-98780-0

von Ahn, L.: Games with a Purpose. IEEE Computer 39(6), 92–94 (2006)

Wang, H.C., Cosley, D., Fussell, S.R.: Idea Expander: Supporting Group Brainstorming with Conversa-tionally Triggered Visual Thinking Stimuli. In: Proceedings of the ACM Conference on Computer Supported Co-operative Work (CSCW), Georgia, USA (2010)

Mining Biological Data on the Cloud – A MapReduce Approach

Zafeiria-Marina Ioannou[1], Nikolaos Nodarakis[1], Spyros Sioutas[2],
Athanasios Tsakalidis[1], and Giannis Tzimas[3]

[1] Computer Engineering and Informatics Department, University of Patras,
26500 Patras, Greece
{ioannouz,nodarakis,tsak}@ceid.upatras.gr
[2] Department of Informatics, Ionian University,
49100 Corfu, Greece
sioutas@ionio.gr
[3] Computer & Informatics Engineering Department,
Technological Educational Institute of Western Greece, 26334 Patras, Greece
tzimas@cti.gr

Abstract. During last decades, bioinformatics has proven to be an emerging field of research leading to the development of a wide variety of applications. The primary goal of bioinformatics is to detect useful knowledge hidden under large volumes biological and biomedical data, gain a greater insight into their relationships and, therefore, enhance the discovery and the comprehension of biological processes. To achieve this, a great number of text mining techniques have been developed that efficiently manage and disclose meaningful patterns and correlations from biological and biomedical data repositories. However, as the volume of data grows rapidly these techniques cannot cope with the computational burden that is produced since they apply only in centralized environments. Consequently, a turn into distributed and parallel solutions is indispensable. In the context of this work, we propose an efficient and scalable solution, in the MapReduce framework, for mining and analyzing biological and biomedical data.

Keywords: Bioinformatics, Data mining, Text mining, Clustering, MapReduce, Hadoop.

1 Introduction

The term *data mining* [9] refers to the process of data analysis to identify interesting and useful information and knowledge from huge collections of data. From a more general and more intuitive aspect, data mining is a research field in computer science that employs a wide diversity of well-established statistical and machine learning techniques, such as neural networks, to derive hidden correlations, patterns and trends from large datasets. Bioinformatics, is a prominent domain, among several others, where the existing data mining algorithms are applicable and enhance the process of knowledge discovery. Especially, as

L. Iliadis et al. (Eds.): AIAI 2014 Workshops, IFIP AICT 437, pp. 96–105, 2014.
© IFIP International Federation for Information Processing 2014

the volume of biological and biomedical data accumulated in large repositories continues to expand at exponential rates, data mining techniques are of critical importance in extracting knowledge efficiently from such datasets. Adding the need to manage heterogeneous data, that in the vast majority are unstructured biomedical text documents (Biomedical Text Mining [1]), and automate the exploration procedure of them, it is easy to understand why data mining plays a fundamental role in bionformatics domain [22].

Despite how efficient a data mining technique might be, as the volume of data collections continues to expand at some point it will be impractical to use due to limits in resources posed by the centralized environment where the algorithm is executed. Typical Biomedical Text Mining tasks include automatic extraction of protein-protein interactions, named entity recognition, text classification and terminology extract. Consider PubMed[1], which is the most widely used biomedical bibliographic text base with millions of records and grows by a rate of 40,000 publications per month. To perform such tasks in an enormous corpus like PubMed is unthinkable. Most existing methods in literature [4,5,10,11,15,16] apply to a few hundreds or thousands of records. As a result, high scalable implementations are required. Cloud computing technologies provide tools and infrastructure to create such solutions and manage the input data in a distributed way among multiple servers. The most popular and notably efficient tool is the *MapReduce* [6] programming model, developed by Google, for processing large scale data.

The method proposed in the context of this work, is under development and is an extension of the work in [13] which proposes an automatic and efficient clustering approach that performs well on multiple types of biological and biomedical data. In [13], two mining tools were developed, Bio Search Engine which is a text mining tool working with biomedical literature acquired from PubMed and Genome-Based Population Clustering tool which extracts knowledge from data acquired from FINDbase [8,20,21]. FINDbase[2] is an online resource documenting frequencies of pathogenic genetic variations leading to inherited disorders in various populations worldwide. In this paper, we take the data mining technique proposed in [13] one step further and adapt it to the needs of big data analysis. We propose a novel and effective data mining technique for extracting valuable knowledge from biological and biomedical data in the cloud. We focus only on Biomedical Text Mining, since the size of PubMed is adequately big and fit the needs of MapReduce model in contrary to the size of existing genetic and mutation databases (like FINDbase). Our approach uses *Hadoop* [18,23], the open source MapReduce implementation, and *Mahout* [19] which is a scalable machine learning library built on top of Hadoop.

The remainder of the paper is organized as follows: Section 2 discusses related work, Section 3 provides a full analysis of the algorithm and proceeds into a detail examination of all of its steps and finally Section 4 concludes the paper and presents future steps.

---

[1] http://www.ncbi.nlm.nih.gov/pubmed/
[2] http://www.findbase.org

2 Related Work

2.1 Biomedical Text Mining

Biomedical Text Mining or BioNLP is the field of research that deals with the automatic processing, retrieval and analysis of scientific texts and more generally literature from the biomedical domain by applying text mining techniques aimed at uncovering previously unknown knowledge. Currently, there has been significant research advances in the area of biomedical text mining, including named entity recognition, text classification, terminology extraction, relationship extraction and hypothesis generation [4].

Text Document clustering, unlike text document classification, is an unsupervised learning process that does not depend on prior knowledge or domain expertise. In particular, document clustering is the task where similar documents are grouped into clusters. Existing biomedical text mining systems that cluster results into topics [15] include GOPubMed[3], ClusterMed[4] and XplorMed[5]. GOPubMed uses both Medical Subject Headings (MeSH) terms and Gene Ontology (GO) in order to organize the search results and, thus, to enhance the user navigation and the search possibilities. It is also capable of sorting results into four categories: "what", "who", "where" and "when". Another prominent example is ClusterMed that also employs clustering in six different ways: i) Title, Abstract and MeSH terms, ii) Title and Abstract, iii) MeSH terms only, iv) Author names, v) Affiliations, vi) Date of publication. XplorMed organizes results by MeSH categories, extracts topic keywords and their co-occurrences and furthermore it provides an interactive navigation through abstracts. For a comprehensive survey of such biomedical text mining systems along with their various characteristics and features, one can consult [4,5,15,16].

It should be noted that besides clustering, there are a handful of other fruitful techniques that have been applied to mine biological text data, that either deviate apart from the bag-of-words model and the tf-idf representation or employ other learning techniques different from clustering. From these approaches, worth to be mentioned are approaches based on the incorporation of semantic information and ontologies in order to correctly disambiguate the meaning of various terms. We could also refer to the employment of second order n-gram Markov models, probabilistic suffix analysis, and named entity recognition. Furthermore special mentioning deserves automatic term recognition techniques, that recognize domain concepts and using automatic term restructuring technique permit the text content organization into knowledge structures (terminologies). More details for these techniques can be found in [1].

2.2 Hadoop and MapReduce Framework

In the context of this work, we focus on the use of *Hadoop* which is an open source framework for managing large-scale data and running computationally heavy

---

[3] http://www.gopubmed.org/web/gopubmed/
[4] http://clustermed.info/
[5] http://www.ogic.ca/projects/xplormed/

tasks in a parallel and distributed manner. Hadoop follows *MapReduce* principles and is designed to scale up from single servers to thousands of machines, each offering local computation and storage. Here we describe in more details how it works.

A task performing a specific computation in Hadoop is called MapReduce *job*. The input data for each job is split and distributed among the nodes consisting the cluster and then is processed by a number of tasks executed on these nodes. The categories in which these tasks belong are defined as *Map* and *Reduce*. Each Map task receives input data and processes them by calling a user-defined *Map* function which outputs a set of intermediate key-value pairs. After that, a Shuffle process groups all intermediate values associated with the same key I and each group is assigned to the corresponding Reduce task. Each Reduce task also calls a user-defined *Reduce* function which accepts an intermediate key I and a set of values for that key and outputs the final set of key-value pairs. The intermediate values are supplied to the Reduce function via an iterator in order for the system to handle lists of values that are too large to fit in memory. Using the MapReduce programming model, the user does not have to worry about how the program is executed over the cluster or designing fault tolerance protocols in case a node fails. The system handles these issues itself thus allowing the user to focus on his own problem exclusively.

Mahout is a scalable machine learning library that consists of a set of data mining algorithms (e.g. k-means clustering, naive Bayes classifier etc.) implemented in the Hadoop MapReduce framework.

3 The Proposed Mining Algorithm

In the following subsections, we present in detail the steps of the proposed data mining technique. The solution consists of a combination of MapReduce jobs that run either on the Hadoop or Mahout framework. The basic steps of the algorithm are: 1) Preprocessing of the data collection in order to represent them in a more processable structure, 2) Latent Semantic Indexing and 3) Coarse Clustering in an attempt to generate an initial partition of data, 4) Agglomerative Hierarchical Clustering [17] in order to reduce the number of the initial clusters (in case that the number is relatively large) and 5) spherical k-means algorithm [7,14] in order to enhance the quality of the proposed clustering. Each of the aforementioned steps is a MapReduce job (or a series of MapReduce jobs) and the output of each MapReduce job feeds the input of the next job in the sequence. Note that we combine hierarchical and non-hierarchical clustering techniques to improve the efficiency and accuracy of clustering [3,24]. An overview of the algorithmic process is depicted in Figure 1.

3.1 Preprocessing the Data Collection

Data preprocessing is an essential step in the data mining process that includes feature selection and representation of data. The data elements have to be transformed into a representation suitable for computational use. For this purpose,

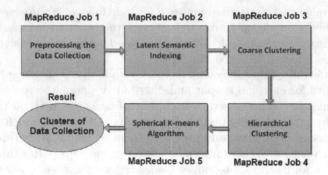

Fig. 1. Overview of MapReduce Clustering Algorithm

we display a method for preprocessing the corpus of PubMed which constitutes the input dataset of our mining method.

The bag-of-words model is the most widely used document representation adopted in the text clustering algorithms. According to the principles of this model, each document is represented as a sequence of terms/words. Initially, we parse each document and distinguish the lexical units (tokens) that constitute it. After this, we proceed two kinds of feature reduction. At first, we remove the stopwords. A stopword is defined as a term which is not thought to convey any meaning as a dimension of vector space. Typically, a compilation of stopwords consists of very common words such as articles, preposition etc. By removing them, we achieve a reduction in the dimensionality of the index by 20-30%. The second step involves Part-Of-Speech Tagging (POS Tagging) and lemmatization. POS Tagging is the process of assigning a particular part of speech (e.g. noun, verb, adjective, etc.) to each term of a document while lemmatization is the process of reducing the words to their basic form (lemma). Consequently, we end up with the unique lemmas of noun words of all documents. To put through the POS Tagging process we use the GENIA Tagger[6] which is specifically tuned for biomedical texts.

Subsequently, according to the vector space model [2], each document is represented as a vector of m dimensions, where m is the number of unique lemmas. For each document, we measure the significance of each lemma in its content using the TF-IDF (Term Frequency - Inverse Document Frequency) scheme. The term frequency is simply the number of times a lemma appears in a document, whereas the inverse document frequency is a measure obtained by dividing the total number of documents by the number of the documents containing the term. So, more formally, we define $TF - IDF(t, d, D) = TF(t, d) \cdot IDF(t, D)$, where $IDF(t, D) = \log \left(\frac{|D|}{|\{d \in D : t \in d\}|} \right)$. Consequently, lemmas that appear frequently in a document but have low document frequency in the whole data collection are given a high value in the TF-IDF scheme.

---

[6] http://www.nactem.ac.uk/tsujii/GENIA/tagger/

In order to preprocess the data in the way we described, we developed a MapReduce Job. The Map function takes as input the data collection of PubMed, proceeds to POS Tagging of the terms and extracts the lemmas of noun words. For each lemma, it outputs a key-value pair in the form <lemma, d, 1>, where lemma is the key and the value composes of number 1 and the name of document that contains the lemma. The Reduce function retrieves the unique lemmas and calculates the TF-IDF value for each document they belong. Next, it outputs key-value pairs in the format <lemma, tfidf, d>, where lemma is the key and the value composes of the TF-IDF weight and the name of the document that is binded with the lemma and the TF-IDF value. The overall outcome of this job is a term-document matrix A of dimension $m \times n$, where m is the number of unique lemmas and n is the number of documents.

3.2 Latent Semantic Indexing

Latent Semantic Indexing (LSI) [2] is a prominent indexing and retrieval method for extracting hidden relationships between terms and concepts contained in large, unstructured collections of documents. The main idea of the method is the projection of document vectors in a new low-dimensional space through the Singular Value Decomposition (SVD) of the term-document matrix A. This can significantly reduce the number of computations needed in the next steps of our data mining algorithm.

The Singular Value Decomposition of an $m \times n$ matrix A of rank r expresses A as a product of three simpler matrices, $A = USV^T$ where $S = diag(\sigma_1, ..., \sigma_r)$ is a diagonal matrix containing the set of singular values, $U = (u_1, ..., u_r)$ is an $m \times r$ matrix whose columns are orthonormal and $V = (v_1, ..., v_r)$ is an $n \times r$ matrix which is also column-orthonormal. LSI omits all but the k largest singular values in the above decomposition, for some appropriate k which will be the dimension of the low-dimensional space referred to in the description above. It should be small enough to enable fast retrieval and large enough to adequately capture the structure of the corpus. Consequently, after SVD and dimension reduction, a matrix $A_k = U_k S_k V^T{}_k$ is created, where $S_k = diag(\sigma_1, ..., \sigma_k)$, $U_k = (u_1, ..., u_k)$ and $V_k = (v_1, ..., v_k)$.

To this end, we utilize the Singular Value Decomposition algorithm implemented in Mahout. More specifically, Mahout implements a version of Lanczos algorithm[7] as a series of MapReduce jobs. We provide as input to the algorithm the matrix A, the number of row and columns and the rank of the output matrix and it produces the V_k and S_k matrices, needed for the next step of our solution.

3.3 Coarse (initial) Clustering

Fuzzy Clustering [12] has been widely used in the area of information retrieval and data mining. In traditional Fuzzy Clustering Methods, data elements can belong to more than one clusters, and each data element is associated to every

---

[7] http://en.wikipedia.org/wiki/Lanczos_algorithm

cluster based on a membership function, that differentiates on the specific cluster. In our approach, we interpret the results of the LSI transformation of the initial matrix A as a form of fuzzy clustering and then we transform the fuzzy clustering to a hard-crisp clustering by assigning each document to exactly one cluster, the cluster where the document has the highest degree of participation. In this way, we produce an initial rough clustering of documents. Later these initial clusters are going to be reduced (Agglomerative Hierarchical Clustering) and furthermore re-organized (spherical k-means algorithm) in order to enhance their quality.

In particular, we rely on the $n \times k$ matrix $V_k S_k$, produced by SVD in the previous step, and consider its k columns as a set of k clusters and the n rows as documents. Each value in position (i, j) of $V_k S_k$ defines the document's i degree of participation to the cluster j. This fuzzy clustering is transformed to a crisp clustering by assigning each document to exactly one cluster, where the document has the highest degree of participation according to the values of the $V_k S_k$.

To perform the fuzzy clustering we run the respective algorithm implemented in the Mahout framework. The algorithm works as follows:

- Each document is assigned to the cluster where it has the highest degree of participation according to the membership values of $V_k S_k$.
- Each initial cluster is represented by a central vector c [17], called centroid, where $c = \frac{1}{|S|} \sum_{x \in S} x$.

More specifically, the Map function reads the probability membership values of documents and assigns them to the corresponding cluster. The records produced by the Map function are key-value pairs where the key is a cluster identifier and the value is a vector that represents a row of the matrix A. The Reduce function receives all the key-value pairs from the map task and produces a centroid for each cluster. The key of the output record is a cluster identifier and the value is the centroid of the cluster.

3.4   Hierarchical Clustering

Hierarchical Clustering [17] is a widely used data analysis method for identifying relatively homogenous clusters of experimental data items based on selected measured characteristics. Commonly, hierarchical clustering techniques generate a set of nested clusters, with a single all-inclusive cluster at the top and single point clusters at the bottom. The result of a hierarchical clustering algorithm can be graphically displayed as tree, called a dendrogram.

Hierarchical clustering algorithms can be divided into two basic approaches: agglomerative (merging) and divisive (splitting). In the context of this work, we adopt the agglomerative approach in an attempt to reduce the number of initial clusters, and develop an iterative algorithm in the Hadoop framework. The agglomerative approach is a bottom-up clustering method that starts with all the data elements as individual clusters and at each step it merges the most similar or closest pair of clusters, based on a cluster similarity or distance measure. The steps are repeated until the desired number of clusters is obtained according to a user defined parameter.

The algorithm consists of an iterative step with two phases. Phase 1 computes similarity between the clusters and phase 2 merges the two most similar clusters. The process repeats until the above prerequisite is fulfilled. The agglomerative approach attempts to shrink the number of initial clusters produced by LSI and fuzzy clustering since the produced number of clusters is equal to the LSI dimension which is generally high. At each step, the algorithm merges the most similar pair of clusters based on the UPGMA (Unweighted Pair Group Method with Arithmetic Mean) scheme:

$$similarity(cluster1, cluster2) = \frac{\sum_{x \in cluster1, y \in cluster2} \cos(x, y)}{size(cluster1) \cdot size(cluster2)}$$

3.5 Spherical k-means Algorithm

The k-means algorithm is an efficient and well known method for clustering large portions of data. Typically, the k-means algorithm determines the distance (similarity) between a data element in a cluster and its cluster centroid by using the squared Euclidean distance measure. The spherical k-means algorithm [14] is a variant of the original method that uses the cosine similarity as distance measure and achieves better results on high-dimensional data such as text documents. Note however that the above mentioned algorithm is significantly sensitive to the initialization procedure meaning that its clustering performance depends heavily on the initial selection of cluster centroids. As a consequence, this selection constitutes a task of crucial importance. In an attempt to overcome this issue and with the goal of enhancing the achieved performance, we initialize the spherical k-means algorithm with the cluster centroids obtained by the previous steps of the proposed technique. Furthermore, the motivation for applying the spherical k-means as the final step of our clustering approach, is to enhance the quality and the precision of the clustering produced by the previous steps of the proposed technique.

More specifically, we utilize a MapReduce version of a refinement algorithm [14] that uses local search in order to refine the clusters generated by spherical k-means. The algorithm alternates between two steps (a) first variation and (b) spherical k-means. The first step moves a single data element from one cluster to another increasing in this way the value of the objective function of the clustering. A sequence of first variation moves allows an escape from a local maximum, so that fresh iterations of spherical k-means (in the second step) can further increase the objective function value. The k-means algorithm is implemented in the Mahout framework and we provide the cosine similarity as input parameter, in order for the method to behave similar to the spherical k-means approach.

4   Conclusions and Future Work

In this paper, we present a novel data mining algorithm for clustering biological and biomedical data in the cloud. The method constitutes of a series of steps/algorithms and is currently under development. We utilize the Hadoop and Mahout frameworks to implement existing centralized data mining algorithms that fit the MapReduce programming model. In the near future, we intend to finish the implementation of our approach and run extensive experiments using the enormous data repository of PubMed. We want to compare the efficiency of our method with existing centralized methods and measure the gain we earn using parallel and distributed solutions instead of centralized ones. Moreover, we plan to explore more existing data mining techniques and create respective MapReduce versions of them to experiment with.

Acknowledgements. This research has been co-financed by the European Union (European Social Fund ESF) and Greek national funds through the Operational Program "Education and Lifelong Learning" of the National Strategic Reference Framework (NSRF) - Research Funding Program: Thales. Investing in knowledge society through the European Social Fund.

References

1. Ananiadou, S., Mcnaught, J.: Text Mining for Biology and Biomedicine. Artech House (2006)
2. Baeza-Yates, R., Ribeiro-Neto, B.: Modern Information Retrieval: The Concepts and Technology behind Search, 2nd edn. ACM Press (2011)
3. Chen, B., Harrison, R., Pan, Y., Tai, P.: Novel Hybrid Hierarchical-K-means Clustering Method (H-K-means) for Microarray Analysis. In: Proceedings of the 2005 IEEE Computational Systems Bioinformatics Conference - Workshops, pp. 105–108. IEEE Computer Society, Washington, DC (2005)
4. Cohen, A.M., Herch, W.R.: A Survey of Current Work in Biomedical Text Mining. Brief Bioinform. 6, 57–71 (2005)
5. Dai, H.J., Lin, J.Y.W., Huang, C.H., Chou, P.H., Tsai, R.T.H., Hsu, W.L.: A Survey of State of the Art Biomedical Text Mining Techniques for Semantic Analysis. In: Proceedings of the IEEE International Conference on Sensor Networks, Ubiquitous and Trustworthy Computing, pp. 410–417 (2008)
6. Dean, J., Ghemawat, S.: MapReduce: Simplified Data Processing on Large Clusters. In: Proceedings of the 6th Symposium on Operating Systems Design and Implementation, pp. 137–150. USENIX Association, Berkeley (2004)
7. Dhillon, I.S., Guan, Y., Kogan, J.: Iterative Clustering of High Dimensional Text Data Augmented by Local Search. In: Proceedings of the 2002 IEEE International Conference on Data Mining, pp. 131–138 (2002)
8. Georgitsi, M., Viennas, E., Gkantouna, V., Christodoulopoulou, E., Zagoriti, Z., Tafrali, C., Ntellos, F., Giannakopoulou, O., Boulakou, A., Vlahopoulou, P., Kyriacou, E., Tsaknakis, J., Tsakalidis, A., Poulas, K., Tzimas, G., Patrinos, G.: Population-Specific Documentation of Pharmacogenomic Markers and their Allelic Frequencies in FINDbase. Pharmacogenomics 12, 49–58 (2011)

9. Han, J., Kamber, M.: Data Mining: Concepts and Techniques, 2nd edn. Morgan Kaufmann Publishers, San Francisco (2006)
10. Ioannou, M., Makris, C., Tzimas, G., Viennas, E.: A Text Mining Approach for Biomedical Documents. In: Proceedings of the 6th Conference of the Hellenic Society for Computational Biology and Bioinformatics, Patras, Greece (2011)
11. Ioannou, M., Patrinos, G.P., Tzimas, G.: Genome-based population clustering: Nuggets of truth buried in a pile of numbers? In: Iliadis, L., Maglogiannis, I., Papadopoulos, H., Karatzas, K., Sioutas, S. (eds.) AIAI 2012, Part II. IFIP AICT, vol. 382, pp. 602–611. Springer, Heidelberg (2012)
12. Inoue, K., Urahama, K.: Fuzzy Clustering Based on Cooccurence Matrix and Its Application to Data Retrieval. Electron. Comm. Jpn. 84(pt. 2), 10–19 (2001)
13. Ioannou, M., Makris, C., Patrinos, G., Tzimas, G.: A Set of Novel Mining Tools for Efficient Biological Knowledge Discovery. In: Artificial Intelligence Review. Springer (2013)
14. Kogan, J.: Introduction to Clustering Large and High-Dimensional Data, pp. 51–72. Cambridge University Press, New York (2007)
15. Lu, Z.: Pubmed and Beyond: A Survey of Web Tools for Searching Biomedical Literature. Database, Oxford (2011)
16. Manconi, A., Vargiu, E., Armano, G., Milanesi, L.: Literature Retrieval and Mining in Bioinformatics: State of the Art and Challenges. In: Adv. Bioinformatics (2012)
17. Steinbach, M., Karypis, G., Kumar, V.: A Comparison of Document Clustering Techniques. In: Proceedings of the KDD Workshop on Text Mining, 6th ACM SIGKDD International Conference on Data Mining (2000)
18. The apache software foundation: Hadoop homepage, http://hadoop.apache.org/
19. The apache software foundation: Mahout homepage, https://mahout.apache.org/
20. Van Baal, S., Kaimakis, P., Phommarinh, M., Koumbi, D., Cuppens, H., Riccardino, F., Macek, M. Jr., Scriver, C.R., Patrinos. G.: FINDbase: A Relational Database Recording Frequencies of Genetic Defects Leading to Inherited Disorders Worldwide. Nucleic Acids Res. 35 (2007)
21. Viennas, E., Gkantouna, V., Ioannou, M., Georgitsi, M., Rigou, M., Poulas, K., Patrinos, G., Tzimas, G.: Population-Ethnic Group Specific Genome Variation Allele Frequency Data: A Querying and Visualization Journey. Genomics 100, 93–101 (2012)
22. Wang, J.T.L., Zaki, M.J., Toivonen, H.T.T., Shasha, D.: Data Mining in Bioinformatics. In: Advanced Information and Knowledge Processing. Springer (2005)
23. White, T.: Hadoop: The Definitive Guide, 3rd edn. O'Reilly Media / Yahoo Press (2012)
24. Zhang, C., Xia, S.: K-means Clustering Algorithm with Improved Initial Center. In: Knowledge Discovery and Data Mining, pp.790–792 (2009)
25. Zhang, T., Ramakrishnan, R., Livny, M.: BIRCH: an Efficient Data Clustering Method for Very Large Databases. In: Proceedings of the 1996 ACM SIGMOD International Conference on Management of Data, Montreal, Quebec, Canada, pp. 103–114 (1996)
26. Zhang, T., Ramakrishnan, R., Livny, M.: BIRCH: a New Data Clustering Algorithm and its Applications. Journal of Data Mining and Knowledge Discovery 1, 141–182 (1997)

On-line Minimum Closed Covers

Costas S. Iliopoulos and Manal Mohamed

Department of Informatics, King's College London,
London WC2R 2LS, United Kingdom
c.iliopoulos@kcl.ac.uk

Abstract. The Minimum Closed Covers problem asks us to compute a
minimum size of a closed cover of given string. In this paper we present an
on-line $O(n)$-time algorithm to calculate the size of a minimum closed
cover for each prefix of a given string w of length n. We also show a
method to recover a minimum closed cover of each prefix of w in greedy
manner from right to left.

Keywords: String, Cover, Closed String, Algorithm.

1 Introduction

The computation of various kinds of "regularities" in a given string has been
of interest for more than a century. Such interest has been initiated with the
computation of periodicities [23] and later extended, in response to applications
arising in data compression and molecular biology, to other concepts of regularities. Apostolico *et al* introduced the idea of *cover* (*quasiperiod*); that is, a
substring u is called a cover of given string w if concatenations of u generates w.
Clearly the notion of a cover is one way to grasp the repetitive structures of w [2].
Several algorithms to compute the covers of a given string were published in the
1990s [5, 22, 14], culminating in an algorithm [19] that in $O(n)$ time computes
a *cover array* specifying all the covers of every prefix of w.

The idea of covers was further extended to consider a set of substrings of w
that covers w rather than a single substring. In particular, aiming for a set of
minimum size, the *k-covers* problem calculates for a particular integer k, a set
of substrings all of length k that covers w [13]. Also, the *λ-covers* problem finds
a minimum set of λ substrings of equal length that covers w with the minimum
error, under a variety of distance models [12]. More recently, the *palindromic
covers* was proposed in which a minimum set of palindromic substrings that
covers w is calculated [24]. In this paper, we consider yet another set of substrings
that cover w, namely the set of *closed* substrings. We chose to investigate the
idea of a cover using set of closed substrings due to the well-studied relation
between palindromic and closed strings[6, 4]. We aim to design an algorithm
that calculates a minimum set of closed substrings of w that covers a given
string w.

More precisely, a string w is called a closed string if it is empty or it has a factor
(different from the string itself) occurring exactly twice in, as a prefix and as a

L. Iliadis et al. (Eds.): AIAI 2014 Workshops, IFIP AICT 437, pp. 106–115, 2014.

suffix, that is, with no internal occurrences. Initially, the notion of closed string has been proposed and characterized by Fici [10] as a way to classify trapezoidal strings. In a more recent work, Badkobeh, Fici and Lipták [4] showed that any string of length n contains at least $n+1$ closed factors. They also investigated and provided a combinatorial characterization of the set of strings with the smallest number of closed factors which they called *C-poor* strings.

The paper is organized as follows: in Section 2, we state the preliminaries used throughout the paper. In Section 3, we define the longest closed cover problem and detail our algorithm.

2   Preliminaries

Throughout the paper w denotes a *string* of length n defined on a finite alphabet Σ. We use $w[i]$, for $i = 1, 2, \ldots, n$, to denote the i-th letter of w, and $w[i..j]$ as a notation for the *factor* (*substring*) $w[i]w[i+1]\cdots w[j]$ of w. The *length* of w, the non-negative integer n, is denoted by $|w|$. The empty string has length zero and is denoted by ϵ. A *power* is a string that is a concatenation of copies of another string.

A *prefix* (resp. a *suffix*) of a string w is any string u such that $w = uz$ (resp. $w = zu$) for some string z. A factor of w is a prefix of a suffix (or equivalently, a suffix of a prefix) of w. A *border* of w is any string $u \neq w$ that is a prefix and a suffix of w. From the definitions, we have that ϵ is a prefix, a suffix and a factor of any string. An *occurrence* of a factor u in a string w is a pair of positions (i, j) such that $1 \leq i, j \leq n$ and $u = w[i..j]$.

We recall the definition of closed string given in [10]

Definition 1. *We say that w is closed if and only if it is empty or has a factor $v \neq w$ occurring exactly twice in w, as a prefix and as a suffix of w.*

The string *aba* is closed, since its factor *a* appears only as a prefix and as a suffix. On the contrary, a string *abaa* is not closed. Note that for any letter $a \in \Sigma$ and for any $n > 0$, the string a^n is closed, as a^{n-1} is a factor that occurs only as a prefix and a suffix in it. More generally, any string that is a power is closed.

The concept of closed string is equivalent to that of *periodic-like* string [7]. A string w is periodic-like if its longest repeated prefix does not appear in w followed by different letters. The concept of closed string is also related to *complete return* to a factor u in w, as considered in [11]. A complete return to u in w is any factor of w having exactly two occurrences of u, one as a prefix and one as a suffix. Therefore, w is a closed string if and only if w is a complete return to its longest repeated prefix.

A factor of w that is closed is called a *closed factor* of w. For any string w of length n, the number of closed factors of w is at least $n + 1$. In particular, a string w of length n is called *C-poor* if it has exactly $n + 1$ closed factors, i.e., if it contains the smallest number of closed factors a string of length n can contain. For example, the string *abca* of length 4 is a C-poor as it has 5 closed factors namely ϵ, a, b, c and $abca$, whereas the string $ababa$ of length 5 is not C-poor as it has 8 closed factors: $\epsilon, a, b, aba, abab, bab, baba$, and $ababa$. Note that there are C-poor strings that are not closed, e.g. ab, and closed strings that are not C-poor, e.g. $abab$.

A set of subintervals $\{[b_1, e_1], \cdots, [b_h, e_h]\}$ is called a *cover* of interval $[1, n]$ if $\bigcup_{i=1}^{h}[b_i, e_i] = [1, n]$. The size of the cover is the number h of subintervals in it. A cover $\{[b_1, e_1], ..., [b_h, e_h]\}$ of $[1, n]$ is said to be a *closed cover* of string w of length n, if $w[b_i..e_i]$ is a closed factor of w for all $1 \leq i \leq h$. A *minimum closed cover* of w is a closed cover of w with the smallest possible size; note that a minimum closed cover is not unique.

2.1 Tools

Suffix Array: The *Suffix Array* of the string w is a data structure used for indexing its content. It is comprised of two arrays:

- The array SA which stores the list of positions of w associated with its suffixes in increasing lexicographic order. Thus SA is indexes by the ranks of suffixes in their sorted list.
- The second array, LCP, which stores the longest common prefixes between consecutive suffixes in the sorted list. That is

$$\text{LCP}[r] = |lcp(\text{SA}[r-1], \text{SA}[r])|,$$

where $lcp(i, j)$ is the longest common prefix of the two suffixes of w starting at positions i and j, i.e. $w[i..n]$ and $w[j..n]$.

The computation of the suffix array can be done in time $O(n \log n)$ in the comparison model [21]. For an integer alphabet, the suffix array can be built in time $O(n)$ [16–18].

Longest Previous Factor: The *Longest Previous Factor* array has been introduced in [8] and gives for each position i in a given string w the length of the longest factor that occurs both at position i and to the left of i in w. The Longest Previous Factor is defined by $\text{LPF}[1] = 0$ and for $1 < i \leq n$, by

$\text{LPF}[i] = \max \{k \mid w[i..i+k-1] = w[j..j+k-1],$ for some $1 \leq j < i\}$.

For example, for string $\underline{abb}a\underline{ba}$, the Longest Previous factor corresponding to position 4 is 2 because ab is the longest factor at position 4 that appears before (at position 1). In the paper we make use of the following known result on LPF from [9]:

Lemma 1. *The Longest Previous Factor array of a string of length n on an integer alphabet can be built from the read-write Suffix Array and Longest Common Prefix array in time $O(n)$ (independently of the alphabet size) with a constant amount of extra memory space.*

The Longest Previous Factor algorithm can compute for no extra cost an array FPOS that stores the positions of the longest previous factor if such factor exists, that is

FPOS[i] = 0 if LPF[i] = 0 and

FPOS[i] = max $\{j \mid w[i..i+$LPF$[i]-1] = w[j..j+$LPF$[i]-1]$, for $1 \leq j < i\}$.

Range Min in Weighted Tree: For any two nodes u and v in the same path of a weighted rooted tree, let $min(u, v)$ be a query that returns a node in the path with minimum weight. The following lemma from [1] states recent result used in this paper:

Lemma 2. *Under a RAM model, a dynamic tree can be maintained in linear space so that a min query and an operation of adding a leaf to the tree are both supported in the worst-case $O(1)$ time.*

3   Minimum Closed Covers

In this section, we present an algorithm that computes Minimum Closed Covers of all prefixes for a given string w of length n. More precisely, we compute array C such that for each position $1 \leq i \leq n$, C[i] stores the size of a minimum closed cover of $w[1..i]$. If $C[i] = h$ then, by definition, there is a closed cover $\{[b_1..e_1], \cdots, [b_h..e_h]\}$ that covers $w[1..i]$ such that $b_1 = 1$, $e_h = i$ and h is minimum.

Obviously there might be several closed covers that are minimum but we are interested on being able to retrieve one of them. In order to achieve this goal, array C_ℓ is maintained such that $C_\ell[i]$ stores the length of the rightmost closed factor ending at position i which is associated with the rightmost subinterval of a minimum closed cover (of size $C[i]$) of $w[1..i]$. A minimum closed cover of w, and indeed of any prefix of w, can be easily computed in a greedy manner from right to left using C_ℓ.

Before presenting our algorithm, we will introduce several known results concerning closed strings; for proof and more details please refer to [4]:

Lemma 3. *For any strings u and v, one has $|CC(u)|+|CC(v)| \leq |CC(uv)|+1$, where $|CC(w)|$ is the size of a set of closed factors of w.*

Corollary 1. *A string w is C-poor if and only if every closed factor of w is a complete return of a single letter.*

By definition empty strings and one-letter strings are closed. Corollary 1 suggests an additional class of closed strings namely complete return of a single letter; that is for any letter α, string $w = \alpha u \alpha$ is closed if α has no occurrence in string u. Additionally, Lemma 3 suggests that if the size of a string increases by one (let say, by concatenating a single letter) then the total number of closed factors of the new strings should increase. In this paper, we are interested with the minimum number of closed factors that cover a given string. Our main observation suggests that if the size of string increases by one letter then the minimum size of closed cover cannot increase by more than one.

Observation 1. *Let $\{[b_1..e_1], \cdots, [b_k..e_k]\}$ be a minimum closed cover $w[1..i-1]$ and $\{[b'_1..e'_1], \cdots, [b'_h..'e_h]\}$ is a minimum closed covers of $w[1..i1]$ then $h \leq k+1$.*

Considering position i, given $C[j]$ for $1 \leq j < i$, while taking $C[i-1]$ into consideration, there are three different possibilities:

1. The size of a minimum closed cover of $w[1..i]$ increases by one. In this case, the one-letter closed factor $w[i]$ becomes the rightmost closed factor of a minimum closed cover.
2. The size of a minimum closed cover of $w[1..i]$ stays the same. This may be achieved if the rightmost closed factor can be extended to the right by letter $w[i]$ and stays closed. If $w[b..i-1]$ is the rightmost closed factor of a minimum closed cover of $w[1..i-1]$ and $w[b..i]$ is closed then $w[b..i]$ is the rightmost closed factor of a minimum closed cover of $w[1..i]$.
3. The size of a minimum closed cover of $w[1..i]$ decreases. This can only be achieved if there exists a position k such that $w[k..i]$ is a closed factor of $w[1..i]$ and $C[k-1] < C[i-1]$. Here $C[i] = C[k-1]+1$ and $w[k..i]$ will be the rightmost closed factor of a minimum closed cover of $w[1..i]$.

Recall that $C[i]$ and $C_\ell[i]$ are the size of a minimum closed cover of $w[1..i]$ and the length of its rightmost closed subinterval, respectively. That is, if $C_\ell[i] = \ell$ then $w[i - \ell + 1..i]$ is the rightmost closed factor associated with interval $[i - \ell + 1..i]$; the rightmost subinterval of a minimum closed cover of $w[1..i]$. For each position i we will store the length of the border of the rightmost closed factor such that $B_\ell[i] = |border(w[i - \ell + 1..i])|$. In this way, array B_ℓ will enable us to check in $O(1)$ time whether the rightmost closed factor can be extended to the right by a single letter and stay closed. We will also maintain an array P that keeps for each position i, the nearest position j to the left of i containing letter $w[i]$ such that $w[j..i]$ is closed as it is a complete return of a single letter (Corollary 1).

Now we are ready to present our Minimum Closed Covers algorithm. The algorithm initializes $C[1] = 1$ and iterates from left to right calculating for each position i the value of $C[i]$ as follows:

$$C[i] = \begin{cases} C[i-1]. & \text{If } w[i - C_\ell[i-1] + B_\ell[i-1]] = w[i]. \\ \min \begin{cases} C[P[i]-1]+1, \\ C[i-1]+1, & \text{Otherwise.} \\ C[k-1]+1. \end{cases} \end{cases}$$

At each iteration i, the algorithms first checks whether the rightmost closed factor can be extended to the right by letter $w[i]$ and stays closed. This can be easily achieved by comparing positions $w[i - C_\ell[i-1] + B_\ell[i-1]]$ and $w[i]$. If the two positions are equal then the size of a minimum closed cover does not increase, as we are still able to cover $w[1..i]$ with a minimum cover of size $C[i-1]$, and since $C[i-1]$ is minimum, then $C[i]$ is also minimum. If the rightmost closed factor cannot be extended, then the algorithm checks all possible closed factors of $w[1..i]$, ending at position i, and chooses the one that minimizes the size of a closed cover. Obviously, there are two closed factors of $w[1..i]$ ending at position i that can be easily checked: factor $w[i]$; a a one-letter closed factor, and factor $w[P[i]..i]$; a complete return of a single letter. These two factors are not the only ones that need to be checked and now we will turn our attention to explain how all other closed factors can be identified and checked efficiently.

Recall that the Longest Previous Factor gives for each position i in w the length of the longest factor that occurs at i and to the left of i in w. We show that the position of the longest previous factor can be calculated at no extra cost. Here, we update the definition of LPF as follows: $LPF[1] = (0,0)$ and

$$LPF[i] = (\ell, j) \text{ such that } \ell = \max\{k \mid w[i..i+k-1] = w[j..j+k-1], 1 \leq j < i\}.$$

Clearly, if $LPF[i] = (\ell, j)$, then $w[j..i+\ell-1]$ is a closed factor with a border (of length ℓ) starting at positions j and i. For the purpose of our algorithm, we want to calculate for each position the maximum closed factor ending at a certain position (not starting). By maximum, we mean non left-expendable. To do so, the longest previous factor array is computed for string $\overleftarrow{w}[1..n] = w[n]w[n-1]..w[1]$; we will call the new array $\overleftarrow{LPF}$. By doing so, we calculate for each position the length of the maximum closed factor ending at each position $n-i+1$ and to the right of $n-i+1$ in w. Thus, if $\overleftarrow{LPF}[i] = (\ell, j)$, then there is a closed factor in w with a left border (of length ℓ), ending at position $n-i+1$, and a right border ending at position $n-j+1$. The relation between closed factors and array $\overleftarrow{LPF}$ has been highlighted in [3].

Example 1. Let $w = aabaaaaabaaaabcdbcd$ be a given string of length 19.

position i	1	2	3	4	5	6	7	8	9	10	11	12	13	14	15	16	17	18	19
$w[i]$	a	a	b	a	a	a	a	a	b	a	a	a	a	b	c	d	b	c	d
$LPF[i](\ell,j)$	0	1	0	2	4	3	7	6	5	5	4	3	②	1	0	0	3	2	1
$LPF[i](\ell,j)$	0	1	0	1	4	5	1	2	3	5	6	7	⑧	9	0	0	14	15	16

For $i = 13$, $LPF[13] = (2,8)$ implies that there is a closed factor $\underline{ab}aaa\underline{ab}$, with a left border starting at position 8, and a right border starting at position 13 in w. The length of the border of this closed factor is 2.
The inverse of string w, $\overleftarrow{w} = dcbdcbaaaabaaaaabaa$.

position i	1	2	3	4	5	6	7	8	9	10	11	12	13	14	15	16	17	18	19
$\overleftarrow{w}[i]$	d	c	b	d	c	b	a	a	a	a	b	a	a	a	a	a	b	a	a
$\overleftarrow{LPF}[i](\ell,j)$	0	0	0	3	2	1	0	3	2	1	⑤	4	7	6	5	4	3	2	1
$\overleftarrow{LPF}[i](\ell,j)$	0	0	0	1	2	3	0	7	8	9	⑥	7	7	8	9	10	11	12	13

For $i = 11$, $\overleftarrow{LPF}[11] = (5,6)$ implies that there is a closed factor $\underline{baaaa}\ \underline{baaaa}$ in $\overleftarrow{w}$, that corresponds to a closed factor $(\underline{aaaab}\ \underline{aaaab})$ in w, with a left border ending at position $9 = 19 - 11 + 1$, and a right border ending at position $14 = 19 - 6 + 1$. This is a maximum closed factor as it cannot be extended to the left.

Note that position 7 appears three times in $\overleftarrow{LPF}$, which implies that there are three different closed factors with a right border ending at position $13 = 19 - 7 + 1$ in w. These three closed factors are different and each is associated with distinct maximum closed factor with respect to the end positions of the left borders:

– left border ending at position $12 = 19 - 8 + 1$ and of length $= 3$: $a\overline{aa}a$,
– left border ending at position $8 = 19 - 12 + 1$ and of length $= 4$: $aaaab\underline{aaaa}$,
– left border ending at position $7 = 19-13+1$ and of length $= 7$: $aabaaa\underline{aa}baaaa$.

From the above example, one should be able to recognize that the number of distinct maximum closed factors ending at a certain position i in w is not always equal to 1. However, the total number of distinct maximum closed factors over all positions i in w is at most n; the length of w.

Now, we can go back and finalize the Minimum Closed Covers algorithm. Recall that in order to calculate $C[i]$, the algorithm checks whether the rightmost factor can be extending with letter $w[i]$ and remain closed. If this not possible, the algorithm finds the rightmost closed factor ending at position i that minimizes $C[i]$.

Lemma 4. *A factor $w[k..i]$ is a closed factor of w ending at position i if and only if $k = i$ or $k = P[i]$ or*

$$k \in \bigcup [j_i - \ell_{j_i} + 1..j_i - \ell_{j_{i+1}}],$$

where $\{(j_1, \ell_{j_1}), (j_2, \ell_{j_2}), \cdots, (j_h, \ell_{j_h})\}$ is the ordered set of all maximal closed factors of w ending at position i, and with left borders ending at positions j_i and of length ℓ_{j_i} $(j_{h+1} = 0)$.

Proof. If $\{(j_1, \ell_{j_1}), (j_2, \ell_{j_2}), \cdots, (j_h, \ell_{j_h})\}$ is the set of closed factors ending at position i in ascending order, i.e. $j_1 < j_2 < ... < j_h$. Then, it should be clear that $\ell_{j_1} > \ell_{j_2} > \cdots > \ell_{j_h}$.

Consider for example positions $j_1 < j_2$, both $w[j_1 - \ell_{j_1} + 1..i]$ and $w[j_2 - \ell_{j_2} + 1..i]$ are closed factors of w ending at position i with borders of length ℓ_{j_1} and ℓ_{j_2}, respectively. By definition, $w[i - \ell_{j_1} + 1i]$ and $w[i - \ell_{j_2} + 1i]$ are the longest suffixes associated with these closed factors. Each suffix appears in the associated factor twice as a suffix and a prefix. If $\ell_{j_1} < \ell_{j_2}$, then $w[i - \ell_{j_1} + 1i]$ is a suffix of $w[i - \ell_{j_2} + 1i]$ and should have internal occurrence in $w[j_1 - \ell_{j_1} + 1..i]$, a contradiction.

Each closed factor $w[j_l - \ell_{j_l} + 1..i]$ is maximum i.e. it cannot be extended to the left by any letter. This is due to the method of construction from LPF. However, in order to calculate all possible positions in w that could start a closed factor ending at position i, non-maximal closed factors need to be considered. This can be done by considering all suffixes of $w[j_l - \ell_{j_l} + 1..i]$ starting at positions $j_l - \ell_{j_l} + 1..j_l - \ell_{j_{l+1}}$. Note that any smaller suffix is not closed due to an internal occurrence. $\square$

Each position k defined by Lemma 4 is a candidate for a start position of a closed factor ending at position i. The closed factor that minimizes $C[i]$ is the one that starts at position k such that $C[k-1]$ is smallest.

Theorem 2. *Given string w of length n, the Minimum Closed Covers algorithm computes $C[i]$ for $1 \le i \le n$ in $O(n)$-time in an online manner.*

Proof. Although for each position i, the number of candidate positions at which a rightmost closed factor (ending at position i) is not constant, the number of intervals that need to be checked is bounded by n. The sum of these intervals over all possible i is linear. According to Lemma 2, calculating for each interval, the position k, with the minimum $C[k]$, can be done in $O(1)$ time.   □

Example 2. For the same string $w = aabaaaaabaaaabcdbcd$.

position i	1	2	3	4	5	6	7	8	9	10	11	12	13	14	15	16	17	18	19
$w[i]$	a	a	b	a	a	a	a	a	b	a	a	a	a	b	c	d	b	c	d
$C[i]$	1	1	2	2	1	2	2	2	1	1	1	1	1	2	3	4	2	2	2
$C_\ell[i]$	1	2	1	3	5	1	2	3	9	10	11	12	13	9	1	1	4	5	6

Consider position $i = 14$, given that $\overline{aabaaaaabaaaa}$ is the rightmost closed factor of a minimum closed cover of $w[1..13]$. Firstly, the Minimum Closed Covers algorithm checks whether this closed factor is extendable to the right by comparing positions $w[14]$ and $w[8]$. Since $w[14] \ne w[8]$, the closed factor cannot be extended to the right and the algorithm proceeds to compute the optimal value for $C[14]$ as follows:

$$
C[14] = \min \begin{cases} C[P[14] - 1] + 1 = C[8] + 1 = 2 + 1 = 3, \\ C[13] + 1 = 1 + 1 = 2, \\ \min\{C[k-1] + 1; \text{ for } k \in [5..9]\} = C[5] + 1 = 1 + 1 = 2. \end{cases}
$$

In *Example1*, we show that there is one maximum closed factor ending at position 14 that starts at position 5 and with a border of length 5. Although, all five suffixes (starting at positions $\in [5..9]$) of this maximum closed factor are also closed and need to be considered, computing position $k \in [5..9]$ with the minimum $C[i-1]$ can be done constant time using a single min query ($k = 6$). Therefore, $C[14] = 2$ is the size of a minimum closed cover of $w[1..14]$; note that there are two minimum closed covers: $\{[1..13], [14..14]\}$ and $\{[1..5], [6..14]\}$, both of size 2.

The size of a minimum closed cover of w calculated by our proposed algorithm is $C[19] = 2$. It consists of two closed factors that can be retrieved using array C_ℓ. In this case, $\{[1..13][14..19]\}$ is a minimum closed cover with two closed factors: $aabaaaaabaaaa$ and $bcdbcd$.

4 Conclusion

We have considered The Minimum Closed Covers problem and proposed an on-line algorithm to calculate such cover in linear time. Our work is an extension to various string cover related problems. Recently, an algorithm to compute the minimum palindromic covers was proposed. Although there are relatively strong relations between palindromic string and closed string, our algorithm is significantly different. Various types of covers are still to be investigated mainly for its theocratical interest. In particular, we are interested in the calculation of the minimum Abelian covers of a string.

References

1. Alstrup, S., Holm, J.: Improved Algorithms for Finding Level Ancestors in Dynamic Trees. In: Welzl, E., Montanari, U., Rolim, J.D.P. (eds.) ICALP 2000. LNCS, vol. 1853, pp. 73–84. Springer, Heidelberg (2000)
2. Apostolico, A., Farach, M., Iliopoulos, C.S.: Optimal Superprimitivity Testing for Strings. Inform. Processing Letter 39, 17–20 (1991)
3. Badkobeh, G., Bannai, H., Goto, K.: I, Tomohiro, Iliopoulos, C. S., Inenaga, S., Puglisi, S.J., Sugimoto, S.: Closed Factorization (to appear)
4. Badkobeh, G., Fici, G., Lipták, Z.: A Note on Words With the Smallest Number of Closed Factors. CoRR abs/1305.6395 (2013)
5. Breslauer, D.: An On-Line String Superprimitivity Test. Information Processing Letters 44, 345–347 (1992)
6. Bucci, M., de Luca, A., De Luca, A.: Rich and Periodic-Like Words. In: Diekert, V., Nowotka, D. (eds.) DLT 2009. LNCS, vol. 5583, pp. 145–155. Springer, Heidelberg (2009)
7. Carpi, A., de Luca, A.: Periodic-Like Words, Periodicity and Boxes. Acta Informatica 37, 597–618 (2001)
8. Crochemore, M., Ilie, L.: Computing Longest Previous Factor in Linear Time and Applications. Inf. Process. Lett. 106(2), 75–80 (2008)
9. Crochemore, M., Ilie, L., Iliopoulos, C.S., Kubica, M., Rytter, W., Walen, T.: Computing the Longest Previous Factor. Eur. J. Comb. 34(1), 15–26 (2013)
10. Fici, G.: A Classification of Trapezoidal Words. In: Ambroz, P., Holub, S., Masakova, Z. (eds.) 8th International Conference on Words, WORDS 2011. Electronic Proceedings in Theoretical Computer Science, vol. 63, pp. 129–137 (2011)
11. Glen, A., Justin, J., Widmer, S., Zamboni, L.Q.: Palindromic Richness. European J. Combin. 30, 510–531 (2009)
12. Guo, Q., Zhang, H., Iliopoulos, C.S.: Computing the λ-Covers of a String. Inf. Sci. 177(19), 3957–3967 (2007)
13. Iliopoulos, C.S., Mohamed, M., Smyth, W.F.: New Complexity Results for the k-Covers Problem. International Journal of Information Sciences 181, 2571–2575 (2011)
14. Iliopoulos, C.S., Mouchard, L.: An $O(n \log n)$ Algorithm for Computing All Maximal Quasiperiodicities in Strings. Theoratical Computer Science 119, 247–265 (1993)
15. Iliopoulos, C.S., Smyth, W.F.: On-Line Algorithms for k-Covering. In: Proc. Ninth Australasian Workshop on Combinatorial Algorithms, pp. 107–116 (1998)

16. Kärkkäinen, J., Sanders, P.: Simpler Linear Work Suffix Array Construction. In: Baeten, J.C.M., Lenstra, J.K., Parrow, J., Woeginger, G.J. (eds.) ICALP 2003. LNCS, vol. 2719, pp. 943–955. Springer, Heidelberg (2003)

17. Kim, D.-K., Sim, J.S., Park, H.-J., Park, K.: Linear-Time Construction of Suffix Arrays. In: Baeza-Yates, R., Chávez, E., Crochemore, M. (eds.) CPM 2003. LNCS, vol. 2676, pp. 186–199. Springer, Heidelberg (2003)

18. Ko, P., Aluru, S.: Space Efficient Linear Time Construction of Suffix Arrays. Journal of Discrete Algorithms 3(24), 143–156 (2005)

19. Li, Y., Smyth, W.F.: Computing the Cover Array in Linear Time. Algorithmica 32, 95–106 (2002)

20. Main, M.G., Lorentz, R.J.: An $O(n \log n)$ Algorithm for Finding All Repetitions in a String. J. Algs. 5, 422–432 (1984)

21. Manber, U., Myers, G.: Suffix Arrays: A New Method for On-Line Search. SIAM J. on Computing, 935–948 (1993)

22. Moore, D., Smyth, W.F.: An Optimal Algorithm to Compute All the Covers of a String. Information Processing Letters 50, 239–246 (1994)

23. Thue, A.: Über Unendliche Zeichenreihen. Norske Vid. Selsk. Skr. I, Mat. Nat. Kl. Christiana 7, 1–22 (1906)

24. Tomohiro, I., Sugimoto, S., Inenaga, S., Bannai, H., Takeda, M.: Computing palindromic factorizations and palindromic covers on-line. In: Kulikov, A.S., Kuznetsov, S.O., Pevzner, P. (eds.) CPM 2014. LNCS, vol. 8486, pp. 150–161. Springer, Heidelberg (2014)

Reciprocal Rank
Using Web Page Popularity

Xenophon Evangelopoulos, Christos Makris, and Yannis Plegas

Computer Engineering and Informatics Department,
University of Patras, Greece
{evangelopo,makri,plegas}@ceid.upatras.gr

Abstract. In recent years, predicting user behavior has drawn much attention in the fields of information retrieval. To that extend, many models and even more evaluation metrics have been proposed, aiming at the accurate evaluation of the information retrieval process. Most of the proposed metrics, including the well-known nDCG and ERR, rely on the assumption that the probability (R) a user finds a document relevant, depends only on its relevance grade. In this paper, we employ the assumption that this probability is a function of a combination of two factors; its relevance grade and its popularity grade. Popularity, as we define it from daily page views, can be considered as users' vote for a document, and by combining this factor in the probability R we can capture user behavior more accurately. We present a new evaluation metric called Reciprocal Rank using Webpage Popularity (RRP) which takes into account not only the document's relevance judgment, but also its popularity, and as a result correlates better with click metrics than the other evaluation metrics do.

Keywords: Information Retrieval, Evaluation, Metrics, User Behavior, User Model.

1 Introduction

Designing evaluation metrics consists an important direction of information retrieval, which has gained tremendous attention over the last few years due to the expeditious evolution of information retrieval systems. Some of the best known evaluation metrics that have been developed over the years are Mean Average Precision (MAP), Precision at k (P@k), normalized Discounted Cumulative Gain (nDCG) [9], Expected Reciprocal Rank (ERR) [2] etc. A good evaluation metric should reflect the rate of relevance of the retrieved results. Furthermore, a proper user model which reflects users' interaction with the retrieval system is of utmost importance.

There are two different types of user behavior models; the *position models* and the *cascade models*. The first [7] assume that a click depends on both relevance and examination. Moreover, the probability of examination depends only on the position. As a result, position models consider each result in a page as

L. Iliadis et al. (Eds.): AIAI 2014 Workshops, IFIP AICT 437, pp. 116–125, 2014.
© IFIP International Federation for Information Processing 2014

independent from other results and thus fail to capture interaction among them in examination probability.

On the other side, cascade models rely on the assumption that users examine all the results sequentially from top to bottom and stop as soon as a relevant document is found and clicked. The examination probability in this case depends on the rank of the document and the relevance of all the previous documents. It is showed by Chapelle and Zhang [3] that cascade models can predict click-through rates mode accurately than position models.

In this paper we induce a novel evaluation metric which incorporates a new concept called *web page popularity*. Our metric is a cascade-based evaluation metric, which means that we assume the user scans all results from top to bottom and stops when she/he finds a relevant document and clicks on it. The difference of our proposed metric lies on the definition of the probability R, that a user finds a document relevant. Unlike previous editorial metrics we assume that the probability a user clicks on a document depends on two factors; not only on its relevance grade, but also on its popularity grade. For each document D_i, we have two values, a relevance grade as it is proposed by experts and popularity grade as it is given by web traffic statistics. We then incorporate these two values in the probability R and evaluate how well our proposed metric captures user behavior by comparing its performance with the performance of click-metrics.

The remainder of this paper is organized as follows. First Section 2 presents some recent related work on the field. Section 3 then describes the cascade model on which our new metric is based. Section 4 explains in detail the notion of web page popularity. Our metric is presented in Section 5 and Section 6 provides evidence of our metric's well-behavior. Finally we conclude the paper in Section 7.

2   Related Work

Evaluating the quality of information retrieval systems constitutes an important task in the research area of information retrieval. A great number of scientific papers have been published trying to best model, how well the search system satisfies users' search needs. As a result, a broad range of evaluation metrics which quantify system performance have been proposed, including Discounted Cumulative Gain (DCG or nDCG as it is mostly used) by Järverlin and Kekäläinen [9], Average Precision (AP) and Expected Reciprocal Rank (ERR) by Chapelle et al. [2] just to name a few. The latter belongs to a group of IR metrics which have an underlying cascade user model, which is the user model our proposed evaluation metric also uses.

The basic concept an information retrieval measures, is the concept of relevance. Most evaluation metrics use a set of relevance judgments for a set of documents as firstly induced by TREC evaluations. Relevance judgments are collected by human experts who are obliged to asses for a document's relevance to a given query and thus capture the notion of user relevance [6] [15]. Instead, we argue that an expert's opinion about a document's relevance is not sufficient enough to account for all users' opinion. Thus, we introduce a second relevance indicator which accounts for a huge amount of real users' opinion. We call this concept web page popularity and is derived from the daily page views a web page gets.

A resultant issue which arises when evaluating information retrieval systems is the test collection incompleteness problem, where a significant amount of relevance judgments is missing from the test collection. Buckley and Voorhees [1], Sakai [14] and Chucklin et al [5] battle this problem either by introducing a new metrics or by alternative solutions. We will also contemplate this problem using the concept of web page popularity.

3   Cascade User Model

One of the first things one should take into consideration when developing an evaluation metric is how well it's user model reflects users' satisfaction [8]. There exist two main categories of user models: position models and cascade models. Position models [7] [13], as their name implies, are trying to express the examination probability as a function of the position. Some of the metrics relying on the position model include the DCG metric and the RBP [10] metric. However, as showed by Chapelle et al. [2] position models appear to face some serious drawbacks due to the fact that they assume that the probability of examination depends only on position.

Our proposed metric is based on the cascade model [3], assuming a linear traversal through the ranking, and that the rest of the results below a clicked document are not examined by the user. Let R_i be the probability of examination of the i-th document. As soon as the user clicks on a result and is satisfied with it, she/he terminates the search and results below are not examined at all, no matter of what their position is.

As showed by Craswell et al. [7] R_i represents the attractiveness of a result. More specifically it measures the probability of a click on a result which can be interpreted as the relevance of the snippet. We expand this by inducing a novel factor contributing in the measurement of examination probability. This factor is called *web page popularity*. A more detailed explanation and definition of web page popularity follows in section 4. In order to develop an evaluation metric which captures user behavior well, one should not only rely on experts' judgments, but also on other factors which capture user behavior. Popularity, as we define it from daily page views, can be considered as users' vote for a document and as result a value which expresses user's behavior. Here, we make the assumption that at each rank, the document's probability relies not only on its relevance, but also on its popularity. This means that when a user examines all documents from top to bottom, at each rank r she/he is expected to click on a result after examining its snippet S_r, which finds relevant **and** its url link L_r, which finds popular. As a result we suggest that probability of clicks is highly correlated both with the relevance of the document and its popularity.

$$R_r = P(C_r | S_r, L_r)$$

The R_i values as we will show in the next section can be set as a function of two factors. The relevance grade of document i and the popularity of the url of document i. For a given set of R_i, the probability of click on the i-th document can thus be expressed as:

$$P(C_r = 1) = \prod_{i=1}^{r-1}(1 - R_i)R_r.$$

4   Web Page Popularity

In the previous section we presented the cascade user model, where we presented a novel concept named as popularity of a web page. But how can one define the popularity of a link or better of a page? Cho et al. [4] in their study define popularity $V(p,\Delta t)$ of a page p as the number of "visits" or "page views" the page gets within a specific time interval Δt. Here, we also use this notation but abstract it generally as Web Page Popularity.

Definition 1 (Web Page Popularity). We define the popularity $P(p,\Delta t)$ of a web page p as the number of page views (pv) that page gets within a specific time interval Δt.

Following from the above definition we construct a popularity grade in accordance with the relevance grade, so that the popularity of each link can be measured and compared properly. Particularly, given a number of page views pv for a link u, we define its popularity grade as follows:

$$p_u = \left\lfloor \frac{\ln pv_u}{5} \right\rfloor \tag{1}$$

Equation (1) was developed based on traffic statistics about every web page. More specifically, pageview values pv ranged from 0 (that is no page views at all) to 500.000.000 (Google's page views) per day. In order to incorporate these large numbers in a metric we decided to get the natural logarithm from each pv value so that we don't lose the amount of information for each website. Moreover we mapped each natural logarithm to a 5-scale ranging from 0 to 4 in order to create a grade-scale similar to that of the relevance grades.

According to equation (1) each website received a 5-scaled grade (unknown, well-known, popular, very popular and famous) according to its daily traffic statistics. Table 1 shows the popularity grade of some websites according to their daily page views.

Table 1. Popularity grade for four sample websites

WebSite	Daily Page views	Popularity Grade
http://google.com	584.640.000	4
http://wikipedia.com	30.451.680	3
http://ceid.upatras.gr	11.228	1
http://sample-site.wordpress.com	11	0

5   Proposed Metric

In this section we will introduce our new proposed evaluation metric. This novel metric is based on the cascade model described in section 3. As stated before our proposed metric is described by two factors of relevance. The relevance grade of the document and the popularity grade of the document. Let g_i be the relevance grade and p_i the popularity grade of the i-th document. Then the relevance probability can be equally defined by g_i and p_i.

$$R_i = \mathcal{R}(g_i, p_i).$$

Here R_i denotes a different value equally defined by two factors and $\mathcal{R}$ consists a mapping function from relevance and popularity grades to probability of relevance. There are many different ways to define $\mathcal{R}$, but here we select it in accordance with the gain function of ERR used in[2]

$$\mathcal{R}(r) = \frac{2^r - 1}{2^{r_{max}}},$$

where

$$r = \frac{g + p}{2}, \qquad r \in \{0, ..., r_{max}\}.$$

This means that the probability of relevance can be defined equally by two factors; its relevance grade and its popularity grade. As we will prove in the next section this is a well-balanced way to define relevance due to the equal contribution of each factor. Thus, when a document is fairly-relevant ($g = 1$), but is very popular ($p = 4$), the probability that a user will click on that result is much higher than if the document was non-popular.

In order to make things more clear, consider the following scenario: a user examines all the results after posing a query in a search machine. She/he discovers two results (regardless of their position) which he considers as very relevant. If the first result is a far more popular web page than the other, then she/he will not examine the latter.

The next step is to define a utility function for our metric. A utility function φ takes as argument the position of each document and should satisfy that at the first position it will take the maximum value 1 and as position increases φ converges to 0. In other words, $\varphi(1) = 1$ and $\varphi(r) \to 0$ as r goes to $+\infty$. In accordance with ERR metric, we select the special case $\varphi(r) = 1/r$. As a result, our metric can be defined as follows:

$$RRP := \sum_{r=1}^{n} \frac{1}{r} \prod_{i=1}^{r-1} (1 - R_i) R_r.$$

Compared to other cascade-based metrics, RRP relies not only on the documents' relevance judgment but also on the popularity grade which consists a real-time "judgment" of users. Thus, our metric captures user behavior more successfully.

6   Evaluation

It is known that the evaluation of a new retrieval metric is not an easy task due to the fact that there is no ground truth to compare with. However, it has been shown [12] that the quality of a retrieval system can be well estimated by click-trough data. A common technique though is comparison with click metrics. That way one can see how well does a metric approximates user behavior and as a result captures user satisfaction. In this section, we try to evaluate our novel metric by computing correlations between click metrics and editorial metrics including our proposed metric.

6.1   Data Collection

In order to collect click through data we developed an information retrieval system based on Indri Search Engine (Lemur Project [11]) using TREC Web Track data. We then asked users from Information Retrieval Class of the Computer Engineering and Informatics Department of Patras to perform an informational search on a number of predefined queries by TREC Web Tracks. Particularly, we used 200 queries from TREC Web Tracks 2009 to 2012 and their document results at depth 20 as returned by Indri. Each user was asked to perform search on 200 predefined queries according to a special informational need. Click-through data from each interaction with the system were collected in log files.

We intersected these click through data with relevance judgments as they were defined by TREC Web Tracks on a five-grade scale ($0 \rightarrow 4$). We also intersected click through data with popularity grades which were computed from daily page views as shown in section 4. Finally, each document was graded according to a five-grade scale from 0 to 4.

The final dataset consisted of relevance and popularity grades for each result-document of each query. For some queries of TREC Web Tracks though, all of their results-documents were graded with zero relevance judgments or with zero popularity grades. We removed these queries in order to simulate a search process which reflects reality, ending with click-through data for 167 different queries.

6.2   Correlation with Click Metrics

Using the dataset described in previous subsection we computed the pearson correlation between a set of click metrics and editorial metrics. Particularly, the editorial metrics used were:

- **nDCG:** Normalized Discounted Cumulative Gain.
- **AP:** Average Precision, where grades perfect, excellent and good are mapped to *relevant* and the rest to *non-relevant*.
- **ERR:** Expected Reciprocal Rank.
- **RRP:** Our proposed metric.

The click metrics used were:

- **PLC** Precision an Lowest Rank as defined by [2].
- **Max, Mean and Min RR** Maximum, Mean and Minimum Reciprocal Ranks of the clicks.
- **UCTR** Binary variable indicating whether there was a click or not in a session.

We did not include the Search Success (SS) metric induced by Chapelle et al. [2], as it uses relevances not only clicks. We also confirmed the findings of [2], that QCTR has negative or close to zero correlation with all the editorial metrics and skipped it as well. Table 2 shows correlations between all the above mentioned metrics.

From table 2 we conclude that our proposed metric shows higher scores in correlation with click metrics than other editorial metrics. Particularly, *RRP* outperforms position-based metrics such as *nDCG* and *Average Precision*; additionally *RRP* correlates better with click metrics even than the cascade-based metric *ERR*.

As click metrics are concerned, we observe that *reciprocal ranks (max,min and mean)* along with *precision at lowest rank* seem to correlate better with *ERR* and our proposed metric *RRP*. This holds due to the fact that they both use as utility function the reciprocal rank $\frac{1}{k}$. *Precision at lowest rank* uses reciprocal rank of the lowest position and as a result shows higher correlation with *ERR* and *RRP* too.

Table 2. Pearson Correlation between editorial and click metrics

	PLC	MeanRR	MinRR	MaxRR	UCTR
nDCG	0.498	0.497	0.503	0.445	-0.024
AP	0.402	0.417	0.395	0.396	-0.004
ERR	0.528	0.512	0.517	0.459	0.064
RRP	0.559	0.554	0.588	0.472	0.041

We can see though, that in most cases our proposed metric appears to score significantly better that other editorial metrics except for *UCTR*. This can be explained as *UCTR* does not account for clicks (rather than their absence) and therefore lacks the source of correlation with click metrics. Moreover, overall results of *UCTR* enhance our assertion as it seems to correlate with editorial metrics much lower (around zero) than other click metrics do.

6.3   Performance on Incomplete Collection

Search engines are using an enormous amount of data, which is constantly changing. Thus it is difficult to maintain complete relevance judgments even for a small proportion of the web corpus. As a result there exists a set of queries which partly or completely lacks relevance judgments.

There has been an effort to confront this problem in various ways over the years. Buckley and Voorhees [1] in their work propose a robust preference-based

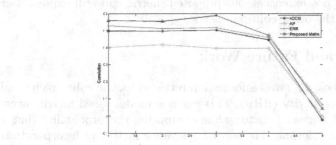

Fig. 1. Correlation between Editorial and Click metrics when no queries are "unjudged"

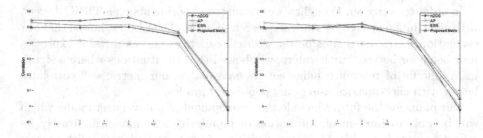

Fig. 2. Correlation between Editorial and Click metrics when 41 and 83 queries are "unjudged" respectively

measure called *bpref* which measures the effectiveness of a system on the basis of judged documents only. On the other side, Sakai [14] proposes an alternative solution which does not need a new metric.

Our proposed metric which accommodates two factors of relevance can battle the missing relevance judgment issue in a complete different way. In situations when there is a document with missing a relevance judgment, we employ the second factor, the popularity grade which accounts for user preference. In order to enhance our assertion we conducted experiments on how correlation between click metrics and our proposed metric was affected when some queries of our dataset lacked relevance grades. Particularly, from the initial set of 167 queries, we created 2 sets where in the first set 41 from 167 queries lacked relevance grades and in the second set 83 from 167 queries lacked relevance grades. We then computed the correlation between click and editorial metrics of each partly "unjudged" dataset.

Figures 1, 2 and 3 show the correlation diagram for the initial dataset and the two partly "unjudged" datasets. In figure 2 where a small set of the dataset is "unjudged", our metric retains the highest scores in correlation with click metrics. As we increase the amount of "unjudged" queries (half of the dataset contains no relevance grades), we observe deterioration of our metric's performance. We can easily conclude that in cases when the amount of relevance judgments absence is not extensive, our proposed metric can still express user behavior better than the other editorial metrics.

7   Conclusions and Future Work

In this paper, we proposed a novel information retrieval metric called reciprocal rank with webpage popularity (RRP). This cascade model-based metric incorporates an additional relevance factor while computing the probability that a user finds a document relevant. The second relevance factor we incorporate is called popularity grade and is calculated by the number of daily page views a document gets. The number of daily page views of a web page, can be viewed as a users' vote for this web page and thus, by combining this factor with experts' relevance judgments, our evaluation metric captures user behavior better.

In order to verify our thoughts, we conducted experiments on a TREC dataset with click data collected by students' search sessions. The results showed that our evaluation metric correlates better with click-metrics and as a result expresses user behavior better. Furthermore, we showed that in situations where a significant amount of relevance judgments is unavailable, our metric still correlates better with click-metrics using only popularity grades.

Our plans for the future include the development of a novel user model which will be able to implement the notion of popularity and model accurately its reflect on user behavior. Moreover, new experiments on a dataset with greater number of queries and sessions is another thought in order to enhance the user model's findings.

Acknowledgments. This research has been co-financed by the European Union (European Social Fund – ESF) and Greek national funds through the Operational Program "Education and Lifelong Learning" of the National Strategic Reference Framework (NSRF) - Research Funding Program: Thales. Investing in knowledge society through the European Social Fund.

References

1. Buckley, C., Voorhees, E.M.: Retrieval evaluation with incomplete information. In: Proceedings of the 27th Annual International ACM SIGIR Conference on Research and Development in Information Retrieval, SIGIR 2004, pp. 25–32. ACM, New York (2004)
2. Chapelle, O., Metlzer, D., Zhang, Y., Grinspan, P.: Expected reciprocal rank for graded relevance. In: Proceedings of the 18th ACM Conference on Information and Knowledge Management, CIKM 2009, pp. 621–630. ACM, New York (2009)
3. Chapelle, O., Zhang, Y.: A dynamic bayesian network click model for web search ranking. In: Proceedings of the 18th International Conference on World Wide Web, WWW 2009, pp. 1–10. ACM, New York (2009)
4. Cho, J., Roy, S., Adams, R.E.: Page quality: In search of an unbiased web ranking. In: Proceedings of the 2005 ACM SIGMOD International Conference on Management of Data, SIGMOD 2005, pp. 551–562. ACM, New York (2005)

5. Chuklin, A., Serdyukov, P., de Rijke, M.: Click model-based information retrieval metrics. In: Proceedings of the 36th International ACM SIGIR Conference on Research and Development in Information Retrieval, SIGIR 2013, pp. 493–502. ACM, New York (2013)
6. Cyril, W.: Cleverdon. The significance of the cranfield tests on index languages. In: Proceedings of the 14th Annual International ACM SIGIR Conference on Research and Development in Information Retrieval, SIGIR 1991, pp. 3–12. ACM, New York (1991)
7. Craswell, N., Zoeter, O., Taylor, M., Ramsey, B.: An experimental comparison of click position-bias models. In: Proceedings of the 2008 International Conference on Web Search and Data Mining, WSDM 2008, pp. 87–94. ACM, New York (2008)
8. Huffman, S.B., Hochster, M.: How well does result relevance predict session satisfaction? In: Proceedings of the 30th Annual International ACM SIGIR Conference on Research and Development in Information Retrieval, SIGIR 2007, pp. 567–574. ACM, New York (2007)
9. Järvelin, K., Kekäläinen, J.: Cumulated gain-based evaluation of ir techniques. ACM Trans. Inf. Syst. 20(4), 422–446 (2002)
10. Moffat, A., Zobel, J.: Rank-biased precision for measurement of retrieval effectiveness. ACM Trans. Inf. Syst., 27(1), 2:1–2:27 (2008)
11. University of Massachusetts and Carnegie Mellon University. The lemur project (January 2014), http://www.lemurproject.org/
12. Radlinski, F., Kurup, M., Joachims, T.: How does clickthrough data reflect retrieval quality? In: Proceedings of the 17th ACM Conference on Information and Knowledge Management, CIKM 2008, pp. 43–52. ACM, New York (2008)
13. Richardson, M., Dominowska, E., Ragno, R.: Predicting clicks: Estimating the click-through rate for new ads. In: Proceedings of the 16th International Conference on World Wide Web, WWW 2007, pp. 521–530. ACM, New York (2007)
14. Sakai, T.: Alternatives to bpref. In: Proceedings of the 30th Annual International ACM SIGIR Conference on Research and Development in Information Retrieval, SIGIR 2007, pp. 71–78. ACM, New York (2007)
15. Sanderson, M., Zobel, J.: Information retrieval system evaluation: Effort, sensitivity, and reliability. In: Proceedings of the 28th Annual International ACM SIGIR Conference on Research and Development in Information Retrieval, SIGIR 2005, pp. 162–169. ACM, New York (2005)

A Survey of Geo-tagged Multimedia Content Analysis within Flickr

Evaggelos Spyrou[1] and Phivos Mylonas[2]

[1] Computational Intelligence Laboratory (CIL),
Institute of Informatics and Telecommunications,
National Center for Scientific Research Demokritos, Athens, Greece
espyrou@iit.demokritos.gr
[2] Department of Informatics
Ionian University, Corfu, Greece
fmylonas@ionio.gr

Abstract. Our survey paper attempts to investigate how recent and undoubted emerge in enriched, geo-tagged social networks' multimedia content sharing works to the benefit of their users and whether it could be handled in a formal way, in order to capture the meaningful semantics rising from this newly introduced user experience. It further specializes its focus by providing an overview of current state-of-the-art techniques with respect to geo-tagged content access, processing and manipulation within the popular Flickr social network. In this manner it explores the role of information retrieval, integration and extraction from the technical point of view, coupled together with human social network activities, like, for instance, localization and recommendations based on pre-processed collaborative geo-tagged photos, resulting into more efficient, optimized search results.

Keywords: Flickr, tags, photos, multimedia, social networks.

1 Introduction

Current digital era is characterized by a single, yet very important observation: an extremely large amount of digital multimedia content is shared online every moment by people interacting within the so-called "social networks". This online social networking explosion shows no signs of abating, with almost twice as many Internet users having an online social profile than two years ago, helping to make Facebook[1] the most viewed website in the European Union, according to recent research efforts. On top of that, more people are using the Internet to create their own multimedia content than ever before, with 73% of online users having a social networking profile [1], compared with 37% in 2008 [2].

Flickr[2] is an image and video hosting website created by a Vancouver-based company named "Ludicorp" back in 2004 and currently owned by Yahoo![3]. What

---

[1] http://www.facebook.com
[2] http://www.flickr.com
[3] http://www.yahoo.com

L. Iliadis et al. (Eds.): AIAI 2014 Workshops, IFIP AICT 437, pp. 126–135, 2014.

makes it special among other social networks is its aspect as an online community, within which users are able to interact by sharing comments about photography and create groups of particular interests. The Verge reported in March 2013 that "Flickr had a total of 87 million registered members and more than 3.5 million new photos uploaded daily"[4]. Each photo may contain metadata added by its photographer, such as tags that describe either its visual content or location, or a free text description. It also contains metadata added by the camera that has been used, such as date taken, camera settings, camera model, etc. Few Global Positioning System (GPS) enhanced cameras automatically geo-tag the photos they take, but in principal this is done manually, by the photographer. The vast majority of images uploaded to Flickr are taken by common users or amateur photographers. The textual metadata associated with the image often serves as a reminder of the context of the image for the photographer and his social circle [3], [4].

In principle, every part of a photo may be tied to a geographic location, but in most typical applications, only the position of the photographer is associated with the entire digital photo. As the reader may imagine, this small detail implicates and significantly burdens most multimedia content search and retrieval tasks. In the most typical example, photos of a landmark may have been taken from very different positions apart and in order to identify all photos of this particular landmark within an image database, all photos taken within a reasonable circular distance from it must be considered. Now, when such geo-tagged photos are uploaded to online multimedia content sharing communities, such as Flickr, Panoramio[5] or Instagram[6], that enable the construction of infinite connections among their users [5], a photo can be placed onto a map to view the location the photo was taken. In this way, social network users can browse photos from a map, search for photos from a given area, and find related photos of the same place from other users; these tasks are considered elementary in order to build additional, ad-hoc value-added digital services on top, like automated route/trip planning or like, to our most recent knowledge, the popular "NOW" app; the latter uses geo-tagged Instagram photos to find nearby events happening now[7].

The act of automatically providing or calculating meaningful photo's geo-tags (the so-called "geo-tagging" process) opens a huge research topic for the researchers' community, mainly to the direction of being able to analyze them, to identify and determine social patterns amongst them. However, issues of credibility on the volunteered user-generated geo-tagging should become of broader research interest in various areas [6], [7], motivating us to further investigate this topic in the following, focusing on the popular Flickr social network. At this point, it should be noted that our work differentiates from previous similar surveys, since they emphasized either on geotagged content without focusing

---

[4] http://en.wikipedia.org/wiki/Flickr#cite_note-4

[5] http://www.panoramio.com

[6] http://www.instagram.com

[7] http://techcrunch.com/2013/01/11/now-app/

specifically on user generated and manually geotagged photos of Flickr [8], or in social media in general [9].

2   Multimedia Content Retrieval

The very first research community dealing with up-to-date multimedia community research challenges is the one depicted by Information Retrieval in general and multimedia content retrieval in particular. Since Flickr is mainly a photo sharing website, the fact that it attracted the interest of the image retrieval community is considered to be rather natural. The main approach followed by researchers is to use either textual metadata or visual properties of photos and often combine them in an effort to improve the accuracy of their respective algorithms. As depicted in the following, such research efforts vary from textual ones, to ones based on low level visual or even hybrid characteristics.

2.1   Aren't Tags Text, After All?

As discussed, the first approach to tackle the problem at hand is based solely on text retrieval, a branch of information retrieval where information is manipulated primarily in the form of text, having each photo represented solely by its textual features, i.e. the manually generated tags. Abbasi et al. [10] identified landmarks using tags and Flickr groups, without exploiting geospatial information. They used SVM classifiers trained on thematical Flickr groups, in order to find relevant landmark-related tags. Ahern et al. [11] analyzed tags associated with geo-referenced Flickr images so as to generate knowledge. This knowledge was a set of the most "representative" tags for an area. They used a TF-IDF approach and presented a visualization tool, namely the *World Explorer*, which allowed users explore their results. Serdyukov et al. [12] adopted a language model which lies on the user collected Flickr metadata and aimed to annotate an image based on these metadata. Their goal was to place photos on a map, i.e. provide an automatic alternative to manual geo-tagging. Venetis et al. [13] examined techniques to create a "tag-cloud", i.e. a set of terms/tags able to provide a brief yet rich description of a large set of terms/tags. They presented and defined certain user models, metrics and algorithms aiming at this goal. Lerman et al. [14] aimed to personalize text-based search results by adding information about users' relations. Finally, Larson et al. [15] tried to detect whether tags correspond to physical objects, and also the scale of these objects, using a natural language approach.

2.2   Shall We Consider Visual Characteristics?

The second approach focuses on the visual aspects of multimedia content analysis. Research efforts in this area discard textual annotations and focus on low-level visual features. Wang et al.[16] proposed a training algorithm and applied it on the problem of image similarity. They worked on a Flickr data set, under the assumption that two images are considered similar if they belong to the same group. Chatzilari et al. [17] used region level annotations and visual features, in an effort to recognize objects with a semi-supervised approach. They started from a set of Flickr photos that contain the same object. Philbin and Zisserman [18] created a graph based on visual features and tried to group similar Flickr photos, from a corpus of 1M photos. Avrithis et al. [19] retrieved similar Flickr photos by using a 2-level clustering processing, both by means of geo-tags and visual features. Yanai et al. [20] focused on the relationship between words and locations. They used visual features and tried to associate them with certain locations, using an entropy based approach. Li et al. [21] used SVMs trained on visual features to classify a 30M data set. They observed that by incorporating temporal information, the accuracy of the results was significantly improved. Joshi and Luo [22] used visual detectors and incorporate bags-of-geotags within a probabilistic framework, in order to detect activities and events in photos. Yu and Luo [23] combined visual context with location information in order to detect concepts in photos. Luo et al. [24] fused information extracted from both a Flickr data set and a set of satellite images, in order to detect events. Batko et al. [25] used MPEG-7 visual features and search into a set of over 50M photos from Flickr. Seah et al. [26] created visual summaries on the results of visual queries on a data set of Flickr images that in contrast to previous works, e.g., the one of [27], they attempted to generate concept-preserving summaries. Finally, Liu et al. [28] incorporated the social aspect of photos, in order to re-rank search results more according to both social and visual relevances.

2.3   A Little Bit of Both – The Hybrid Approach!

Since the visual content of images may provide a powerful description, many research efforts try to combine visual descriptions with textual metadata. Barrios et al. [29] presented an image retrieval system that combines textual and visual content. They downloaded and stored locally images from Flickr and used simple color and texture visual descriptors, along with the title, description and tags, for each image. Crandall et al [30] used visual, temporal and geospatial information to automatically identify places and/or events in city and landmark level. They also added temporal metadata information to improve classification performance. With the same motivation, Quack et al. [31] divided the area of interest into non-overlapping, square tiles, then extracted and used visual, textual and geospatial features. They handled tags by a modified TF-IDF ranking and linked their results to Wikipedia[8]. Gammeter et al. [32] overlaid a geospatial grid over earth and matched pairwise retrieved photos of each tile using visual features. Then they clustered photos into groups of images depicting the same scene. The metadata were used to label these clusters automatically, using a TF-IDF scheme. Moëllic et al [27] aimed to extract meaningful and representative clusters from large-scale image collections. They proposed a method based on a shared nearest neighbors approach that treats both visual features and tags. Li

---

[8] http://www.wikipedia.org

et al [33] proposed an algorithm that learns tag relevance by voting from visually similar neighbors. They did not use geospatial data, nor limited their approach on landmarks/places of interest and aimed to retrieve semantically similar images. Moxley et al. [34] classified mined geo-referenced tags as places, by extending [35], landmarks by clustering image datasets considering mutual information and prior knowledge from Wikipedia and visual terms using the mutual information between visual descriptors and tags. Ulges et al. [36] adopted a context- based approach, assuming that users place semantically similar photos in Flickr groups. Fan et al. [37] proposed a system, namely *JustClick* which exploits both visual and textual information and after a search and retrieval process, it recommends photos using an interactive interface. Simon et al. [38] created visual summaries of large image data set based mainly on visual features, but also exploiting tags. Kennedy and Naaman [39] used visual features and tags, in order to extract the most representative tags and views for landmarks, working on a corpus of 110K Flickr photos from San Fransisco. Finally, Liu et al. [40] were the first to consider user uploading patterns, geotagging behaviors, and the relationship between the temporal and the spatial gap of two photos from the same user.

3   Automatic Tag/Geo-tag Generation

In a slightly different approach, special attention has been given to methods exploiting the automatic generation of tags, a process often called "(tag-)recommendation", as well as the prediction of geo-tags, i.e., of the geographic coordinates where a photo has been taken, a process often referred to as "localization". In the following, we briefly present both approaches, summarizing most important research works in the fields.

3.1   Tag Recommendation

Initially and as expected, tag recommendation approaches often adopt traditional tag processing techniques. In this manner, Chen et al. [41] proposed a system that automatically recommends tags for photos and also for adding photos into appropriate popular groups. For the latter case, they used SVM predictors in order to identify concepts and used these results so as to search for groups. Then, they used these groups to harvest more tags and attach them to their photos. Anderson et al. [42] presented a system, namely *TagEz* which combined both textual and visual features, so as to recommend tags. Their results indicated that the use of textual metadata outperformed both visual and combined feautures. Chaundry et al. [43] presented an approach for tag assignment to geographic areas, using a TF-IDF scheme and logistic regression, for various levels of detail. Hsieh and Hsu [44] exploited visual similarity and after a tag expansion process, aim to automatically annotate photos. Kennedy et al. [45] selected representative tags from urban areas using a multimodal approach. Their results indicate that the use of visual features can drastically improve precision. Sigurbjörnsson and Van Zwol [46] extracted tag co-occurence statistics and tag aggregation algorithms, in order to recommend tags by investigating and evaluating four different strategies. Furthermore, they introduced a "promotion" function, whose role was to promote the most descriptive tags. Garg and Weber [47], [48] presented a system that while users tagged their photos, it dynamically suggested related tags by considering similar groups to user's preferences. Moxley et al [49] presented *SpiritTagger* tool, in order to recommend tags for Flickr photos of urban regions, which is unaware of the user's tags and lies on visual properties and geographic distance, in order to select similar photos. Popescu and Moëllic [50] presented *Monuanno*, a system that uses visual features to automatically annotate georeferenced landmark images. Kleban et al. [51] presented a world scale system for tag recommendation, based on geotags and visual features. Finally, Chen and Shin [52] used both textual and social features of tags and a machine learning approach, in order to extract representative tags that can be related to the users favorite topics.

3.2   Content Localization

On the other hand, automatic geo-tag generation has gained huge research interest, mainly due to the vast available Flickr database of geo-tagged photos. Kelm et al. [53] adopted a hierarchical approach and tried to automatically predict geo-tags for Flickr videos. Their technique lies on both textual and visual features and also uses external resources, such as Geonames[9] and Wikipedia. Van Laere et al. [54] trained naive Bayes classifiers at different spatial resolutions. They used only textual features and worked at various spatial resolutions, for a set of 55 european cities. De Rouck et al [55] used language probabilistic models that have been trained on Flickr photos, in order to geo-tag Wikipedia pages. Their approach outperformed Yahoo! Placemaker[10] and their results indicated that the increasing growth of tagged content in Flickr would continuously improve their accuracy. Friedland et al. [56] combined textual and visual features and worked on a the MediaEval 2010 data set. They concluded that solely visual information proves inadequate for accurate geo-localization, but when combined with textual it can assist on the improvement of the accuracy. Hauff and Houben [57] added information considering user's activities in Twitter[11]. However, even if their results were promising, the median location error was still far from usable. Van Laere et al. [58] divided areas into disjoint regions and then used statistics and a Naive Bayes classifier. Van Laere et al. [59] proposed a tag-based approach that uses language models and similarity search, in order to estimate geo-tags based on a training set. Friedland et al. [60] worked on Flickr videos and used both textual and visual metadata. Their results may seem poor, however they were superior to all other contributions of MediaEval 2010 [12]. Kalantidis et al. [61] presented *Viral*, a system that aims to localize photos uploads by users,

---

[9] http://www.geonames.org/

[10] http://www.programmableweb.com/api/yahoo-placemaker

[11] http://www.twitter.com

[12] http://www.multimediaeval.org/mediaeval2010/

based on the visual similarity to geo-tagged images. They used a database of more than 2M photos taken from 40 cities. In previous work [62] we suggested a probabilistic framework which aimed to place Flickr data on a map based on their tags.

Joshi et al. [63] proposed a probabilistic framework for tag-based localization. Their work was extended by Gallagher et al. [64] who used a large geotagged corpus from Flickr, extracted several visual features and used location probability maps for tags. They integrated them and tried to localize photos. Hays and Efros [65] present *IM2GPS*, a system for image localization using visual features. It should be noted that they provided a probability distribution over the Earth. Kalogerakis et al. [66] extended this work by adding temporal information, in an effort to extract information about image sequences. O' Hare and Murdock [67] presented a statistical language modeling approach, in order to identifying locations in arbitrary text. They investigated several ways to estimate models, based on the term and the user frequencies. To this goal. they used a set of public, geo-tagged photos in Flickr as ground truth. Hare et al. [68] estimated a continuous probability density function (PDF) over the Earth and combined textual with a number of weighted visual features. Their approach on tags differs from the others as they do not filter any of the tags, but they rather use them for evidence, i.e. certain words may be associated with certain countries. Finally, Li et al [69] removed "noisy" photos, i.e. photos that cannot contribute sufficiently to location estimation. They extracted both local and global features and instead of using the whole dataset, they performed clustering and use the resulting centroids, instead.

4   Conclusions

In this position paper we attempted to conduct a detailed survey and provide a brief summarization of current state-of-the-art techniques regarding geo-tagged Flickr content access, processing and manipulation issues. In this manner, we explored related research efforts mostly focused on information retrieval tasks. Our intention was to identify the trends in the surveyed area and organize them in a novel way that would integrate and add understanding to the work in the field with respect to the Flickr social network, so as for fellow researchers to be able to seek and reference related information efficiently. Among our future work is the extension of this survey to other Flickr application domains, other popular social networks and even other content types, such as text snippets according to the social network under investigation.

References

1. Dugan, M., Smith, A.: Social media update 2013 (2013),
   http://pewinternet.org/Reports/2013/Social-Media-Update.aspx
2. Lenhart, A., Purcell, K., Smith, A., Zickuhr, K.: Social media & mobile internet use among teens and young adults (2010),
   http://pewinternet.org/Reports/2010/Social-Media-and-Young-Adults.aspx

3. Nov, O., Naaman, M., Ye, C.: Analysis of participation in an online photo-sharing community: A multidimensional perspective. Journal of the American Society for Information Science and Technology 61(3), 555–566 (2010)
4. Van House, N.A.: Flickr and public image-sharing: distant closeness and photo exhibition. In: Proc. of ACM CHI (2007)
5. Van Dijck, J.: Flickr and the culture of connectivity: Sharing views, experiences, memories. Memory Studies 4(4), 401–415 (2011)
6. Flanagin, A.J., Metzger, M.J.: The credibility of volunteered geographic information. Geo Journal 72(3-4), 137–148 (2008)
7. Spielman, S.E.: Spatial collective intelligence? credibility, accuracy, and volunteered geographic information. Cartography and Geographic Information Science (ahead-of-print), 1–10 (2014)
8. Luo, J., Joshi, D., Yu, J., Gallagher, A.: Geotagging in multimedia and computer vision a survey. Multimedia Tools and Applications 51(1), 187–211 (2011)
9. Wu, J., Sun, H., Tan, Y.: Social media research: A review. Journal of Systems Science and Systems Engineering 22(3), 257–282 (2013)
10. Abbasi, R., Chernov, S., Nejdl, W., Paiu, R., Staab, S.: Exploiting flickr tags and groups for finding landmark photos. In: Boughanem, M., Berrut, C., Mothe, J., Soule-Dupuy, C. (eds.) ECIR 2009. LNCS, vol. 5478, pp. 654–661. Springer, Heidelberg (2009)
11. Ahern, S., Naaman, M., Nair, R., Yang, J.H.-I.: World explorer: visualizing aggregate data from unstructured text in geo-referenced collections. In: Proc. of the ACM/IEEE-CS JCDL (2007)
12. Serdyukov, P., Murdock, V., Zwol, R.V.: Placing flickr photos on a map. In: Proc. of ACM SIGIR (2009)
13. Venetis, P., Koutrika, G., Garcia-Molina, H.: On the selection of tags for tag clouds. In: Proc. of ACM WSDM (2011)
14. Lerman, K., Plangprasopchok, A., Wong, C.: Personalizing image search results on flickr (2007)
15. Larson, M., Kofler, C., Hanjalic, A.: Reading between the tags to predict real-world size-class for visually depicted objects in images. In: Proc. of ACM MM, pp. 273–282 (2011)
16. Wang, G., Hoiem, D., Forsyth, D.: Learning image similarity from flickr groups using stochastic intersection kernel machines. In: Proc. of IEEE ICCV (2009)
17. Chatzilari, E., Nikolopoulos, S., Papadopoulos, S., Zigkolis, C., Kompatsiaris, Y.: Semi-supervised object recognition using flickr images. In: Proc. of IEEE CBMI (2011)
18. Philbin, J., Zisserman, A.: Object mining using a matching graph on very large image collections. In: Proc. of IEEE ICVGIP (2008)
19. Avrithis, Y., Kalantidis, Y., Tolias, G., Spyrou, E.: Retrieving landmark and non-landmark images from community photo collections. In: Proc. of ACM MM (2010)
20. Yanai, K., Kawakubo, H., Qiu, B.: A visual analysis of the relationship between word concepts and geographical locations. In: Proc. of ACM ICIVP (2009)
21. Li, Y., Crandall, D.J., Huttenlocher, D.P.: Landmark classification in large-scale image collections. In: Proc. of IEEE ICCV (2009)
22. Joshi, D., Luo, J.: Inferring generic activities and events from image content and bags of geo-tags. In: Proc. of ACM CBIVR (2008)
23. Yu, J., Luo, J.: Leveraging probabilistic season and location context models for scene understanding. In: Proc. of ACM CBIVR (2008)
24. Luo, J., Yu, J., Joshi, D., Hao, W.: Event recognition: viewing the world with a third eye. In: Proc. of ACM MM (2008)

25. Batko, M., Falchi, F., Lucchese, C., Novak, D., Perego, R., Rabitti, F., Sedmidub-sky, J., Zezula, P.: Building a web-scale image similarity search system. Multimedia Tools and Applications 47(3), 599–629 (2010)
26. Seah, B.-S., Bhowmick, S.S., Sun, A.: Summarizing social image search results. In: Proc. of ACM WWW Companion (2014)
27. Moëllic, P.-A., Haugeard, J.E., Pitel, G.: Image clustering based on a shared nearest neighbors approach for tagged collections. In: Proc. of ACM CIVR (2008)
28. Liu, S., Cui, P., Luan, H., Zhu, W., Yang, S., Tian, Q.: Social-oriented visual image search. Computer Vision and Image Understanding 118, 30–39 (2014)
29. Barrios, J.M., Dıaz-Espinoza, D., Bustos, B.: Text-based and content-based image retrieval on flickr: Demo
30. Crandall, D.J., Backstrom, L., Huttenlocher, D., Kleinberg, J.: Mapping the world's photos. In: Proc. of ACM WWW (2009)
31. Quack, T., Leibe, B., Van Gool, L.: World-scale mining of objects and events from community photo collections. In: Proc. of ACM CIVR (2008)
32. Gammeter, S., Bossard, L., Quack, T., Gool, L.V.: I know what you did last sum-mer: object-level auto-annotation of holiday snaps. In: Proc. of IEEE ICCV (2009)
33. Li, X., Snoek, C.G., Worring, M.: Learning tag relevance by neighbor voting for social image retrieval. In: Proc of ACM MIR (2008)
34. Moxley, E., Kleban, J., Xu, J., Manjunath, B.S.: Not all tags are created equal: Learning flickr tag semantics for global annotation. In: Proc. of IEEE ICME (2009)
35. Rattenbury, T., Good, N., Naaman, M.: Towards automatic extraction of event and place semantics from flickr tags. In: Proc. of ACM SIGIR (2007)
36. Ulges, A., Worring, M., Breuel, T.: Learning visual contexts for image annotation from flickr groups. IEEE Trans. on Multimedia 13(2), 330–341 (2011)
37. Fan, J., Keim, D.A., Gao, Y., Luo, H., Li, Z.: Justclick: Personalized image recom-mendation via exploratory search from large-scale flickr images. IEEE Trans. on Circuits and Systems for Video Technology 19(2), 273–288 (2009)
38. Simon, I., Snavely, N., Seitz, S.M.: Scene summarization for online image collec-tions. In: Proc. of ICCV (2007)
39. Kennedy, L.S., Naaman, M.: Generating diverse and representative image search results for landmarks. In: Proc. of ACM WWW (2008)
40. Liu, B., Yuan, Q., Cong, G., Xu, D.: Where your photo is taken: Geolocation prediction for social images. Journal of the Association for Information Science and Technology (2014)
41. Chen, H.M., Chang, M.H., Chang, P.C., Tien, M.C., Hsu, W.H., Wu, J.L.: Sheep-dog: group and tag recommendation for flickr photos by automatic search-based learning. In: Proc. of ACM MM (2008)
42. Anderson, A., Ranghunathan, K., Vogel, A.: Tagez: Flickr tag recommendation. Association for the Advancement of Artificial Intelligence (2008)
43. Chaudhry, O., Mackaness, W.: Automated extraction and geographical structuring of flickr tags. In: Proc. of GEOProcessing (2012)
44. Hsieh, L.C., Hsu, W.H.: Search-based automatic image annotation via flickr photos using tag expansion. In: Proc. of IEEE ICASSP (2010)
45. Kennedy, L., Naaman, M., Ahern, S., Nair, R., Rattenbury, T.: How flickr helps us make sense of the world: context and content in community-contributed media collections. In: Proc. of ACM MM (2007)
46. Sigurbjörnsson, B., Van Zwol, R.: Flickr tag recommendation based on collective knowledge. In: Proc. of ACM WWW (2008)
47. Garg, N., Weber, I.: Personalized tag suggestion for flickr. In: Proc. of ACM WWW (2008)

48. Garg, N., Weber, I.: Personalized, interactive tag recommendation for flickr. In: Proc. of ACM RecSys (2008)
49. Moxley, E., Kleban, J., Manjunath, B.S.: Spirittagger: a geo-aware tag suggestion tool mined from flickr. In: Proc. of ACM MIR (2008)
50. Popescu, A., Moëllic, P.-A.: Monuanno: automatic annotation of georeferenced landmarks images. In: Proc. of ACM CIVR (2009)
51. Kleban, J., Moxley, E., Xu, J., Manjunath, B.S.: Global annotation on georeferenced photographs. In: Proc. of ACM CIVR (2009)
52. Chen, X., Shin, H.: Tag recommendation by machine learning with textual and social features. Journal of Intelligent Information Systems 40(2), 261–282 (2013)
53. Kelm, P., Schmiedeke, S., Sikora, T.: A hierarchical, multi-modal approach for placing videos on the map using millions of flickr photographs. In: Proc. of ACM SBNMA (2011)
54. Van Laere, O., Schockaert, S., Dhoedt, B.: Combining multi-resolution evidence for georeferencing flickr images. In: Deshpande, A., Hunter, A. (eds.) SUM 2010. LNCS, vol. 6379, pp. 347–360. Springer, Heidelberg (2010)
55. De Rouck, C., Van Laere, O., Schockaert, S., Dhoedt, B.: Georeferencing wikipedia pages using language models from flickr (2011)
56. Friedland, G., Choi, J., Lei, H., Janin, A.: Multimodal location estimation on flickr videos. In: Proc. of ACM WSM (2011)
57. Hauff, C., Houben, G.-J.: Geo-location estimation of flickr images: Social web based enrichment. In: Baeza-Yates, R., de Vries, A.P., Zaragoza, H., Cambazoglu, B.B., Murdock, V., Lempel, R., Silvestri, F. (eds.) ECIR 2012. LNCS, vol. 7224, pp. 85–96. Springer, Heidelberg (2012)
58. Van Laere, O., Schockaert, S., Dhoedt, B.: Towards automated georeferencing of flickr photos. In: Proc. of ACM GIR (2010)
59. Van Laere, O., Schockaert, S., Dhoedt, B.: Finding locations of flickr resources using language models and similarity search. In: Proc. of ACM ICMR (2011)
60. Friedland, G., Choi, J., Janin, A.: Video2gps: a demo of multimodal location estimation on flickr videos. In: Proc. of ACM MM (2011)
61. Kalantidis, Y., Tolias, G., Avrithis, Y., Phinikettos, M., Spyrou, E., Mylonas, P., Kollias, S.: Viral: Visual image retrieval and localization. Multimedia Tools and Applications 51(2), 555–592 (2011)
62. Spyrou, E., Mylonas, P.: Placing user-generated photo metadata on a map. In: Proc. of IEEE SMAP (2011)
63. Joshi, D., Gallagher, A., Yu, J., Luo, J.: Exploring user image tags for geo-location inference. In: Proc. of IEEE ICASSP (2010)
64. Gallagher, A., Joshi, D., Yu, J., Luo, J.: Geo-location inference from image content and user tags. In: Proc. of IEEE CVPR (2009)
65. Hays, J., Efros, A.A.: Im2gps: estimating geographic information from a single image. In: Proc. of IEEE CVPR (2008)
66. Kalogerakis, E., Vesselova, O., Hays, J., Efros, A.A., Hertzmann, A.: Image sequence geolocation with human travel priors. In: Proc. of IEEE ICCV (2009)
67. O'Hare, N., Murdock, V.: Modeling locations with social media. Information Retrieval 16(1), 30–62 (2013)
68. Hare, J., Davies, J., Samangooei, S., Lewis, P.H.: Placing photos with a multimodal probability density function
69. Li, J., Qian, X., Tang, Y.Y., Yang, L., Mei, T.: Gps estimation for places of interest from social users' uploaded photos. IEEE Transactions on Multimedia 15(8), 2058–2071 (2013)

Algebraic Interpretations
Towards Clustering Protein Homology Data

Fotis E. Psomopoulos[1,*] and Pericles A. Mitkas[2]

[1] Center for Research and Technology Hellas,
GR570 01, Thessaloniki, Greece
[2] Dept. of Electrical and Computer Engineering,
Aristotle University of Thessaloniki,
GR541 24, Thessaloniki, Greece
`fpsom@issel.ee.auth.gr`

Abstract. The identification of meaningful groups of proteins has always been a principal goal in structural and functional genomics. A successful protein clustering can lead to significant insight, both in the evolutionary history of the respective molecules and in the identification of potential functions and interactions of novel sequences. In this work we propose a novel metric for distance evaluation, when applied to protein homology data. The metric is based on a matrix manipulation approach, defining the homology matrix as a form of block diagonal matrix. A first exploratory implementation of the overall process is shown to produce interesting results when using a well explored reference set of genomes. Near future steps include a thorough theoretical validation and comparison against similar approaches.

1 Introduction

In the era of Big Data, the quest for identifying hidden patterns and relationships is becoming an ever increasingly demanding objective, but at the same time, an imperative goal for most researchers. This situation holds particularly true in the fields of structural and functional genomics, where the need to assign potential functions and interactions to a rapidly expanding number of novel protein sequences is increasingly evident [1] [2]. There exist several algorithms in literature that address the issue of protein data clustering, ranging from generally applicable approaches [3] [4] [5], to highly specialized algorithms tailored for specific studies (i.e. studies focused on particular species [6], sets of genomes [7] or groups of molecules [8]).

A common concept in the vast majority of the clustering algorithms is "protein homology", i.e. the inherent degree of similarity that is assigned to a pair of protein sequences after application of a pair-wise comparison algorithm such as BLAST [9]. This similarity metric is consequently used to define new measures of distance, in order to produce the necessary data partitioning, and therefore, insight into the intrinsic organization of the data involved.

---

[*] Corresponding author.

L. Iliadis et al. (Eds.): AIAI 2014 Workshops, IFIP AICT 437, pp. 136–145, 2014.

A second key issue in any given clustering algorithm is the number of partitions created. As is often the case with big data, the actual number of "correct" and meaningful clusters is either unknown or hard to evaluate. Therefore there are two main approaches towards this issue; either approximate the number of clusters through external algorithms or parameters (such as in the case of k-means), or allow the clustering algorithm to construct an arbitrary number of clusters, based on its inner design (such as in the case of the popular MCL algorithm [3]).

In this work we propose a new algorithm, where the distance metric and an estimate of the clusters to be constructed is directly interpreted from the protein homology data. The rest of the paper is organized as follows: first we define the key concepts and techniques used within the context of the clustering process. The next section outlines the proposed algorithm and formally defines the metrics used. We conclude with the application of the novel metric on a well studied set of target genomes.

2   Problem Outline and Definitions

Attempting to formally define the protein clustering problem, we first need to provide the definitions of the concepts involved.

Given a set of n protein sequences, the homology matrix, a key concept in any protein clustering algorithm, can be defined as an $n \times n$ matrix H as follows:

$$
H = \begin{bmatrix} h_{1,1} & h_{1,2} & \cdots & h_{1,n} \\ h_{2,1} & h_{2,2} & & h_{2,n} \\ \vdots & & \ddots & \vdots \\ h_{n,1} & h_{n,2} & \cdots & h_{n,n} \end{bmatrix} \tag{1}
$$

where $h_{i,j}$ is the expect value of the pair-wise sequence comparison of protein sequence i and j, using the BLAST algorithm. The expect-value, or e-value, is defined as the number of hits (correct alignments) expected to emerge by chance when searching a database of a certain size. At this point it is important to note that matrix H is square but not necessarily symmetrical, i.e. $h_{i,j} \neq h_{j,i}$.

A second key aspect of the homology matrix is that it is inherently sparse, i.e $|\{h_{i,j} \neq 0, \forall i, j\}| \ll |\{h_{i,j} = 0, \forall i, j\}|$. This qualitative characteristic is quantified through the sparsity metric of the matrix, defined as:

$$
s = \frac{|\{h_{i,j} \neq 0, \forall i, j\}|}{|\{h_{i,j} = 0, \forall i, j\}|} \tag{2}
$$

A third aspect of a homology matrix emerges when protein sequences across k genomes (where $k > 1$) are included in the data set. The sparsity pattern of the matrix H reveals a structure reminiscent of a block diagonal matrix. By definition, a block diagonal matrix is a square diagonal matrix in which the diagonal elements are square matrices of any size (possibly even 1×1), and the off-diagonal elements are zero matrices, i.e.:

$$A = \begin{bmatrix} A_1 & 0 & \cdots & 0 \\ 0 & A_2 & & 0 \\ \vdots & & \ddots & \vdots \\ 0 & 0 & \cdots & A_l \end{bmatrix} \quad \text{where each } A_i \text{ is a square matrix} \qquad (3)$$

However, the formal definition of a block diagonal matrix differs with regard to the case of a homology matrix produced from sequences across k genomes. The difference lies in the fact that the off-diagonal elements are not zero, but exhibit significantly higher sparsity percentage s compared to the diagonal elements.

With the above definitions, the problem we are attempting to address can be formally defined as follows: given a homology matrix H, define an appropriate distance metric m applied directly on H, which will be consequently used within an agglomerative clustering process in order to produce a segmentation of H based on the $h_{i,j}$ values.

3   Methods

The algorithm comprises three distinct stages. The first pre-processing stage transforms the pair-wise comparison data into a full homology matrix. This is a necessary step as the standard output of the BLAST algorithm cannot be readily used in matrix manipulations, mainly due to the missing values and the scoring system employed. The second stage uses the constructed matrix in order to identify the target number of clusters. Finally, the proposed distance metric is applied through a standard agglomerative clustering process.

At this point it must be noted that the proposed method does not aim to replace the BLAST algorithm or to provide similar functionality. Instead, by directly utilizing the output of BLAST, we aim to construct a singular analysis method that evaluates this information in the form of clusters.

3.1   Pre-processing Stage

The pair-wise alignment algorithm BLAST, employs a scoring system based on the expect value (e-value). Therefore, by definition, given two sequences seq_i and seq_j, the respective e-value would be:

$$e_{i,j} = \begin{cases} \text{missing value} & \text{, when } seq_i \text{ cannot be aligned with } seq_j \\ a > 0 & \text{, when there exist a valid alignment} \\ 0 & \text{, when the two sequences are identical} \end{cases} \qquad (4)$$

In order to produce the homology matrix H, we apply the following transformation:

$$h_{i,j} = \begin{cases} 0 & \text{, when } e_{i,j} = NaN \\ -\log_{10}(e_{i,j}) & \text{, when } e_{i,j} > 0 \\ \text{large constant c} & \text{, when } e_{i,j} = 0 \end{cases} \quad (5)$$

This linear transformation in essence changes only the range of values that appear in matrix H without affecting the attributes and characteristics of the data involved. With regard to the large constant, in our case we have set $c = 1000$, but any sufficiently large number can be used

3.2   Cluster Number Estimation

As stated earlier, one of the key issues in data clustering is the definition of the number of clusters to be produced. In our case, this number is estimated directly from the characteristics of the homology matrix H.

In linear algebra terms, the use of a block matrix corresponds to having a linear mapping thought of in terms of corresponding sets of basis vectors. This can be further viewed as having separate direct sum decompositions of both the domain and the range of the matrix. By completeness purposes, for any arbitrary matrices $A_{m \times n}$ and $B_{p \times q}$, the direct sum of A and B is denoted by $A \oplus B$ and defined as:

$$A \oplus B = \begin{bmatrix} a_{1,1} & \cdots & a_{1,n} & 0 & \cdots & 0 \\ \vdots & \ddots & \vdots & \vdots & \ddots & \vdots \\ a_{m,1} & \cdots & a_{m,n} & 0 & \cdots & 0 \\ 0 & \cdots & 0 & b_{1,1} & \cdots & b_{1,q} \\ \vdots & \ddots & \vdots & \vdots & \ddots & \vdots \\ 0 & \cdots & 0 & b_{p,1} & \cdots & b_{p,q} \end{bmatrix} \quad (6)$$

Given the fact that the matrix H is square, we can also interpret the mappings as an endomorphism of an n-dimensional space V, i.e. a linear map such as $f: V \rightarrow V$. In that regard, the block structure is of importance as it corresponds to having a single direct sum decomposition on V.

However, we must take under consideration the fact that the homology matrix H is an approximation of a block diagonal matrix. In order to evaluate the potential number of blocks existent within H, we employ the λ eigenvalues of the matrix as the functional characteristic degree for the final segmentation.

3.3   Distance Metric

The final stage in the clustering process requires the definition of an adequate distance metric. There are two key points that should be taken under consideration:

— Both dimensions of the homology matrix directly correspond to an ordered list of protein sequences. For the purposes of this work, the ordering of the protein sequences is based on the relative position of the respective genes on the genome chromosome.
— As can be also surmised from Equation (5), the distribution of values of matrix H is heavily biased on two ends, corresponding to the two cases of protein sequence alignment; no similarity (hence the sparsity of matrix) and complete identity.

Therefore, special care has be taken to include those attributes into the proposed metric, as seen in the equation below:

$$dist\left(h_{i_1,j_1}, h_{i_2,j_2}\right) =$$

$$\begin{cases} \sqrt[3]{(h_{i_1,j_1} - h_{i_2,j_2})^2} & , \sqrt[2]{\Delta i^2 + \Delta j^2} \leq c_1 \,\&\&\, \Delta h \leq c_2 \\ (h_{i_1,j_1} - h_{i_2,j_2}) + \sqrt[2]{\Delta i^2 + \Delta j^2}, & \sqrt[2]{\Delta i^2 + \Delta j^2} > c_1 \,\&\&\, \Delta h \leq c_2 \\ \sqrt[2]{\Delta i^2 + \Delta j^2 + \Delta h^2} & , \sqrt[2]{\Delta i^2 + \Delta j^2} > c_1 \,\&\&\, \Delta h > c_2 \\ \sqrt[2]{|\Delta h^2 + (\Delta i^2 - \Delta j^2)|} & , \sqrt[2]{\Delta i^2 + \Delta j^2} \leq c_1 \,\&\&\, \Delta h > c_2 \end{cases} \quad (7)$$

where $c_1 = \frac{n}{|\lambda|}$ and c_2 an arbitrary constant satisfying $\left(\overline{\frac{\sum h_{i,j}}{|\{h_{i,j} \neq \{0,c\}, \forall i,j\}|}}\right) < c_2 < c.$

The metric clearly defines four distinct states in the homology matrix H:

1. closely located genes within the same range of similarity. In this case, as the genes are expected to be linked at some level, the distance takes into account only the homology values but with a bias towards smaller distance.
2. closely located genes with significant difference in homology. In this case, the respective genes are expected to belong to different functional groups, and therefore the distance is biased towards larger values.
3. distant genes within the same range of similarity. This is a very interesting case, as it should contain genes across different species that exhibit a high level of similarity.
4. distant genes with significant difference in homology. This is the most distant case of gene similarity, therefore the maximum distance is assigned.

The function shown in Equation (7) is used in the clustering process in order to produce the distance matrix and, consequently, the required clusters.

4   Results

In order to evaluate the effectiveness of the proposed metric, we employed a well-studied group containing the following five genomes ([10], [11]):

1. *Mycoplasma genitalium, G-37* [12] (Bacteria; Firmicutes; Mollicutes; Mycoplasmatales) 479 genes, COGENT code: MGEN-G37-01.
2. *Ureaplasma urealyticum, serovar 3* [13] (Bacteria; Firmicutes; Mollicutes; Mycoplasmatales) 613 genes, COGENT code: UURE-SV3-01.
3. *Streptococcus pyogenes M1, SF370* [14] (Bacteria; Firmicutes; Bacilli; Lactobacillales) 1696 genes, COGENT code: SPYO-SF3-01.
4. *Buchnera aphidicola, SG* [15] (Bacteria; Proteobacteria; Gammaproteobacteria; Enterobacteriales) 545 genes, COGENT code: BAPH-XSG-01.
5. *Nanoarchaeum equitans, Kin4-M* [16] (Archaea; Nanoarchaeota) 563 genes, COGENT code: NEQU-N4M-01.

The phylogenetic relationships of the species is represented by the dendrogram in Fig. 1.

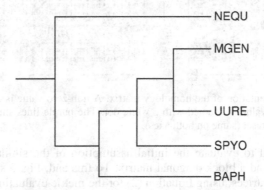

Fig. 1. Simplified dendrogram of the five species in the input dataset. The COGENT codes are used for genome representation.

In order to produce the initial matrix containing the expect value of the sequence similarities, we employed the BLAST algorithm with the default parameters. The homology matrix H is produced directly from the e-values, through the pre-processing step (Equation 5, where $c = 1000$). A visual representation of the final matrix is shown in Fig. 2 below.

It must be noted that Fig. 2 showcases only the sparsity of the homology matrix, and does not take into account the actual values of the non zero elements. However, even this simplified representation is sufficient to evaluate the patterns that emerge from the genome-wide sequence comparison.

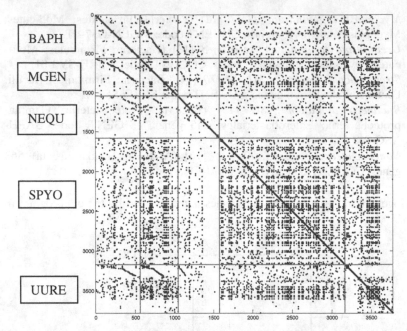

Fig. 2. Visual representation of the homology matrix. A non-zero value is denoted with a blue dot, whereas a zero value is denoted with a white dot. The purple lines show the limits of the five species in the dataset (same on both axes).

It is also critical to evaluate the initial assumption of the similarity between the homology matrix and a block diagonal matrix. To this end, Fig. 3 shows the sparsity value of the sub-matrices, using Equation (2) for the metric evaluation.

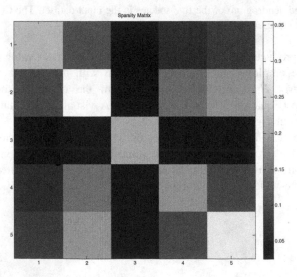

Fig. 3. Sparsity value of the homology matrix. Each sub-matrix corresponds to a pair of genomes.

Before applying the distance metric for the clustering process, we calculate the eigenvalues of the input matrix in order to set the number of clusters to be produced. Finally, inserting the proposed distance metric into the clustering process, we can produce a clustering of the protein sequences as shown in the following figures (Fig. 4. and Fig. 5.).

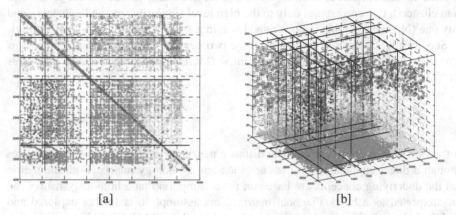

[a]   [b]

Fig. 4. Clustering of the homology matrix across the three dimensions, where x and y correspond to the ordered protein sequences and z corresponds to the homology value of the respective sequence comparison. [a] shows the clustering on the two dimensions (i.e. ignoring the z axis) and [b] shows the clustering on all three dimensions.

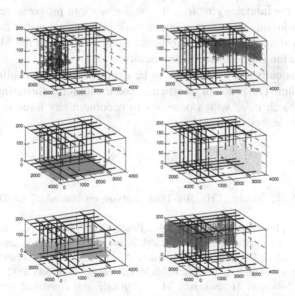

Fig. 5. Visualization of the six clusters the comprise Fig. 4 employing the same color coding. It is evident that there is significant segmentation of the data points, both regarding the homology value and the different species the sequences belong to.

There are mainly two observations that can be made from the produced clusters; (i) the proposed distance metric allows for the discrimination of the different levels of homology, and (ii) the differentiation of the pair-wise genome comparisons (i.e. the different "cells" of the matrix) is not following closely the expected relationships.

Regarding the first observation, we can further infer from the results that there exist three distinct clusters containing the low value homology scores (green, yellow and cyan clusters), two correspond only to the high level similarity (red and magenta), and only one (blue cluster) that is constrained within a specific area of mid-level similarity. Specifically, this area corresponds to the two genome comparisons; *Streptococcus pyogenes* (SPYO) – *Mycoplasma genitalium* (MGEN) and *Streptococcus pyogenes* (SPYO) – *Buchnera aphidicola* (BAPH).

5 Discussion

In this work we propose an alternate distance metric for clustering protein sequences through a direct application on the corresponding homology matrix. Both the metric and the underlying concepts are based on the assumption that a homology matrix can be interpreted as a block diagonal matrix. This assumption is further explored and exploited through the estimate of the expected clusters and the definition of the distance metric.

We have provided some preliminary results by applying the proposed method in order to cluster data from a well-studied set of protein sequences. Although the results are very encouraging, by showing significant differentiation of the various levels evident within the homology matrix, it is still a work in progress, requiring rigorous testing and validation. To this end, future work includes extensive comparison of the proposed method against similar algorithms, both from within the Machine Learning community and the Bioinformatics community.

Finally, an equally important issue to be explored is the scalability of the implemented algorithm. The problem of efficient and meaningful clustering of protein data is an open research issue which promises to become a key issue in next generation sequencing data analysis.

References

1. Williams, S.M., Moore, J.H.: Big Data analysis on autopilot? BioData Min. 6(1), 22 (2013)
2. Overbeek, R., Fonstein, M., D'Souza, M., Pusch, G.D., Maltsev, N.: The use of gene clusters to infer functional coupling. Proc. Natl. Acad. Sci. U.S.A. 96(6), 2896–2901 (1999)
3. Enright, A.J., Van Dongen, S., Ouzounis, C.A.: An efficient algorithm for large-scale detection of protein families. Nucleic Acids Res. 30(7), 1575–1584 (2002)
4. Sarkar, A., Soueidan, H., Nikolski, M.: Identification of conserved gene clusters in multiple genomes based on synteny and homology. BMC Bioinformatics 12(suppl.9), S18 (2011)
5. Miele, V., Penel, S., Duret, L.: Ultra-fast sequence clustering from similarity networks with SiLiX. BMC Bioinformatics 12(1), 116 (2011)

6. Röttger, R., Kalaghatgi, P., Sun, P., Soares, S.D.C., Azevedo, V., Wittkop, T., Baumbach, J.: Density parameter estimation for finding clusters of homologous proteins–tracing actinobacterial pathogenicity lifestyles. Bioinformatics 29(2), 215–222 (2013)
7. Fouts, D.E., Brinkac, L., Beck, E., Inman, J., Sutton, G.: PanOCT: automated clustering of orthologs using conserved gene neighborhood for pan-genomic analysis of bacterial strains and closely related species. Nucleic Acids Res. 40(22), e172 (2012)
8. Bonet, J., Planas-Iglesias, J., Garcia-Garcia, J., Marín-López, M.A., Fernandez-Fuentes, N., Oliva, B.: ArchDB 2014: structural classification of loops in proteins. Nucleic Acids Res. 42(database issue), D315–D319 (2014)
9. Altschul, S.F., Gish, W., Miller, W., Myers, E.W., Lipman, D.J.: Basic local alignment search tool. J. Mol. Biol. 215(3), 403–410 (1990)
10. Freilich, S., Goldovsky, L., Gottlieb, A., Blanc, E., Tsoka, S., Ouzounis, C.A.: Stratification of co-evolving genomic groups using ranked phylogenetic profiles. BMC Bioinformatics 10, 355 (2009)
11. Psomopoulos, F.E., Mitkas, P.A., Ouzounis, C.A.: Detection of genomic idiosyncrasies using fuzzy phylogenetic profiles. PLoS One 8(1), e52854 (2013)
12. Fraser, C.M., Gocayne, J.D., White, O., Adams, M.D., Clayton, R.A., et al.: The minimal gene complement of Mycoplasma genitalium. Science 270, 397–403 (1995), doi:10.1126/science.270.5235.397
13. Glass, J.I., Lefkowitz, E.J., Glass, J.S., Heiner, C.R., Chen, E.Y., et al.: The complete sequence of the mucosal pathogen Ureaplasma urealyticum. Nature 407, 757–762 (2000), doi:10.1038/35037619
14. Ferretti, J.J., McShan, W.M., Ajdic, D., Savic, D.J., Savic, G., et al.: Complete genome sequence of an M1 strain of Streptococcus pyogenes. Proc. Natl. Acad. Sci. U.S.A. 98, 4658–4663 (2001), doi:10.1073/pnas.071559398
15. Shigenobu, S., Watanabe, H., Hattori, M., Sakaki, Y., Ishikawa, H.: Genome sequence of the endocellular bacterial symbiont of aphids Buchnera sp. APS. Nature 407, 81–86 (2000)
16. Waters, E., Hohn, M.J., Ahel, I., Graham, D.E., Adams, M.D., et al.: The genome of Nanoarchaeum equitans: insights into early archaeal evolution and derived parasitism. Proc. Natl. Acad. Sci. U.S.A. 100, 12984–12988 (2003), doi:10.1073/pnas.1735403100

Comparative Evaluation of Feature Extraction Methods for Human Motion Detection

Olga Politi, Iosif Mporas, and Vasileios Megalooikonomou

Multidimensional Data Analysis and Knowledge Management Laboratory
Dept. of Computer Engineering and Informatics, University of Patras
26500 Rion-Patras, Greece
{politi,vasilis}@ceid.upatras.gr, imporas@upatras.gr

Abstract. In this article we conduct an evaluation of feature extraction methods for the problem of human motion detection based on 3-dimensional inertial sensor data. For the purpose of this study, different preprocessing methods are used, and statistical as well as physical features are extracted from the motion signals. At each step, state-of-the-art methods are applied, and the produced results are finally compared in order to evaluate the importance of the applied feature extraction and preprocessing combinations, for the human activity recognition task.

Keywords: Accelerometers, movement classification, human motion recognition.

1 Introduction

One of the most important tasks in pervasive computing is to provide accurate and opportune information on people's activities and behaviors. Applications in medicine, security and entertainment constitute examples of this effort. For instance, patients with obesity, diabetes or heart disease, are often required to fulfill a program of physical exercise that is integrated within their daily activities [1]. In computer vision, complex sensors such as cameras have been used to recognize human activities but their accuracy falls under a real-home setting, due to the high-level activities that take place in the natural environments, as well as the variable lighting or clutter [8]. As a result, body-attached accelerometers are commonly used as an alternative in order to assess variable daily living activities.

The human motion detection problem using accelerometers is an emerging area of research. Sensors embedded in objects or attached on the body are generally chosen to study movement patterns or human behavior. Accelerometers have been used extensively due to their low-power requirements, small size, non-intrusiveness and ability to provide data regarding human motion. In a common scenario, these data can be processed using signal processing and pattern recognition methods, in order to obtain a real-time recognition of human motion.

Several human activity recognition systems have been proposed in the past, which include the use of accelerometers. Some of them analyze and classify different kinds

L. Iliadis et al. (Eds.): AIAI 2014 Workshops, IFIP AICT 437, pp. 146–154, 2014.

of activity using acceleration signals [2], [3], while others apply them for recognizing a wide set of daily physical activities [4], or describe a human activity recognition framework based on feature selection techniques [5]. Bernecker et al [6], proposed a reclassification step that increases accuracy of motion recognition. Karantonis et al. [7] introduced an on-board processing technique for a real-time classification system, yielding results that demonstrate the feasibility of implementing an accelerometer-based, real-time movement classifier using embedded intelligence.

Khan et al. [8] propose a system that uses a hierarchical recognition scheme, i.e., the state recognition at the lower level using statistical features and the activity recognition at the upper level using the augmented feature vector followed by linear discriminant analysis. Several powerful machine learning algorithms have been proposed in the literature for the detection of human motion. The most widely used are the artificial neural networks [6, 8, 10], the naïve-Bayes [4, 14] and the support vector machines [5, 14].

In this paper, a comparative evaluation of feature extraction methods for human motion detection is presented. The main contribution of the paper is the proposal of the best combination between preprocessing and feature extraction, for the human activity recognition task. For this purpose, we choose to evaluate different preprocessing and feature extraction combinations, formed after thoroughly examining the existing state-of-the-art methods. Although this study may seem quite simple, there is little known in bibliography regarding official comparison between methods that include preprocessing combined with feature extraction. Since most studies concentrate on preprocessing or feature extraction separately, focusing on the above combinations seems important not only for the acquisition of better results, but also for discovering meaningful data interpretations and features that "characterize" human motion.

The rest of this paper is organized as follows: In Section 2 we present the framework constructed for human motion detection. Section 3 offers details about the experimental setup and in Section 4 we present the achieved experimental results. Finally in Section 5 we conclude this work.

2   Framework for Comparative Evaluation

In the present framework for comparative evaluation, we assume that the input to the framework consists of 3-dimensional $\{x, y, z\}$ signal streams, as illustrated in Fig. 1. Each stream represents one movement direction in the sense of moving forward/backward, up/down and left/right. Preprocessing$_i$ consists of applying a sliding window W to the incoming streams, of constant length, resulting to W_i frames, where $1 \leq i \leq I$. The time-shift between two successive frames is also constant and can result to frame sequences with or without overlap. After applying the sliding window, each signal frame W_i is either led directly to the feature extraction stage(Non-preprocessing method) or is previously processed using two different techniques.

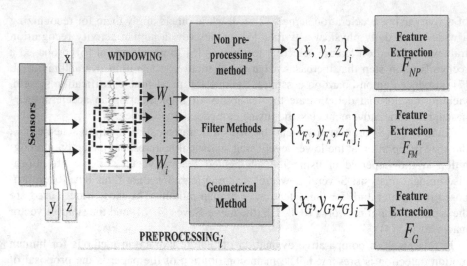

Fig. 1. Block diagram of the preprocessing stage of the motion detection framework

Specifically, the framework uses filtering methods F_n, with $1 \leq n \leq N$, that produce n signal frame representations per W_i frame and b) a geometrical representation method G. Each one of the previously mentioned methods yields different 3-dimensional data representations per frame W_i, namely $\{x, y, z\}$ for non-preprocessed data, $\{x_{F_n}, y_{F_n}, z_{F_n}\}$ for n chosen filters, $1 \leq n \leq N$, and $\{x_G, y_G, z_G\}$ for the geometrical representation. Each representation produced by the preprocessing stage is led to the feature extraction stage.

In this way, we have feature extraction carried out in parallel for the non-preprocessing representation, F_{NP}, for n filter methods representation, $F_{FM}{}^n$, and for the geometrical representation, F_G, per signal frame W_i. The block diagram of the feature extraction and classification stage of the motion detection framework is depicted in Fig. 2. In the feature extraction stage, every input is processed by statistical and physical feature extraction algorithms. The statistical algorithms can briefly be divided to time, frequency and time-frequency domain methods.

The utilization of these methods ensures that we get as much possible information from the data retrieved. In detail, each incoming representation of a signal frame is processed in parallel by each one of the feature extraction modules shown in Fig. 2. The estimated feature vectors could be mathematically represented as follows: the time-domain features $F_T^i \in \mathbb{R}^{|F_T^i|}$, the frequency-domain features $F_F^i \in \mathbb{R}^{|F_F^i|}$, the time-frequency domain features $F_{TF}^i \in \mathbb{R}^{|F_{TF}^i|}$ and the physical features $F_P^i \in \mathbb{R}^{|F_P^i|}$.

After the decomposition of the different incoming frame representations to the described feature vectors, an individual classifier is build per feature vector. Specifically, in the classification stage, for time-domain features we have C_T classifier, for frequency domain features C_F classifier, for time-frequency C_{TF} and finally for physical features C_P classifier.

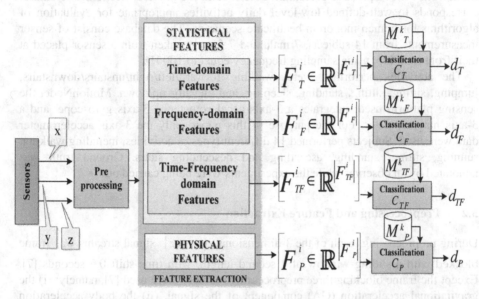

Fig. 2. Block diagram of the feature extraction and classification stage of the motion detection framework

During the training phase a set of motion data (training data) with known labels, i.e. with a-priori annotated motion labels, is used to estimate models, $M^k{}_f$, with $1 \le k \le K$ and $f = \{T, F, TF, P\}$, for each human motion k of interest and for each feature vector f. At the test phase the unknown motion data (test data) will be preprocessed and decomposed to feature vectors as in the training phase. The classification algorithms C_f will compare each vector $F^j{}_f$, with $f = \{T, F, TF, P\}$ and $1 \le j \le J$ against each motion model $M^k{}_f$ in order to decide the corresponding motion class, i.e. $d_j = \arg\max_k \left\{ C_f \left(F^j{}_f, M^k{}_f \right) \right\}$, where d_j is the motion class label assigned to the j-th test frame of the sensor data. After classification a post-processing algorithm could be applied on the automatically labeled frames of the test recording in order to fine-tune the detected human motion classes, but this is left for further studying.

3   Experimental Setup

3.1   Sensor Data Description

The USC-HAD database [9], which is a freely available dataset provided by the University of Southern California was chosen in order to conduct experiments using the framework for comparative evaluation described in the previous section. The dataset corresponds to well-defined low-level daily activities appropriate for evaluation of algorithms for human motion in healthcare scenarios. The database consist of sensor measurements from 14 subjects, 7 male and 7 female, taken from a sensor placed at their front right hip, using sampling frequency equal to 100 Hz.

The activities performed were: walking forward/left/right/upstairs/downstairs, jumping, sitting, sitting, standing, sleeping, and elevator up/down. MotionNode, the sensing platform used, integrates a 3-axis accelerometer, a 3-axis gyroscope, and a 3-axis magnetometer. For the purpose of this paper, only the 3-axis accelerometer data were used. Subjects performed 14 different types of activities, including walking, running, sitting, standing, ascending and descending stairs. Ground truth was annotated by an observer while the experiments were being carried out.

3.2   Preprocessing and Feature Extraction

During preprocessing, each of the 3-dimensional $\{x, y, z\}$ signal streams was frame blocked with a sliding window of 1 second length, with time-shift 0.5 seconds [7]. Except the frame blocking three preprocessing methods were used [7], namely: (i) the gravitational acceleration (GA) component of the signal, (ii) the body acceleration (BA) component of the signal and (iii) the Tilt sensing, defined as the angular position with respect to gravity per axis. Together with the baseline preprocessing (non-preprocessing method), four preprocessing setups were created.

For each of the four preprocessed outputs per frame, statistical and physical features were extracted [5]. As far as feature analysis is concerned, statistical features are briefly divided into *time-domain, frequency domain* and *time-frequency domain* [11]. Time-domain features include mean, median, variance, root mean square (RMS), standard deviation, skewness, kurtosis and interquartile range (25%, 50%, 75%). Frequency-domain features, which mainly represent the periodic structure of the signal, are the Fourier transform, the spectral entropy, the spectral energy and the 3rd order autoregressive-filter (AR) coefficients.

Time-frequency domain features are used to investigate both time and frequency characteristics of complex signals and in general employ wavelet techniques, such as wavelet coefficients or energy of wavelet coefficients. The physical features [5] include the movement intensity (MI), the eigenvalues of dominant directions (EVA), correlation of acceleration along gravity and heading directions (CAGH) and the averaged velocity along each direction and the averaged acceleration energy (AAE).

3.3   Classification

For each of the four preprocessing methods the estimated feature vector has dimensionality equal to 77. Thus, the final feature vector per frame is of dimensionality 4×77=308. For the classification of the estimated feature vectors, we relied on the support vector machines (SVMs) implemented with the sequential minimal optimization method [13] using the polynomial kernel function (poly). The choice of the kernel-based algorithm is owed to the fact that SVMs do not suffer from the curse of dimensionality phenomenon [12]. The classifier was implemented using the WEKA machine learning toolkit software [13].

4   Experimental Results

The human motion detection framework presented in Section 2 was evaluated using the experimental setup described in Section 3. To avoid overlap between the training and test datasets we followed a 10-fold cross validation protocol. The experimental results, in percentages, for the four feature extraction methods and for each of the 14 subjects of the dataset are presented in Table 1. The best accuracy achieved for each subject is indicated in bold. The classification was carried out using individual classifiers, under a subject-dependent scenario.

Table 1. Motion detection accuracy (%) per subject and feature extraction method

Sub.	NP	BA	GA	Tilt
01	91,80	57,66	**91,88**	89,72
02	**90,55**	62,93	90,29	89,01
03	**90,74**	61,92	90,63	89,80
04	**85,93**	55,33	85,42	83,94
05	86,23	59,00	**86,74**	85,10
06	91,99	**92,51**	56,28	90,42
07	92,46	60,56	**92,72**	90,55
08	**91,79**	65,61	91,35	90,10
09	**92,07**	69,24	91,24	91,08
10	**91,72**	64,56	91,42	88,27
11	**93,91**	58,54	93,89	91,37
12	93,34	63,44	**93,51**	92,81
13	**92,57**	57,61	92,33	90,81
14	**87,45**	53,98	86,67	83,48

From Table 1 we can see that the best accuracy achieved is approximately 94%, in the non-preprocessed data representation and for Subject 11. The BA method is the one that performs the worst among feature extraction methods. Both the non prepro-cessed method and the tilt representation achieve very good results, with GA feature extraction method following shortly after the first two methods.

The experimental results may show that the non preprocessed method performs the best, so not using preprocessing can yield good accuracy in motion detection, but the interesting finding is that the Tilt method also performs as well. This is very important, due to the fact that by using Tilt sensing, we achieve to add a geometrical aspect to the previously linear acceleration signals. As a result, when it comes to motions that are characterized by changes in angular velocity or torque, i.e. turning or clockwise movements that may be followed by movements that are linearly translated, the Tilt sensing method is preferable.

Another observation is that the GA method also performs well, apart from the case of Subject 6. This outcome may be due to the fact that the database used for the experiments includes many motions that have a strong relation to gravity acceleration. To be more specific, five out of fourteen activities (walking upstairs/downstairs, elevator up/down and jumping) could be translated to changes only in the gravity dimension, not to mention the three static activities (lying, standing and sitting) which are also "characterized" by gravity acceleration.

As a result, to face the problem of variability in human motion, it seems more appropriate to deal with data preprocessing that involves mathematical or geometrical representation, rather than focus on the idea of applying filter-based signal processing. In addition, we investigated the importance of features used in the overall experimental process, in terms of feature influence in the classification process. The feature ranking was performed using the ReliefF algorithm [15], and the top-5 features ranked by ReliefF are presented in Table 2.

Table 2. Top-5 features ranked by the ReliefF algorithm

Ranking	Features
1	Spectral entropy
2	CAGH of x axis (gravity)
3	Correlation xz
4	Median
5	Mean

As can be seen in Table 2, within the most discriminative motion features for the task of automatic motion recognition are spectral entropy, the correlation between acceleration along gravity direction (CAGH), the correlation between gravity and heading direction, mean and median. At this point, it is evident that frequency domain features play an important role, as we meet spectral entropy and correlation at the top five ranked features. A physical feature (CAGH) is second in ranking, hence the importance of the addition of physical features in the feature extraction process. Attempting to interpret these ranking results, we could initially observe that spectral entropy is justifiably ranked first, since it helps differentiate between signals that have similar energy values but correspond to different activity patterns [16]. The USC-HAD database includes movements with similar energy, i.e. sitting compared to standing and lying, or walking in different patterns (forward, left, right, upstairs and downstairs).

Following in rank is the CAGH feature that improves differentiation between ac-tivities that involve translation (in terms of acceleration intensity) in a single di-men-sion, compared to the total acceleration intensity of the other two dimensions. Move-ments such as walking upstairs/ downstairs and walking left/right/forward fall in the previous category, hence the importance of using the CAGH feature. Correlation xz (between gravity and heading direction) also seems to play an important role in the classification results. Its interpretation is similar to CAGH, only in this case the com-parison is inner-axial. Motions that have a strong xz correlation in the database are jumping, elevator up/down as well as static movements (lying, sitting, and standing).

Finally, statistical features like mean and median, despite their simplicity, could not be missing from the top-5 ranking, since they are known for the robustness and stability they provide to the classification process.

5   Conclusion

In the present work we present a comparative evaluation of feature extraction methods, based on 3-dimensional data acquired from inertial sensors worn by humans. The framework uses four different preprocessing methods and the motion signals are parameterized by statistical and physical features. The experimental results indicated an average performance of 90.89% for the best combination of preprocessing and feature extraction methods, for 14 everyday human motion activities. The highest performance achieved was approximately 94%. We deem the application of experiments constructed under a subject-independent scenario, in order to see if the present study responds well in intra-class errors, namely the errors introduces in the classification process when different people execute the same movement.

Acknowledgement. The reported research was partially supported by the BioMed-Mine project (Thalis – University Of Patras – Mining Biomedical Data And Images: Development Of Algorithms and Applications), as well as the ARMOR Project (FP7-ICT-2011-5.1 - 287720) "Advanced multi-paRametric Monitoring and analysis for diagnosis and Optimal management of epilepsy and Related brain disorders".

References

1. Jia, Y.: Diatetic and exercise therapy against diabetes mellitus. In: Second International Conference on Intelligent Networks and Intelligent Systems, pp. 693–696 (2009)
2. Mantyjarvi, J., Himberg, J., Seppanen, T.: Recognizing human motion with multiple acceleration sensors. In: IEEE Int. Conf. Syst., Man, Cybern., vol. 2, pp. 747–752 (2001)

3. Sekine, M., Tamura, T., Akay, M., Fujimoto, T., Togawa, T., Fukui, Y.: Discrimination of walking patterns using wavelet-based fractal analysis. IEEE Trans. Neural Syst. Rehabil. Eng. 10(3), 188–196 (2002)
4. Ermes, M., Parkka, J., Mantyjarvi, J., Korhonen, I.: Frequent Bit Pattern Mining Over Triaxial Accelerometer Data Streams for Recognizing Human Activities and Detecting Fall. Procedia Computer Science 19, 56–63 (2013)
5. Zhang, M., Sawchuk, A.: A Feature Selection-Based Framework for Human Activity Recognition Using Wearable Multimodal Sensors. In: BodyNets 2011, University of Southern California, Los Angeles (2011)
6. Bernecker, T., Graf, F., Kriegel, H., Moennig, C.: Activity Recognition on 3D Accelerometer Data. Technical Report (2012)
7. Karantonis, D.M., Narayanan, M.R., Mathie, M., Lovell, N.H., Celler, B.G.: Implementation of a Real-Time Human Movement Classifier Using a Triaxial Accelerometer for Ambulatory Monitoring. IEEE on Information Technology in Biomedicine 10 (2006)
8. Khan, A., Lee, Y., Lee, S.Y., Kim, T.: Triaxial Accelerometer-Based Physical-Activity Recognition via Augmented-Signal Features and a Hierarchical Recognizer. IEEE transactions on information technology in biomedicine 14(5) (2010)
9. Zhang, M., Sawchuk, A.: USC-HAD: A Daily Activity Dataset for Ubiquitous Activity Recognition Using Wearable Sensors. In: UbiComp 2012, USA (2012)
10. Khan, A.M., Lee, Y.K., Lee, S.Y.: Accelerometer's Position Free Human Activity Recognition Using A Hierarchical Recognition Model. In: IEEE HealthCom (2010)
11. Avci, A., Bosch, S., Marin-Perianu, M., Marin-Perianu, R., Havinga, P.: Activity Recognition Using Inertial Sensing for Healthcare, Wellbeing and Sports Applications: A Survey, University of Twente, The Netherlands (2010)
12. Burges, C.: A tutorial on Support Vector Machines for Pattern Recognition. In: Data Mining and Knowledge Discovery, vol. 2(2), pp. 121–167. Kluwer Academic Publishers (1998)
13. Keerthi, S.S., Shevade, S.K., Bhattacharyya, C., Murthy, K.R.K.: Improvements to Platt's SMO Algorithm for SVM Classifier Design. Neural Computation 13(3), 637–649 (2001)
14. Ravi, N., Dandekar, N., Mysore, P., Littman, M.L.: Activity Recognition from Accelerometer Data, Department of Computer Science, Rutgers University (2005)
15. Robnik-Sikonja, M., Kononenko, I.: An adaptation of Relief for attribute estimation in regression. In: ICML 1997, pp. 296–304 (1997)
16. Brezmes, T., Gorricho, J.-L., Cotrina, J.: Activity recognition from accelerometer data on a mobile phone. In: Omatu, S., Rocha, M.P., Bravo, J., Fernández, F., Corchado, E., Bustillo, A., Corchado, J.M. (eds.) IWANN 2009, Part II. LNCS, vol. 5518, pp. 796–799. Springer, Heidelberg (2009)

Eye Blink Artifact Removal in EEG Using Tensor Decomposition

Dimitrios Triantafyllopoulos and Vasileios Megalooikonomou

Multidimensional Data Analysis and Knowledge Management Laboratory
Dept. of Computer Engineering and Informatics, University of Patras
26500 Rion-Patras, Greece
{dtriantafy,vasilis}@ceid.upatras.gr

Abstract. EEG data are usually contaminated with signals related to subject's activities, the so called artifacts, which degrade the information contained in recordings. The removal of this additional information is essential to the improvement of EEG signals' interpretation. The proposed method is based on the analysis, using Tucker decomposition, of a tensor constructed using continuous wavelet transform. Our contribution is an automatic method which processes simultaneously spatial, temporal and frequency information contained in EEG recordings in order to remove eye blink related information. The proposed method is compared with a matrix based removal method and shows promising results regarding reconstruction error and retaining the texture of the artifact free signal.

Keywords: eye blink, Tucker decomposition, wavelet transform, EEG.

1 Introduction

Electroencephalographic (EEG) data are used in several cases where the brain's functionality needs to be analyzed. Their relative ease of use, their low power consumption, compact size and convenience as well as their temporal resolution, are some of the reasons that EEG recordings are preferred in several cases from their alternatives, such as functional Magnetic Resonance Imaging. Some of the applications where EEG data are used are Brain Computer Interaction (BCI) applications as well as medical tasks that require the supervision of a patient's brain functionality, like seizure monitoring.

Unfortunately, during EEG recordings except the underlying brain related activity, there are several signals captured also that degrade the information gathered. These signals are either related to external or internal reasons and are generally referred to as artifacts. External artifacts are often occurring due to technology's shortcomings such as electrical lines introducing a 50 Hz component, affecting the data recorded. The surrounding environment, such as walls and electrical devices, may cause additional noise. Those artifacts are usually counteracted with filtering techniques, for example the application of a Notch filter at 50 Hz.

Internal artifacts are related to the signals captured due to the subject's physical actions. For example eye blinking causes capturing of additional information in the

L. Iliadis et al. (Eds.): AIAI 2014 Workshops, IFIP AICT 437, pp. 155–164, 2014.

frontal electrodes. Eye movement, as well as muscle movement, generate also undesired signals that are, of course, recorded. Avoiding these artifacts is one solution, but as showed in several studies [1,2] , instructing subjects to avoid movement or any other activity that can cause an artifact affects the recorded data. On the other hand, the treatment of these artifacts is a challenging procedure and requires two steps. First the detection of their occurrence and then the handling of the detected events aim in the manipulation of these phenomena.

The detection of artifacts can be performed with visual inspection from a medical expert. But this procedure is, as expected, time consuming. Alternatively there are several automatic techniques that can be applied during or after recordings and detect the occurring artifacts. There are techniques, which are based in thresholds applied in time or frequency domain like in [3]. Some others are based in a supervised learning method using statistical [4] or autoregressive (AR) [5] features. Other methods, like [6], use a combination of feature extraction and data driven thresholds.

Once the artifacted epochs are detected, there are two ways of manipulating these periods. The first one is rejecting the epoch that contains the artifacted signals. This procedure may solve the problem of artifacted epochs but results in substantial information loss. The second methodology is the removal of artifact related information and conservation of brain activity.

Removal methods are separated in two main categories. The first one aims in the removal of the EOG captured signal, which is captured with a separate sensor, with the use of linear combination and regression techniques [6]. The second category of removal methods aims in the clarification of the components-sources that compose the recorded data and afterwards the identification of the artifact related ones. The most used decompositions are Principal Component Analysis (PCA), Canonical Correlation Analysis (CCA) and Independent Component Analysis (ICA). In [7] and [8] CCA is performed in order to remove muscle artifacts. Several methods report that ICA based removal methods outperform the methods based in other decompositions. In [9] an ICA based method discriminates the artifact related independent components, using an SVM classifier. In [10] using high order statistics as criteria, wavelet decomposition in order to divide the recorded signal in frequency bands and ICA as the decomposition method, the proposed technique removes eye blink or muscle movement related artifacts from a pre specified epoch. In [11] the proposed ICA based method by using correlation coefficients between independent sources, specifies the independent component that most likely contains the artifact's signature. In [12] a combination of ICA and Wiener filtering supress eye blink related artifacts.

In this study the effectiveness of a different decomposition method is examined. The decompositions in the methods mentioned so far are based on two dimensional matrices, whose dimensions correspond to electrodes and time samples respectively. In order to examine, when necessary, the frequency information encapsulated in recordings the aforementioned methods, first extract spectral features and then classify the components as artifacts or not. By constructing a three-way array (tensor) whose first two modes correspond to electrode and time domain, while the third encloses the frequency information, it is possible to examine the three domains simultaneously. This concept was used previously in EEG signals in several studies [13, 14, 15].

In [16] the same model was used in order to examine it's abilities in seizure localization as well as in artifact extraction. Regarding the latter the proposed method suggests multilinear subspace analysis with applying Tucker decomposition in the tensor created, in favor of removing the information that is correlated with the artifact. The main aim of our work is to examine the capability of tensors and more specifically Tucker decomposition's as a model to discriminate artifact related information from brain related activity. Our contribution is an automatic artifact removal method based on a similar tensor model with the one in [13] and [16]. The proposed method is compared with an ICA method, similar to the one proposed in [10]. The result of the proposed modeling is a procedure which automatically removes artifact related information, while encapsulating simultaneously the spatial, temporal and frequency underlying structure of the recorded data.

In the following sections the theoretical background, the methodology followed as well as the experimental scheme is presented. In the end the conclusions and future work are discussed.

2   Background

2.1   Multilinear Arrays

Multilinear arrays or tensors, as more than often are referred to, are a multidimensional generalization of vectors. A N-th order tensor is a product on N vector spaces having their own coordinate system [17]. A matrix is a special case of tensors of order 2. It should be noted that the order of a tensor is the number of its dimensions. A general N-th order real tensor is notated as $\mathcal{X} \in \mathbb{R}^{I_1 \times I_2 \times \dots \times I_N}$. As in the 2-dimensional case, where there exist several decomposition methods, like SVD, QR or NMF, there are high order generalizations of those methods that can be applied in high order tensors.

The most important of those decompositions are PARAFAC and Tucker decomposition. The former, introduced in [18] and in [19] independently, decomposes the tensor in its rank-1 tensor components. The latter proposed in [20] is considered as the higher order generalization of SVD.

2.2   High Order Decomposition

PARAFAC. Introduced in [18] and in [19] independently, is based in the proposition of rank-1 decomposition of tensors in [21]. In [22] the same model was augmented, by introducing the parallel proportional profiles, whose aim was the estimation of sources ("source traits") that fit simultaneously in many profiles. After its third introduction, in 1970 by the two teams, the model is considered a very important tool in multilinear analysis. Its computation in [23], the toolbox used in this research, is based in an Alternating least squares (ALS) algorithm. The 3-rd order tensor $\mathcal{X} \in \mathbb{R}^{I_1 \times I_2 \times I_3}$ using PARAFAC is decomposed in R rank-1 tensors as follows:

$$\mathcal{X} = \sum_{i=1}^{R} a_i \circ b_i \circ c_i + \mathcal{E} \tag{1}$$

where $a_i \in \mathbb{R}^{I_1}$, $b_i \in \mathbb{R}^{I_2}$, $c_i \in \mathbb{R}^{I_3}$, $\mathcal{E} \in \mathbb{R}^{I_1 \times I_2 \times I_3}$ is the residual tensor and $\circ$ denotes the outer product.

Tucker. It is considered as the generalization of SVD, hence there are several cases where it is referred to as High Order SVD (HOSVD). A three way example of this decomposition is parted from three projection matrixes corresponding in each of the modes of the tensor, and one core tensor capturing the correlation of the different components. Since this interaction is allowed Tucker is considered as a more flexible model than PARAFAC. In [23] Tucker is computed using a truncated SVD like method as an initialization step and then an ALS algorithm is followed. The Tucker decomposition of a 3-rd order tensor $\mathcal{X} \in \mathbb{R}^{I_1 \times I_2 \times I_3}$, with R components in every mode is considered as follows:

$$\mathcal{X} \approx \mathcal{C} \times_1 A \times_2 B \times_3 C \tag{2}$$

where $\mathcal{C} \in \mathbb{R}^{R \times R \times R}$ is the core tensor, $A \in \mathbb{R}^{I_1 \times R}$, $B \in \mathbb{R}^{I_2 \times R}$, $C \in \mathbb{R}^{I_3 \times R}$ are the projection matrices and $\times_i$ denotes the i-mode product (for more detail refer to [17]).

Figure 1 shows the Tucker decomposition of the above scheme:

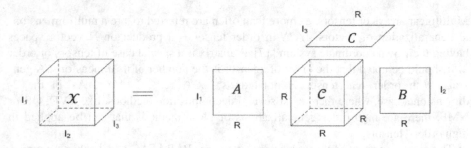

Fig. 1. Tucker decomposition. PARAFAC is a special case of the Tucker decomposition where the core tensor is diagonal and the number of components in each mode are the same.

3   Methodology

In the proposed method the higher order decomposition used is Tucker decomposition. As described in [16] and verified by our experiments, PARAFAC was unable to distinguish the artifact related information without removing also a large amount of useful information. This is due to the fact that generally artifact-related components correspond to components with large variance, therefore the removed information contains also great amounts of useful information. The more relaxed model of Tucker decomposition, due to the core tensor's non-diagonal structure, is more flexible to distinguish the artifact from clean signals' components.

The tensor constructed in the proposed method is the outcome of the application of continuous wavelet transform (CWT) in an EEG matrix with a variety of scales. The resulting tensor's modes are therefore *electrode × samples × scales*, capturing the spatial, temporal and frequency information of the recorded data. The mother wavelet used in our experiments is Morlet wavelet due to its ability to reconstruct better an EEG signal. Before the application of CWT each channel is normalized using min-max normalization.

Once the decomposition of the tensor takes place the projection matrix corresponding to the second mode, which is the time domain, is examined for artifact related information. The criteria used in the identification of the artifact related components are the same as in [10]. For each temporal signature kurtosis and entropy is extracted. The kurtosis criteria aims in the identification of peaky distributions, expected from an eye blink signature, while the entropy aims in artifacts of a noisy background. Then these features are separately normalized in order to have zero mean and unit variance. A threshold is applied in order to decide which signatures are artifact related. The latter ones are then removed and the reconstruction of the signal follows. Fig. 2 summarizes the process followed by the proposed method.

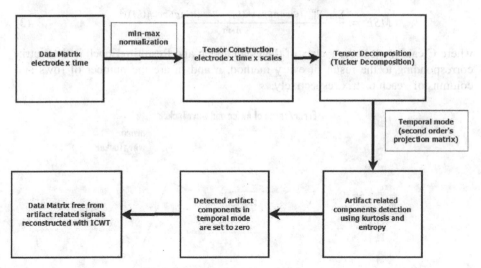

Fig. 2. Flowchart of the proposed method

4   Experiments

The data used in our experiments are from [7]. The recorded data contain intentionally generated artifacts from a healthy 27-year-old male. From the EOG related recording an unmarked epoch was extracted as reference point of artifact free EEG period. A marked as an eye blink period was used as the reference point for artifact related information. The whole set of artifacted period's channels were added to the clean epoch's corresponding ones in order to create a test set with a ground truth data set.

Afterwards the ICA based method, as well as the tensor based one were used in order to remove the artifact related information. It should be mentioned that it is not the same as performing the removal methods in the artifact's period on it's own, since the artifact data contain their clean information also, which will be added to the clean epoch used as references. For better comparison of the two methods we plan to apply the two methods on the artifact epoch alone and then with the help of a medical expert having the results examined.

Once the clean epoch was extracted, one of the marked artifacts was added to the clean epoch's center. The clean epoch's duration was twice the size of the one of the artifact's epoch. The two methods were compared with different thresholds.

Fig.3 shows the average mean square error for both of these techniques. For every method and threshold value combination (thresholds belong in the set (1.2,1.3,1.4...1.8) five experiments of the removal procedure were performed. This error evaluation procedure was followed due to the fact that both methods rely on a random initialization. In more detail the ICA initializes the unmixing matrix randomly, while the Tucker decomposition initializes randomly the projection matrices. The mean square error is calculated based on the equation:

$$MSE = \frac{\sum_{i=1}^{n}\sum_{j=1}^{m}(CleanEpoch(i,j) - ResultingEpoch(i,j))^2}{n*m} \tag{3}$$

where CleanEpoch is the matrix of the clean epoch and ResultingEpoch is the matrix corresponding to the result of every method, n and m are the number of rows and columns of each matrix respectively.

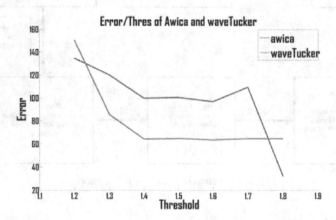

Fig. 3. Tensor based (waveTucker) in comparison with ICA based (awica) for several threshold values.

It is shown that the tensor based technique is more stable than the ICA based and has the lowest error for most of the thresholds. After a certain value, the ICA based method is clearly a better performer but the artifact signal in those thresholds is not cleared completely. It should be noted that for thresholds greater than 1.9 the suspected, for artifact containment, independent components were less than two. Therefore it was not possible to apply the ICA algorithm, leading in not removing the artifact at all. As noted in [10] the preferred value of the threshold for the ICA method is 1.5. Applying the same threshold in the tensor based method had satisfactory results, as proven by our experinments, since the artifact was sufficiently removed without removing a great amount of the rest information in the signal.

Fig.4 and Fig.5 show the two best results of the two methods. Each electrode's signal was normalized using min-max normalization, in order to show the relative amount removed from each method. The channels shown in this figure are part of the whole set of electrodes and are chosen to show each methods drawbacks.

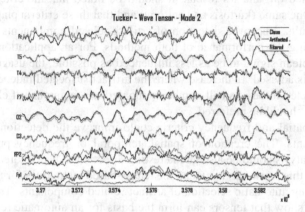

Fig. 4. The result of the tensor-based method. Every channel is affected, since the method is in essence a higher order PCA, but the texture of the reference signal is retained.

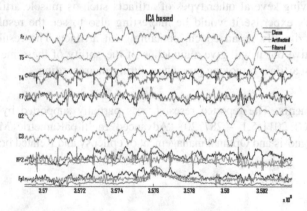

Fig. 5. The ICA-based method. Most of the electrode signals correctly remain intact but the artifact related electrodes, as well as the ones containing some sort of components that fulfilled the removal criteria, have lost the texture of the reference signal.

It should be noted that the artifact exists only in the FP1 and FP2 channels, in the figures showed above. Any other information loss is unacceptable. The ICA based method retains most of the channels unfiltered. However, in the ones where a component was rejected, no texture was retained. By texture we mean the morphology of the clean signals. On the other hand the Tucker based method rejected information from all channels. But the texture of the signals is retained with a certain detail in almost every channel.

5   Discussion

Each method's usage has to be chosen depending on the application where the cleared data will be used on, and it's results. It should be noted that the criteria in the two methods were the same (kurtosis and entropy) and that these criteria play an important role in the results. Improved detection of artifact related components will certainly result in boosting the performance of both methods. For an application requiring the texture of the clean signal to be as less influenced as possible, the usage of the tensor based method is advised. In the case where the artifact suspected independent components are expected to be of small number compared to the amount of electrodes, ICA based methods should be preferred.

Applying spatial and frequency criteria could improve the detection of artifact related components. The creation of spatial, temporal or frequency profiles derived from tensor analysis could be also an interesting direction to be explored.

The goal of this work was to extend the tensor-based techniques for artifact removal by introducing automatic selection of artifact-related components. The results were promising and show that tensors can form the basis for an automatic removal method obtaining comparable results to the ICA based methods.

For future directions we aim to explore the possibilities of success of these methods on removing several other types of artifacts such as muscle artifacts. Using a medical expert's expertise it would be interesting also to see the result of applying these methods directly in an epoch, containing an artifact. The result will in this case be subjective but it will give a guideline of the ability of these methods to work efficiently in an artifact removing procedure during the recording.

Acknowledgements. The reported research was partially supported by the ARMOR Project (FP7-ICT-2011-5.1 - 287720) "Advanced multi-paRametric Monitoring and analysis for diagnosis and Optimal management of epilepsy and Related brain disorders".

References

1. Ochoa, C.J., Polich, J.: P300 and blink instructions. Clinical Neurophysiology: Official Journal of the International Federation of Clinical Neurophysiology 111(1), 93–98 (2000), http://www.ncbi.nlm.nih.gov/pubmed/10656515 (retrieved)
2. Verleger, R.: The instruction to refrain from blinking affects auditory P3 and N1 amplitudes. Electroencephalography and Clinical Neurophysiology 78(3), 240–251 (1991), http://www.ncbi.nlm.nih.gov/pubmed/1707797 (retrieved)
3. Delorme, A., Sejnowski, T., Makeig, S.: Enhanced detection of artifacts in EEG data using higher-order statistics and independent component analysis. NeuroImage 34(4), 1443–1449 (2007), doi:10.1016/j.neuroimage.2006.11.004

4. Chadwick, N., McMeekin, D., Tele, T.: Classifying eye and head movement artifacts in EEG Signals. In: Digital Ecosystems and Technologies Conference (DEST), vol. 5, pp. 285–291 (2011),
   http://ieeexplore.ieee.org/xpls/abs_all.jsp?arnumber=5936640 (retrieved)
5. Lawhern, V., Hairston, W.D., McDowell, K., Westerfield, M., Robbins, K.: Detection and classification of subject-generated artifacts in EEG signals using autoregressive models. Journal of Neuroscience Methods 208(2), 181–189 (2012), doi:10.1016/j.jneumeth(2012).05.017
6. Wu, J., Zhang, J., Yao, L.: An automated detection and correction method of EOG artifacts in EEG-based BCI. Complex Medical Engineering 100875, 1–5 (2009), doi:10.1109/ICCME(2009).4906624
7. De Clercq, W., Vergult, A.: Canonical correlation analysis applied to remove muscle artifacts from the electroencephalogram. Biomedical Engineering 53(5), 2583–2587 (2006), doi:10.1109/TBME.2006.879459
8. Vergult, A., De Clercq, W., Palmini, A., Vanrumste, B., Dupont, P., Van Huffel, S., Van Paesschen, W.: Improving the interpretation of ictal scalp EEG: BSS-CCA algorithm for muscle artifact removal. Epilepsia 48(5), 950–958 (2007), doi:10.1111/j.1528-1167.2007.01031.x
9. Halder, S., Bensch, M., Mellinger, J., Bogdan, M., Kübler, A., Birbaumer, N., Rosenstiel, W.: Online artifact removal for brain-computer interfaces using support vector machines and blind source separation. Computational Intelligence and Neuroscience 10 (2007), doi:10.1155/2007/82069
10. Mammone, N., la Foresta, F., Marabito, F.C.: Automatic artifact rejection from multichannel scalp EEG by wavelet ICA. IEEE Sensors Journal 12(3), 533–542 (2012), doi:10.1109/JSEN.2011.2115236
11. Sheng, H., Tian, H.: A Novel Method to Remove Eye-Blink Artifacts Based on Correlation Using ICA. In: Proceedings of the 2013 IEEE International Conference on Green Computing and Communications and IEEE Internet of Things and IEEE Cyber, Physical and Social Computing, pp. 1946–1948. IEEE Computer Society Press (2013), doi:10.1109/GreenCom-iThings-CPSCom.2013.362
12. Heute, U., Guzman, A.: Removing "cleaned" eye-blinking artifacts from EEG measurements. Signal Processing and Integrated ..., 576–580 (2014),
    http://ieeexplore.ieee.org/xpls/abs_all.jsp?arnumber=6777020 (retrieved)
13. Miwakeichi, F., Martínez-Montes, E., Valdés-Sosa, P.A., Nishiyama, N., Mizuhara, H., Yamaguchi, Y.: Decomposing EEG data into space-time-frequency components using Parallel Factor Analysis. NeuroImage 22(3), 1035–1045 (2004), doi:10.1016/j.neuroimage.2004.03.039
14. Mørup, M., Hansen, L.K., Herrmann, C.S., Parnas, J., Arnfred, S.M.: Parallel Factor Analysis as an exploratory tool for wavelet transformed event-related EEG. NeuroImage 29(3), 938–947 (2006), doi:10.1016/j.neuroimage.2005.08.005
15. Acar, E., Bingöl, C.A., Bingöl, H., Yener, B.: Computational analysis of epileptic focus localization. In: Ruggiero, C. (ed.) Proceedings of the 24th IASTED International Conference on Biomedical Engineering (BioMed 2006), pp. 317–322. ACTA Press, Anaheim (2006)
16. Acar, E., Aykut-Bingol, C., Bingol, H., Bro, R., Yener, B.: Multiway analysis of epilepsy tensors. Bioinformatics s 23(13), i10-8 (2007), doi:10.1093/bioinformatics/btm210

17. Kolda, T.G., Bader, B.W.: Tensor Decompositions and Applications. SIAM Review 51(3), 455–500 (2009), doi:10.1137/07070111X
18. Harshman, R.: Foundations of the PARAFAC procedure: Models and conditions for an "explanatory" multimodal factor analysis (10), 1–84 (1970)
19. Carroll, J., Chang, J.: Analysis of individual differences in multidimensional scaling via an N-way generalization of "Eckart-Young" decomposition. Psychometrika 35 (1970), http://link.springer.com/article/10.1007/BF02310791 (retreived)
20. Tucker, L.R.: Implications of factor analysis of three-way matrices for measurement of change. In: Harris, C.W. (ed.) Problems in Measuring Change, pp. 122–137. University of Wisconsin Press, Madison (1963)
21. Hitchcock, F.L.: The expression of a tensor or a polyadic as a sum of products (1927)
22. Cattell, R.B.: Parallel proportional profiles and other principles for determining the choice of factors by rotation. Psychometrika 9(4), 267–283 (1944)
23. Bader, B.W., Kolda, T.G.: MATLAB Tensor Toolbox Version 2.5., http://www.sandia.gov/~tgkolda/TensorToolbox/ (accessed)

The Greek Audio Dataset

Dimos Makris, Katia Lida Kermanidis, and Ioannis Karydis

Dept. of Informatics, Ionian University, Kerkyra 49100, Greece
{c12makr,kerman,karydis}@ionio.gr

Abstract. The Greek Audio Dataset (GAD), is a freely available collection of audio features and metadata for a thousand popular Greek tracks. In this work, the creation process of the dataset is described together with its contents. Following the methodology of existing datasets, the GAD dataset does not include the audio content of the respective data due to intellectual property rights but it includes MIR important features extracted directly from the content in addition to lyrics and manually annotated genre and mood for each audio track. Moreover, for each track a link to available audio content in YouTube is provided in order to support researchers that require the extraction of new feature-sets, not included in the GAD. The selection of the features extracted has been based on the Million Song Dataset in order to ensure that researchers do not require new programming interfaces in order to take advantage of the GAD.

Keywords: music, information retrieval, audio, Greek music, dataset.

1 Introduction

Music Information Retrieval (MIR) research, in other words research on methods for Information Retrieval and Data Mining on musical data, has a number of requirements. Central to these requirements is the necessity to experiment with the methods on real musical data. This experimentation mainly achieves testing on the efficiency and effectiveness of the methods, as well as comparison of existing existing methods in order to show improvement.

In MIR research, the term musical data refers to sound recordings, sheet music as well as associated information to the musical content (i.e. metadata, social tags, etc). The process of accumulating a set of musical data (dataset) is very important for most scientific processed, including MIR. Datasets allow researchers to scientifically to compare and contrast their methods by testing these on commonly available collection of musical works.

Accordingly, MIR requires data for all kinds of music as results are not always intuitive due to the highly artistic nature of music. Although a number of widely used datasets do exist for the purposes of MIR research, most of these are collections of mainstream English language music. Local music has numerous differences from the these musical collections such us different instruments and rhythms. To the best of our knowledge, a standardized Greek audio dataset is not currently available.

L. Iliadis et al. (Eds.): AIAI 2014 Workshops, IFIP AICT 437, pp. 165–173, 2014.

1.1 Motivation and Contribution

To address the aforementioned requirements, we introduce the Greek Audio Dataset, a freely-available collection of Greek audio data for the purposes of MIR that, for each song it contains, offers:

- audio features for immediate use in MIR tasks,
- the lyrics of the song,
- manually annotated mood and genre labels,
- a link to YouTube for further feature extraction, on the basis of MIR research processes.

The rest of the paper is organised as follows: Section 2 presents the related work on MIR available datasets, while Section 3 discuses the dataset, its creation processes as well as a detailed analysis of its content. Next, Section 4 details future directions concerning the dataset that could ameliorate is usability and further support MIR research. Finally the paper is concluded in Section 5.

2 Related Research

Since the early years of the MIR research, there have been numerous efforts to build datasets facilitating the work of MIR researchers.

The dataset titled RWC [4] is a copyright-cleared music database that is available to researchers as a common foundation for research. RWC was one of the first large-scale music database containing six original collections on different genres.

CAL500 [19] is another widely used dataset, a corpus of 500 tracks of Western popular music each of which has been manually annotated by at least three human labelers with a total number of 1700 human-generated musical annotations.

Tzanetakis and Cook [20], introduced a dataset under the name GTZAN Genre Collection. The dataset consists of 1000 audio tracks, each 30 seconds long. It also contains 10 genres, each represented by 100 tracks and its size is approximately 1.2 Gb. The dataset was collected gradually with no titles from a variety of sources including personal CD collections or radio and microphone recordings, while it doesn't contain any music information or metadata.

USPOP2002 [1] was introduced in order to perform comparison between different acoustic-based similarity measurements. 400 artists were chosen for popularity and for representation, while overall, the dataset contains 706 albums and 8764 tracks.

The Swat10k [18] data set contains 10,870 songs that are weakly-labeled using a tag vocabulary of 475 acoustic tags and 153 genre tags. These tags have all been harvested from Pandora [13] result from song annotations performed by expert musicologists involved with the Music Genome Project.

The Magnatagatune dataset features human annotations collected by Edith Law's TagATune game [9] and corresponding sound clips encoded in 16 kHz/ 32kbps /mono/mp3 format. It also contains a detailed analysis from The Echo

Nest of the track's structure and musical content, including rhythm, pitch and timbre.

Schedl et al. [15], presented the MusiClef dataset, a multimodal data set of professionally annotated music. MusiClef contains editorial meta-data, audio features (generic and MIR oriented) are provided. Additionally MusiClef includes annotations such as collaboratively generated user tags, web pages about artists and albums, and the annotation labels provided by music experts.

Finally the Million Song Dataset (MSD) [2] stands out as the largest currently available for researchers with 1 million songs, 44,745 unique artists, 280 GB of data, 7,643 unique terms containing acoustic features such as pitch, timbre and loudness and much more.

Table 1 (adapted from [2]) collectively presents the aforementioned datasets by comparison of size and samples/audio availability.

Table 1. Comparison of existing datasets: size and samples/audio availability

dataset	# songs	includes samples/audio
CAL500	500	No
Greek Audio Dataset (GAD)	1000	No (YouTube links available)
GZTAN genre	1000	Yes
Magnatagatune	25863	Yes
Million Song Dataset (MSD)	1000000	No
MusiCLEF	1355	Yes
RWC	465	Yes
Swat10K	10870	No
USPOP	8752	No

3 The Dataset

The musical tradition of Greece is diverse and celebrated through its history. Greek music can be separated into two major categories: Greek traditional music and Byzantine music, with more eastern sounds [7]. Greek traditional (folk) music or "δημοτική μουσική", as it is most commonly referred to, is a combination of songs, tempos and rhythms from a litany of Greek regions. These compositions, in their vast majority, have been created by unknown authors, are more than a century old and are the basis for the Modern Greek traditional music scene. GAD contains many Greek traditional music songs, starting from the 60's and 70's as well as some current variants that are very popular nowadays.

Greek music contains special characteristics that are not easily to be found on the existing datasets. The unique instruments (bouzouki, lyra, laouto, etc) and the unique rhythms reflect on the audio feature extraction while the complexity of the Greek language can be an area of study for linguistic experts on lyric feature extraction.

3.1 Creation Process

The audio data collection covers the whole range of Greek music, from traditional to modern. We made every effort to make the song and genre selection balanced in GAD as much as possible. The selection of the music tracks was made from personal CD collections, while some of the songs' recordings were available in live performances. The audio feature extraction process was applied on CD quality wave files (44,1KHz, 16 bit) using jAudio [11] and AudioFeatureExtraction, a processing tool introduced by the MIR research team of Vienna University of Technology[21].

For each song, the GAD additionally includes its lyrics. The lyrics included in the dataset have been retrieved among various sources, such as the web site stixoi.info [6], and are used within for the purposes of academic and scientific research of MIR researchers. It should be noted that copyright of the lyrics remains with their respective copyright owners.

As far as genre classification is concerned, the following Greek musical culture oriented tags where used: Rembetiko, Laiko, Entexno, Modern Laiko, Rock, Hip Hop/R & B, Pop, Enallaktiko, that are described in the sequel (Section 3.2). Because of the different music styles that artists may adopt during their career, we performed listening tests to every song before choosing the appropriate genre tag. For mood annotation, the Thayer model [17] has been adopted. To record the mood categories, we invited 5 annotators to listen and read the lyrics for one couple and refrain for each song. Then we computed a standard F-measure as a measure of the inter-annotator agreement [3] and we found a value of 0.8 approximately. For clusters of mood with smaller F-measure, a discussion between the annotators was taken part in order to reduce controversy.

The dataset additionally includes a YouTube [22] link for each song that allows for access to the audio content. In order to identify the best match of the YouTube available alternative songs, for every song we attempted to identify the YouTube URL using the following criteria: number of views, number of responses, best audio quality and as close as possible performance as in the wave library where the features were extracted from.

The GAD is available as a download from the webpage of the Informatics in Humanistic and Social Sciences Lab of the Ionian University[1].

3.2 The Content

The GAD contains audio and lyrics features and metadata for selected Greek Songs:

- 1000 songs
- 1000 txt files with the lyrics
- 1000 best YouTube links
- 277 unique artists

---

[1] http://di.ionio.gr/hilab/gad

– 8 Genres. We give the necessary explanations to distinguish the unique Greek genres.

- Ρεμπέτικο: 65 tracks. In urban centers with strong Greek presence, among other genres, it appeared a kind of folk called "Rembetiko". Major schools of Rembetiko were: Smyrnaiiki and the Piraeus school classic Rembetiko [12]. "Zeibekiko", "karsilamas" and "hasapiko" are a few characteristic rhythms.
- Λαικό: 186 tracks. As an evolution of Rembetiko are considered Greek Folk songs of decades of 1950-1960, which continues to evolve and sound today [10]. The transition to the so-called "Laiko" music is evident in the imposition of European instrument tuning, rhythms and now that the author can write songs with "harmony".
- Εντεχνo: 195 tracks. A complex musical work of art that combines Modern Greek music with poetry [16]. It differs from the Laiko mainly in verse, but also in music (instrumentation, style).
- Μοντέρνο Λαικό: 175 tracks. The "Modern Laiko" is considered as the current evolution of popular music. Several pop elements, electronic sounds can be identified and is now the most common music heard in Greek live stages. The style and the rhythms have not changed much and the theme of the songs is adapted to the current daily problems.
- Rock: 195 tracks. It also includes 80s Pop-Rock.
- Hip Hop/R & B: 60 tracks.
- Pop: 63 tracks. It also includes Dance-Club music style and older Greek disco hits.
- Εναλλακτικό: 60 tracks. Although with the term "Enallaktiko" it is usually considered as "Alternative Rock" [14], we include tracks fusing Greek modern kinds of music. Pop Rock and Entexno elements can be found in this class.

– 16 Mood taxonomies: 2 dimensions, valence and arousal, which divide a 2-dimensional emotive plane into 4 parts by having positive/high and negative/low values respectively. Arousal and valence are linked to energy and tension, respectively. Arousal values correspond to moods such as "angry" and "exciting" to "tired" and "serene" from 1-4. Valence values correspond to moods such as "sad" and "upset" to "happy" and "content" from A-D. Figure 1 shows the distribution of tracks by mood and we observe that the tracks with positive emotions (valence C, D), have variations in arousal and either are very calm (arousal 1) or intense stress (arousal 4). We also include all the lyrics for further feature extraction. The accumulated lyrics contain:

- 32024 lines
- 143003 words
- 1397244 characters

The data is available in two formats, HDF5 and CSV.

HDF5 is efficient for handling the heterogeneous types of information such as audio features in variable array lengths, names as strings, and easy for adding

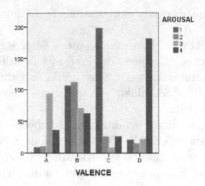

Fig. 1. Valence and Arousal on the GAD

new types of features. Every song is described by a single file the contents of which are as shown in Figure 2. Each file has 2 groups: the folder "Audio Features" containing the three most important types of features, such as Timbral, Rhythm and Pitch [20] and the "Metadata" containing all the other information that is related to the song (i.e. title, artist, mood, genre and YouTube link).

CSV is compatible for processing with Weka [5], RapidMiner [8] and other similar data mining platforms. The GAD provides the commonly used, on the discipline of MIR, audio feature sets in separate CSV files. Weka, besides the well-known options for preprocessing and classification, additionally offers attribute visualization. Figure 3 shows one such mapping between two timbral characteristics with sorting class the genre and more specifically, Rock and Entexno.

The main acoustic features are Timbral, Rhythm and Pitch. Most of the features were extracted using jAudio which also calculates derived features. These features are created by applying metafeatures to primary features.

- Timbral Texture Features: features used to differentiate mixture of sounds based on their instrumental compositions when the melody and the pitch components are similar. (251 features)
  - FFT: Spectral Centroid, Flux, Rolloff, Low Energy, Zero Crossings, Compactness, Spectral Variability, Root Mean Square. 16 features
  - MFCC: 26 features
  - Spectrum: Power & Magnitum Spectrum. 199 features
  - MoM: Method Of Moments. 10 features
- Rhythm Features: Rhythmic content features are used to characterize the regularity of the rhythm, the beat, the tempo, and the time signature. These features are calculated by extracting periodic changes from the beat histogram. (63 features)
  - Beat and Freq: Strongest beat, sum, strength + Strongest FREQ, Zero spectral, FFT + Peak Finder. 13 features

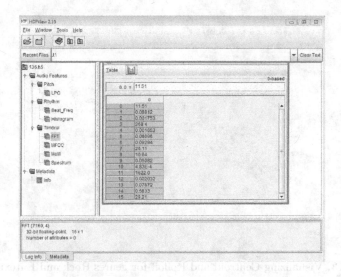

Fig. 2. View of a single track using HDFView

- Beat Histogram. 60 features
- Pitch Content Features: The pitch content features describe the distribution of pitches. (18 features)
  - LPC: Linear Predictive Coding for voice recognition.

4   Future Direction of the Dataset

The GAD is not without issues that can be ameliorated in future versions. One of these issues pertains to the balance of the distributions of genre types in the data. In addition, classification of Greek music into genres can be confusing when use of only one tag is assumed. An extension of the dataset by addition of more data as well as a multi-label version of the dataset will eliminate those problems and it is already under development.

Some of the main future actions that would greatly enhance the GAD, to name a few, are:

- the inclusion of user generated tags (from tagging games or web-services),
- the collection of labels for mood and genre based on more users,
- the expansion of the number of songs (i.e. include latest top-chart songs),
- the refinement of genres by adding more detailed labels with descriptions,
- the balancing of moods and/or genres,
- the inclusion of scores for each song,
- the development of programming language wrappers.

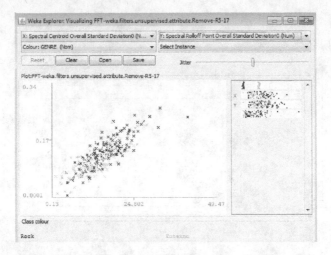

Fig. 3. Visualizing Centroid and Rolloff for genres Rock and Entexno

5 Conclusion

The Greek Audio Dataset is, to the best of the authors' knowledge, the first complete attempt to create an annotated Greek audio dataset. Although its size is relatively small compared to latest datasets, it constitutes a good start for researchers who want to study how the Greek music reflects in the context of MIR. Following the methodology of the, ubiquitous in MIR research, Million Song Dataset, the GAD does not include the audio content and it uses the HDF5 format to store all the information. This makes it easy for any future enrichment with more features and musical information as well as for the utilisation of existing methodologies in order to access the data. In order to enhance further the dataset, it also contains lyrics information so as to provide for linguistic methods on MIR. In addition, the dataset is also available in CSV format that makes it very easy to explore it in data mining experiments using widely available software as WEKA and RapidMiner.

References

1. Berenzweig, A., Logan, B., Ellis, D.P.W., Whitman, B.P.W.: A large-scale evaluation of acoustic and subjective music-similarity measures. Comput. Music J. 28(2), 63–76 (2004)
2. Bertin-Mahieux, T., Ellis, D.P., Whitman, B., Lamere, P.: The million song dataset. In: Proceedings of the 12th International Conference on Music Information Retrieval, ISMIR 2011 (2011)
3. Boisen, S., Crystal, M., Schwartz, R.M., Stone, R., Weischedel, R.M.: Annotating resources for information extraction. In: LREC (2000)

4. Goto, M., Hashiguchi, H., Nishimura, T., Oka, R.: Rwc music database: Popular, classical, and jazz music databases. In: Proceedings of 3rd International Conference on Music Information Retrieval, pp. 287–288 (2002)
5. Holmes, G., Donkin, A., Witten, I.H.: Weka: a machine learning workbench, pp. 357–361 (1994)
6. stixoi info: Greek lyrics for songs and poetry, http://www.stixoi.info/
7. Institute for research on music and acoustics: Greek traditional music, http://www.musicportal.gr/greek_traditional_music/
8. Jungermann, F.: Information extraction with rapidminer. In: Proceedings of the GSCL Symposium 'Sprachtechnologie und eHumanities', pp. 50–61 (2009)
9. Law, E., von Ahn, L.: Input-agreement: A new mechanism for collecting data using human computation games. In: Proceedings of the SIGCHI Conference on Human Factors in Computing Systems, pp. 1197–1206 (2009)
10. Liavas, L.: The greek song: from 1821 to the 1950s. Emporiki Bank Of Greece (2009)
11. Mcennis, D., Mckay, C., Fujinaga, I., Depalle, P.: jaudio: An feature extraction library. In: Proceedings of the 6th International Conference on Music Information Retrieval (2005)
12. Ordoulidis, N.: The greek laiko (popular) rhythms: Some problematic issues. In: Proceedings 2nd Annual International Conference on Visual and Performing Arts (2011)
13. Pandora: A free personalized internet radio, http://www.pandora.com/
14. di Perna, A.: Brave noise-the history of alternative rock guitar. Guitar World (1995)
15. Schedl, M., Liem, C.C., Peeters, G., Orio, N.: A Professionally Annotated and Enriched Multimodal Data Set on Popular Music. In: Proceedings of the 4th ACM Multimedia Systems Conference (2013)
16. Sideras, A.: The sung poetry. Musicology 3, 89–106 (1985)
17. Thayer, R.: The biopsychology of mood and arousal. Oxford University Press, USA (1989)
18. Tingle, D., Kim, Y.E., Turnbull, D.: Exploring automatic music annotation with "acoustically-objective" tags. In: Proceedings of the International Conference on Multimedia Information Retrieval, pp. 55–62 (2010)
19. Turnbull, D., Barrington, L., Torres, D., Lanckriet, G.: Towards musical query-by-semantic-description using the cal500 data set. In: Proceedings of the 30th Annual International ACM SIGIR Conference on Research and Development in Information Retrieval, pp. 439–446 (2007)
20. Tzanetakis, G., Cook, P.: Musical genre classification of audio signals. IEEE Transactions on Speech and Audio Processing 10(5), 293–302 (2002)
21. Vienna University of Technology: Audio feature extraction web service, http://www.ifs.tuwien.ac.at/mir/webservice/
22. YouTube: Share your videos with friends, family and the world, http://www.youtube.com/

Affect in Complex Decision-Making Systems: From Psychology to Computer Science Perspectives

Amine Chohra, Aziza Chohra, and Kurosh Madani

Images, Signals, and Intelligent System Laboratory (LISSI / EA 3956), Paris-East University (UPEC), Senart Institute of Technology, Avenue Pierre Point, Lieusaint 77127, France
{chohra,madani}@u-pec.fr
http://www.lissi.fr
http://chohra.webs.com

Abstract. The increasing progresses in both psychology and computer science allow continually to deal with more and more complex systems and closer to real-world applications in order to solve particularly the decision-making process with uncertain and incomplete information. The aim of this research work is to highlight the irrationality (affect) and to understand some different ways in which the irrationality enter into the decision-making from psychology and computer science perspectives. The goal is also to present some of such integrations of affect in decision-making which emerged in computer science particularly in negotiation and robotics. Thus, the role of the affect in decision-making is developed and discussed. Afterwards, an overview is given on the influence of the affect on an individual level emphasizing the idea that irrationality also has strong social components which can influence the interactions on a group (collective) level. Then, the Emotional Intelligence (EI) is discussed as a first step towards an attempt to answer to how to regulate/control the irrationality part for 'positive' decision-making, and consequently its effects on the though and action. Finally, some developments of computational models are presented. Open questions and challenges are discussed in conclusion, particularly the importance of the personality and its links to the affect.

Keywords: Complex systems, decision-making, uncertain and incomplete information, affect (emotions and moods), personality, negotiation and robotics.

1 Introduction

Affect (or feelings) is the overarching category, which incorporates moods and emotions [1]. Emotions are specific (e.g., anger, happiness, fear, ...) and brief subjective feelings, which usually have an identifiable cause (e.g., I'm angry with ...) [1]. Moods, on the other hand, are usually more diffuse (e.g., good or bad mood) and weaker feeling states, and tend to lack source identification (e.g., I'm in a bad mood).

Throughout recorded human intellectual history, there has been active debate about the nature of the role of emotions or "passions" in human behavior [2], with the dominant view being that passions are a negative force in human behavior [3]. By contrast, some of the latest research has been characterized by a new appreciation of the

L. Iliadis et al. (Eds.): AIAI 2014 Workshops, IFIP AICT 437, pp. 174–183, 2014.

positive functions served by emotions [4]. An appreciation for the positive functions is not entirely new in behavioral science. Darwin, in 1872, was one of the first to hypothesize the adaptive mechanisms through which emotion might guide human behavior [5].

The increasing progress in computer science and information technology allowed to deal with more complex systems and closer to real-world applications in order to solve particularly the decision-making problems. Traditionally, the decision-making was viewed as a rational process where reason calculates the best way to achieve the goal. Investigations from different areas of cognitive science have shown that human decisions and actions are much more influenced by intuition and emotional responses than it was previously thought [6], [7], [8], [9].

Emotions are one of the important subconscious mechanisms that influence human behaviors, attentions, and decision making. The emotion process helps to determine how humans perceive their internal status and needs in order to form consciousness of an individual. Emotions have been studied from multidisciplinary perspectives and covered a wide range of empirical and psychological topics, such as understanding the emotional processes, creating cognitive and computational models of emotions, and applications in computational intelligence.

This paper deals with a review on the role of the affect (emotions and moods) on decision-making from psychology perspective, and the awareness to such role from computer science perspective. The goal is also to emphasize some central topics emerged from such integrations of the irrationality part (emotions and moods, personality) in the decision-making of complex systems with uncertain and incomplete information, particularly in computer science research fields such as in negotiation and robotics.

Thus, the role of the affect in decision-making is developed and discussed in Sect. 2. Afterwards, an overview is given on the influence of the affect on an individual level, in Sect. 3, emphasizing the idea that irrationality also has strong social components which can influence the interactions on a group (collective) level. Then, the Emotional Intelligence (EI) is discussed in Sect. 4, as a first step towards an attempt to answer to how to regulate/control the irrationality part for 'positive' decision-making, and consequently its effects on the though and action. Finally, some developments of computational models are presented in Sect. 5. Open questions and challenges are discussed in Sect. 6, particularly the importance of the personality and its links to the affect.

2   The Role of the Affect in Decision-Making

Until middle of 1990s, the affect (emotions and moods) attracted little attention from researchers on decision-making. In fact, the decision-making was viewed, in general, as a cognitive process: a matter of estimating which of various alternative actions would yield the most positive consequences [2]. The research community which emerged in the late 1960s under the heading of behavioral decision theory largely adhered to this cognitive perspective.

The main thrust of behavioral decision theory [2] has been to identify:
- cognitive errors that people make when they judge the likelihood of future consequences,
- simplifying heuristics that people use to cope with the complexity of decision-making.

The great interest and large investigations in decision-making research associated with the emergence of behavioral decision theory, then, largely ignored the role played by the irrationality part, related to affect in general, in decision-making [2].

However, with the research developments particularly psychology-related fields from 1990s, a great interest have been oriented towards the role of the irrationality part related to emotions in decision-making. Research has shown that:
- even incidental affect (emotion and mood), affect that is unrelated to the decision at hand, can have a significant impact on judgment and choice [10],
- emotional deficits, whether innate [4] or experimentally induced, can degrade the quality of decision-making,
- incorporating affect (emotion and mood) in models of decision-making can greatly increase their explanatory power.

From the perspective of neuroscience and psychology, Antonio Damasio [4] states that an important part of the decision-making process consists of the comparison of potential alternatives with emotions and feelings from similar past situations. Furthermore, the process also involves the estimation of results brought about by these past events and potential rewards or punishments that might have been gained during such events. This procedure enables us to simulate potential future outcomes based on our past experiences and then opt for a move that will lead to the best possible solution.

Thus, contemporary decision-making research is characterized by an intense focus on the irrationality part, related to affect in general [2].

When studying the effects of affect (emotion) on judgment, decision, and behavior, two types of affective (emotional) phenomena should be distinguished: *incidental emotional states* and *integral emotional responses* [11], [12], [6].

Loewenstein and Lerner [2] classify according to their place along the time course of a decision process, beginning with a deliberation phase leading to a choice, then implementing the choice, and, eventually, experiencing the outcomes. They distinguish between anticipated emotions and immediate emotions, with immediate emotions further classified into *incidental* and *anticipatory* emotions. Anticipated emotions are beliefs about one's future emotional states that might ensue when the outcomes are obtained. Immediate emotions, in contrast, are actually experienced when making a decision, thereby exerting an effect on the mental processes involved in making a choice; for similar distinctions see [13]. Immediate emotions come in two variants, either as *incidental* emotions caused by factors which are not related to the decision problem at hand, and as *anticipatory* or integral emotions, which are caused by the decision problem itself.

There is ample evidence that these kinds of emotion frequently do influence the judgments and choices people make. Lerner and Keltner [10] demonstrated the effects of incidental fear and anger on risk judgments. The influence of immediate anticipatory emotions in intertemporal choice has been examined by Loewenstein 1996, see [2], [14]. The importance of anticipated emotions such as anticipated regret and disappointment in decision making has been demonstrated by Zeelenberg, van Dijk, Manstead, and van der Pligt 2000, see [Pfister2008].

Recently, Peters, Västfjäll, Gärling, and Slovic, in 2006 proposed another classification of the roles that affect plays in decision making, see [14]. Affect is loosely defined as experienced feelings about a stimulus, either integral or incidental. Four roles are identified:

- First, affect plays a role as information, especially via the affect as information mechanism, Schwarz and Clore 1988. These feelings, possibly misattributed to the stimulus, act as good-versus-bad information to guide choices, according to the affect heuristic proposed by Slovic, Finucane, Peters, and MacGregor 2002.

- The second role played by affect is as a spotlight, focusing the decision maker's attention on certain kinds of new information and making certain kinds of knowledge more accessible for further information processing. This role is reminiscent of mood-congruent memory as studied by Bower 1991.

- Third, affect operates as a motivator, influencing approach-avoidance tendencies as well as efforts to process information Frijda, 1986 ; Zeelenberg and Pieters 2006.

- Finally, a fourth role of affect is to serve as a common currency in judgments and decisions Cabanac 1992. Just as money does for goods, affect provides a common currency for experiences. Following Cabanac 1992, Peters claims that affective reactions enable people to compare disparate events and complex arguments on a common underlying dimension, see [14].

Another interesting four-fold classification of emotional mechanisms is given in [14] by Pfister which is similar to the Peters approach in some respects, and where the commonalities and differences are discussed.

Thus, several other researchers investigate the role of affect in decision making from different point of views [15], [16], [17], [3], [18], [19], [20], [9], [21].

By another way, until the end of 1990s, works from computer science research fields implying Artificial Intelligence and particularly robotics have been limited and mostly oriented towards synthetic agents, and while some robot systems incorporate emotions, they focus too much on emotional expression, leaving behind perhaps some of the most important functions and aspects of emotional processing, such as their influence in behavior selection, coherence, relevance, and learning.

Thus, a transition has been done in order to deal with more sensing abilities and a larger repertoire of behaviors (than navigation, obstacle avoidance, localization, …). In fact, such transition emerged from many research efforts which have shifted from behavior-based approaches that deal with insect-level competence, to those that try to build robot systems that exhibit behaviors comparable in complexity to those of humans, as it is the case in friendly robotics and cognobotics [22].

As Brooks 1997 has eloquently argued, the robotic systems must manipulate the world and interact in it in non-trivial ways. This requires a much richer set of abilities to obtain sensory information and coordinate motor control, as well as an increased and more complex behavioral repertoire that includes, among other competences, the ability to interact socially. However, robots that exhibit robust and adaptive behavior, and which are responsive to social interactions, must deal with issues such as motivation and emotion which have not been essential in behavior-based robotics, see [22].

Thus, Juan D. Velasquez proposed in [23], [22] an emotion-based approach to robotics that relies on the use of computational frameworks for the emotion-based control: control of autonomous agents that relies on, and arises from, emotional processing. In this work, six different types of affect programs have been identified and created in explicit models: *anger*, *fear*, *distress/sorrow*, *joy/happiness*, *disgust*, and *surprise* [22].

In negotiation, Li developed a strategic emotion in negotiation implying emotion, cognition and culture [24]. A one to one bargaining process, in which a buyer agent and a seller agent negotiate over single issue (price), is developed, where the basic behaviors based on time and personality aspects (conciliatory, neutral, and aggressive) have been suggested [25].

3   Emotion on Individual and Group (Collective) Levels

There have been three major contexts within which researchers have studied the effects of affective states on intergroup perception and behavior. Two of the domains have to do with affect that is elicited by the group itself and the social situations within which the group is experienced (termed "integral" affect by Bodenhausen, 1993). Research on chronic integral affect has examined the impact of enduring affective reactions to the social group on attitudes and behavior toward the group and its members. Research on episodic integral affect has examined the impact of affective reactions that are situationally created in intergroup settings, which may in principle be quite different from more chronic feelings about the group (as when one has a pleasant interaction with a member of an otherwise disliked group). The final domain involves affective states that arise for reasons having nothing to do with the intergroup context itself, but which are carried over from other events into an intergroup setting (termed "incidental affect" by Bodenhausen, 1993), see [11], [12].

By another way, Treur developed agent models from social neuroscience concepts and discussed how such neurological concepts can be used to obtain emergence of shared understanding and collective power of groups of agents, both in a cognitive and affective sense [26]. A generic contagion model is then developed emphasizing the idea that irrationality also has strong social components which can influence the interactions on a group (collective) level, see also [27].

Also, Becker Asano and Wachsmuth designed the "WASABI" affect simulation architecture, which uses a three-dimensional emotion space called PAD (Pleasure-Arousal-Dominance) space [28]. In this study, social robots generate and express their emotions in human-robot interaction.

4   Emotional Intelligence (EI)

"All learning has an emotional base." from the greek philosopher Plato. Emotional Intelligence (EI) refers to the ability to perceive, control and evaluate emotions. Some researchers suggest that EI can be learned and strengthened, while others claim it is an inborn characteristic.

Since 1990, Peter Salovey and John D. Mayer [29] have been the leading researchers on EI. In their influential article "Emotional Intelligence, in 1990" they defined EI as, "the subset of social intelligence that involves the ability to monitor one's own and others' feelings and emotions, to discriminate among them and to use this information to guide one's thinking and actions".

Thus, several research works have been developed on EI [30], [31], [32].

5   Computational Models

An excellent review is given on the development of computational models of emotions for autonomous agents in [33]. Particularly, the comprehensive survey where five design aspects that influence the development process of computational models are investigated: *theoretical foundations*, *operating cycle*, *cognition and emotion interaction*, *architecture design*, and *role in cognitive agent architectures*.

A comprehensive survey of cognitive and computational models of emotions resulted from multidisciplinary studies is given in [34]. It explores how cognitive models serve as the theoretical basis of computational models of emotions. The mechanisms underlying affective behaviors are examined as important elements in the design of these computational models. A comparative analysis of current approaches is elaborated based on recent advances towards a coherent cognitive computational model of emotions, which leads to the machine simulated emotions for cognitive robots and autonomous agent systems in cognitive informatics and cognitive computing.

Because of the multiple facets and components underlying the process of human emotions, it can be approached from a diversity of perspectives. Moreover, due to the nature of this process and its applications, emotions are currently the focus of study in multiple disciplines such as psychology, neuroscience, philosophy, computer science, cognitive sciences, and cognitive informatics [34]. This multidisciplinary inquiry has provided evidence that shows the significance of emotions not only to the rational behavior of individuals, but to achieve more believable and human-like behaviors in intelligent systems. In particular, fields such as psychology and neuroscience have contributed a number of theories and models that explain the diversity of the emotion process. These theories are focused on revealing the mechanisms underlying the process by which humans transform external stimuli into emotional perspectives. Similarly, in fields such as computer science, cognitive informatics, computational intelligence, and artificial intelligence, researchers are interested in the design of formal and computational models of emotions that help improve Artificial Intelligent systems used for cognitive robots, autonomous agents, and human-computer interactions [34]. In this dual approach, computational modeling technologies are used for testing and refining psychological, biological, and cognitive models, which are further used to support the design of computational models of emotions.

Thus, several computational models have been developed [35], [36], [37], [38], [39].

6   Discussion and Conclusion

In this paper, the role of the affect in decision-making has been developed and discussed. Afterwards, an overview is given on the influence of the affect on an individual level emphasizing the idea that irrationality also has strong social components which can influence the interactions on a group (collective) level. Then, the Emotional Intelligence (EI) has been discussed as a first step towards an attempt to answer to how to regulate/control the irrationality part for 'positive' decision-making, and consequently its effects on the though and action. Finally, some developments of computational models have been presented.

Throughout the paper, the goal was also to present some of such integrations of affect in decision-making which emerged in computer science particularly in negotiation and robotics.

In the following, some open questions and challenges are discussed.

First, it is very important to discuss the importance of the personality and its links to the affect. A brief definition would be that personality is made up of the characteristic patterns of thoughts, feelings and behaviors that make a person unique. In addition to this, personality arises from within the individual and remains fairly consistent throughout life. "Personality refers to individuals' characteristic patterns of thought, emotion, and behavior, together with the psychological mechanisms -- hidden or not -- behind those patterns. This definition means that among their colleagues in other subfields of psychology, those psychologists who study personality have a unique mandate: to explain whole persons." Funder 1997.

"Although no single definition is acceptable to all personality theorists, we can say that personality is a pattern of relatively permanent traits and unique characteristics that give both consistency and individuality to a person's behavior." Feist 2009.

"Having closed in on a sense of what personality is, it may be helpful to compare the concept to others with related meanings. Two concepts that quickly come to mind are 'temperament' and 'character.' In everyday language these terms are sometimes used more or less interchangeably with 'personality,' and historically they have often been used in contexts where, in more recent times, 'personality' would be employed. Within psychology, however, they have somewhat distinct meanings. Temperament usually refers to those aspects of psychological individuality that are present at birth or at least very early on in child development, are related to emotional expression, and are presumed to have a biological basis... Character, on the other hand, usually refers to those personal attributes that are relevant to moral conduct, self-mastery, will-power, and integrity." Haslam 2007.

A very well known model of the personality is the five-factor model in personality developed in [40]. An example of negotiation based on such five-factor model has been developed in [41].

Thus, several research works have been developed using the affect and personality [42], [43].

Another open question is related to the philosophers Solomon and Stone 2002, see [8], recently reviewed the emotion literature and concluded that: The analysis of emotions in terms of "valence," while it recognizes something essential about emotions...

is an idea that we should abandon and leave behind. It serves no purpose but confusion and perpetrates the worst old stereotypes about emotion, that these are simple phenomena unworthy of serious research and analysis. The idea is that if we really want to understand emotion and emotion's impact we have to go beyond mere valence [8].

Another interesting alternative for future consists to investigate the research work initiated in [44]. The interest in this issue is that Authors not used a predefined human emotional model but tried to create an agent (robot) specific emotional architecture under the assumption that emotions are unique to an agent's possible interactions with its world.

References

1. Andrade, E.B., Ariely, D.: The Enduring Impact of Transient Emotions on Decision-Making; Organizational Behavior and Human Decision Processes, vol. 109, pp. 01–08. Elsevier (2009)
2. Loewenstein, G., Lerner, J.S.: The Role of Affect in Decision-Making. In: Davidson, R.J., Scherer, K.R., Goldsmith, H.H. (eds.) Handbook of Affective Sciences, pp. 619–642. Oxford University Press (2003)
3. Elster, J.: Alchemies of the Mind: Rationality and the Emotions. Cambridge University Press, Cambridge (1999)
4. Damasio, A.R.: Descartes' Error: Emotion, reason, and the human brain. Putnam, New York (1994)
5. Darwin, C.: The Expression of the Emotions in Man and Animals, 3rd edn. Oxford University Press, New York (1998); Original Work Published in 1872 and Produced by Charles Keller and David Widger
6. Pham, M.T.: Emotion and rationality: a critical review and interpretation of empirical evidence. Review of General Psychology, American Psychological Association 11(2), 155–178 (2007)
7. Seo, M.-G., Barrett, L.F.: Being emotional during decision making – good or bad? an empirical investigation. Academy of Management Journal 50(4), 923–940 (2007)
8. Zeelenberg, M., Nelissen, R.M.A., Breugelmans, S.M., Pieters, R.: On emotion specificity in decision making: why feeling is for doing. Judgment and Decision Making 3(1), 18–27 (2008)
9. Markic, O.: Rationality and emotions in decision-making. Interdisciplinary Description of Complex Systems 7(2), 54–64 (2009)
10. Lerner, J.S., Keltner, D.: Beyond valence: Toward a model of emotion-specific influences on judgement and choice. Cognition and Emotion 14(4), 473–493 (2000b)
11. Bodenhausen, G.V., Mussweiler, T., Gabriel, S., Moreno, K.N.: Affective influences on stereotyping and intergroup relations. In: Forgas, J.P. (ed.) Handbook of Affect and Social Cognition. Lawrence Erlbaum Associates Publishers, Mahwah (2001)
12. Forgas, J.P. (ed.): Handbook of Affect and Social Cognition. Lawrence Erlbaum Associates Publishers, Mahwah (2001)
13. Kahneman, D.: Experienced utility and objective happiness: A moment-based approach. In: Kahneman, D., Tversky, A. (eds.) Choices, Values, and Frames, pp. 673–692. Cambridge University Press, Cambridge

14. Pfister, H.-R., Böhm, G.: The multiplicity of emotions: a framework of emotional functions in decision making. Judgment and Decision Making 3(1), 5–17 (2008)
15. Kahneman, D., Tversky, A.: Prospect theory: an analysis of decision under risk. Econometrica 47(2), 263–292 (1979)
16. Keeney, R.L.: Decision Analysis: An Overview. Operations Research 30(5), 803–838 (1982)
17. Hammond, J.S., Keeney, R.L., Raiffa, H.: The Hidden Traps in Decision Making. Harvard Business Review, 01–11 (September-October 1998)
18. Raiffa, H.: Decision Analysis: a personal account of how it got started and evolved. Operations Research 50(1), 179–185 (2002)
19. Naqvi, N., Shiv, B., Bechara, A.: The role of emotion in decision-making: a cognitive neuroscience perspective. Association for Psychological Science 15(5), 260–264 (2006)
20. Niedenthal, P.M.: Embodying emotion. Science 316(5827), 1002–1005 (2007)
21. Ahn, H.-I.: Modeling and Analysis of Affective Influences on Human Experience, Prediction, Decision Making, and Behavior. PhD Thesis, Massachusetts Institute of Technology (2010)
22. Velasquez, J.D.: An emotion-based approach to robotics. In: Proc. of the IEEE/RSJ Int. Conf. on Intelligent Robots and Systems, vol. 1, pp. 235–240 (1999)
23. Velasquez, J.D.: When robots weep: emotional memories and decision-making. In: Proc. of the American Association for Artificial Intelligence (1998)
24. Li, S., Roloff, M.E.: Strategic emotion in negotiation: cognition, emotion, and culture. In: Riva, G., Anguera, M.T., Wiederhold, B.K., Mantovani, F. (eds.) From Communication to Presence: Cognition, Emotions and Culture Towards the Ultimate Communicative Experience, pp. 169–188. IOS Press, Amsterdam (2006)
25. Chohra, A., Bahrammirzaee, A., Kanzari, D., Madani, K.: Personality aspects and fuzzy logic for bilateral negotiation behaviors with incomplete information. In: Lim, M.K. (ed.) Bidding: Types, Strategies and the Impact of Irrationality, ch. 12, pp. 251–276. Nova Science Publishers, New York (2013)
26. Treur, J.: From mirroring to the emergence of shared understanding and collective power. In: Jędrzejowicz, P., Nguyen, N.T., Hoang, K. (eds.) ICCCI 2011, Part I. LNCS (LNAI), vol. 6922, pp. 1–16. Springer, Heidelberg (2011)
27. Duell, R., Treur, J.: A Computational Analysis of Joint Decision Making Processes. In: Aberer, K., Flache, A., Jager, W., Liu, L., Tang, J., Guéret, C. (eds.) SocInfo 2012. LNCS, vol. 7710, pp. 292–308. Springer, Heidelberg (2012)
28. Becker-Asano, C., Wachsmuth, I.: Affect simulation with primary and secondary emotions. In: Prendinger, H., Lester, J.C., Ishizuka, M. (eds.) IVA 2008. LNCS (LNAI), vol. 5208, pp. 15–28. Springer, Heidelberg (2008)
29. Salovey, P., Mayer, J.: Emotional intelligence. Imagination, Cognition, and Personality 9(3), 185–211 (1990)
30. Mayer, J.D., Salovey, P.: The intelligence of emotional intelligence. Intelligence 17, 433–442 (1993)
31. Elfenbein, H.A., Marsh, A.A., Ambady, N.: Emotional intelligence and the recognition of emotion from facial expressions. In: Barret, L.F., Salovey, P. (eds.) The Wisdom of Feelings: Processes Underlying Emotional Intelligence, pp. 01–019 (2003)
32. Barsade, S.G., Gibson, D.E.: Why does affect matter in organizations? In: Academy of Management of Perspectives, pp. 36–59 (February 2007)
33. Rodriguez, L.-F., Ramos, F.: Development of computational models of emotions for autonomous agents: A review. Cognitive Computation (January 2014), doi:10.1007/s 12559-013-9244-x

34. Wang, Y., Rodriguez, L.-F., Ramos, F.: Cognitive computational models of emotions and affective behaviors. Int. J. of Software Science and Computational Intelligence 4(2), 41–63 (2012)

35. Gratch, J., Marsella, S.: Evaluating a computational model of emotion. Autonomous Agents and Multi-Agent Systems (2005)

36. Ochsner, K.N., Gross, J.J.: The cognitive control of emotion. Trends in Cognitive Sciences 9(5), 242–249 (2005)

37. Marsella, S., Gratch, J., Petta, P.: Computational models of emotion. In: Scherer, K.R., Banziger, T., Roesch, E. (eds.) A Blueprint for an Affectively Competent Agent: Cross-Fertilization Between Emotion, Psychology, Affective Neuroscience, and Affective Computing. Oxford University Press (2010)

38. Lin, J., Spraragen, M., Blythe, J., Zyda, M.: EmoCog: computational integration of emotion and cognitive architecture. Association for the Advancement of Artificial Intelligence (2011)

39. Lin, J., Spraragen, M., Zyda, M.: Computational models of emotion and cognition. Advances in Cognitive Systems 2, 59–76 (2012)

40. McAdams, D.P.: The five-factor model in personality: a critical appraisal. Journal of Personality 60(2), 329–361 (1992)

41. Chohra, A., Bahrammirzaee, A., Madani, K.: Time and personality based behaviors under cognitive approach to control the negotiation process with incomplete information. In: Nguyen, N.T. (ed.) Transactions on Computational Collective Intelligence XII. LNCS, vol. 8240, pp. 69–88. Springer, Heidelberg (2013)

42. Gmytrasiewicz, P.J., Lisetti, C.L.: Emotions and personality in agent design and modeling. In: Meyer, J.-J.C., Tambe, M. (eds.) Intelligent Agents VIII. LNCS (LNAI), vol. 2333, pp. 21–31. Springer, Heidelberg (2002)

43. Ahn, H.S.: Designing of a Personality Based Emotional Decision Model for Generating Various Emotional Behavior of Social Robots. Advances in Human-Computer Interaction, Hindawi Publishing Corporation, Article ID 630808, 01–14 (2014)

44. Mohammad, Y., Nishida, T.: Modelling interaction dynamics during face-to-face interactions. In: Nishida, T., Jain, L.C., Faucher, C. (eds.) Modeling Machine Emotions for Realizing Intelligence. SIST, vol. 1, pp. 53–87. Springer, Heidelberg (2010)

45. Peters, E.: The functions of affect in the construction of preferences. In: Lichtenstein, S., Slovic, P. (eds.) The Construction of Preferences. Cambridge University Press, New York (2006)

InfluenceTracker: Rating the Impact of a Twitter Account

Gerasimos Razis and Ioannis Anagnostopoulos

Computer Science and Biomedical Informatics Dpt., University of Thessaly, Lamia, Greece
{razis,janag}@dib.uth.gr

Abstract. This paper describes a methodology for rating the influence of a Twitter account in this famous microblogging service. Then it is evaluated over real accounts, under the belief that influence is not only a matter of quantity (amount of followers), but also a mixture of quality measures that reflect interaction, awareness, and visibility in the social sphere. The authors of this paper have created "InfluenceTracker", a publicly available website[1] where anyone can rate and compare the recent activity of any Twitter account.

Keywords: Twitter, Influence, Information Diffusion.

1   Introduction

Microblogging is a form of Online Social Network (OSN) which attracts millions of users on daily basis. Twitter is one of these microblog services. Their users vary from citizens to political persons and from news agencies to huge multinational companies. Independent of the type of the user and of the degree of influence on other users, all share the same need; to spread their messages to as many users as possible.

The messages, which are regarded as pieces of information, can be spread in two ways, either directly or indirectly. A case of direct message is when a company reveals information about a new product to its followers. When such a follower decides to share it among his or her own followers, then that is a case of indirect information dissemination. In this paper, we propose a methodology for calculating the importance and the influence of a user in Twitter, as well as a framework, which describes the maximization of diffusion of information in such a network.

The remainder of this paper is organized as follows. In the next section, we provide an overview over the related work on discovering influential users and on information diffusion in OSNs. Then, in Section 3 we describe the proposed methodology and the basic steps of the framework we use. In Section 4 real case scenarios are presented, in order to clearly show how we calculate the dissemination of information in Twittersphere, while in parallel we present the results along with their assessment. Finally, Section 5 provides the conclusions of our work by summarizing the derived outcomes, while providing considerations on our future directions.

2   Related Work

The calculation of the impact a user has on social networks, as well as the discovery of influencers in them is not a new topic. It covers a wide range of sciences, ranging

---

[1] http://www.influencetracker.com

L. Iliadis et al. (Eds.): AIAI 2014 Workshops, IFIP AICT 437, pp. 184–195, 2014.
© IFIP International Federation for Information Processing 2014

from sociology to viral marketing and from oral interactions to Online Social Networks (OSNs). In the related literature the term "influence" has several meanings and it is differently considered most of the times.

Romero et al. (Romero et al., 2011) utilize a large number of tweets containing at least one URL, their authors and their followers in order to calculate how influential or passive, in terms of activity, the Twitter users are. The produced influence metric depends on the "Follower-Following" relations of the users as well as their retweeting behavior. The authors state that the number of followers a user has is a relatively weak predictor of the maximum number of views a URL can achieve. As our work has shown (Section 3.2), through the retweet functionality information can be diffused to audience not targeted.

Cha et al. (Cha et al., 2010) introduce for each Twitter user three types of influence, namely "Indegree" (number of followers), "Retweet" (number of user generated tweets that have been reweeted) and "Mention" influence (number of times the user is mentioned in other users' tweets). A necessary condition for the computation of these influence types is the creation of at least ten tweets per user. The authors claim that "Retweet" and "Mention" influence correlate well with each other, while the "Indgree" does not. Therefore they suggest that users with high such influence type are not necessarily influential.

A topic oriented study on the calculation of influence in OSNs is presented in (Weng et al. 2010). The authors propose an algorithm which takes into consideration both the topical similarity between users and their link structure. It is claimed that due to homophily, which is the tendency of individuals to associate and bond with others having similar interests, most of the "Follower-Following" relations appear. This work also suggests that the active users are not necessarily influential.

Another approach which defines influence in terms of copying what the directly related do is presented by Goyal et al. (Goyal et al., 2010). In this work, the authors propose an influenceability score, which represents how easily a user is influenced by others or by external events. It is built on the hypothesis that a very active user performs actions without getting influenced by anyone. That type of users are regarded as responsible for the overall information dissemination in the network.

Yang and Counts (Yang and Counts, 2010) attempt to measure how topics propagate in Twitter. They constructed a diffusion network based on user mentioning, with constrains on topical similarities in the tweets. The authors state that given the lack of explicit threading in Twitter, this is the optimal approach of a network diffusing information about a specific topic. Moreover, the mention rate of the person tweeting is a strong predictor regarding the information diffusion in Twitter.

Lerman and Ghosh (Lerman and Ghosh, 2010) studied the diffusion of information in two social networks, Digg and Twitter. They conclude that the structure of these networks affects the dynamics of information flow and spread. According to the authors, information in denser and highly interconnected networks, such as of Digg's, reaches nodes faster compared to sparser networks, such as of Twitter's. Because of this structure information is spread slower, but it continues spreading at the same rate as time passes and penetrates the network further.

The authors in (Yang and Leskovec, 2010) propose a Linear Influence Model by calculating the influence of a node on the rate of diffusion through the network. For each influenced node an influence function quantifies how many subsequent ones can be by that node. This is based on the assumption that the number of newly influenced nodes depends on which other nodes were influenced before. The study concludes that the diffusion of information is governed by the influence of individual nodes.

Another influence-oriented study is the one presented by Kimura and Saito (Kimura and Saito, 2006). It is based on the Independent Cascade Model. This model is a stochastic process in which information propagates from a node to its neighboring ones according to a probabilistic rule. Similarly to the previous study, the problem lies in discovering influential nodes in a social network based on the computation of the expected number of influenced nodes. Two information diffusion models are proposed for the efficient calculation.

The study of (Bakshy et al., 2012) examines the information diffusion regarding exposure to signals about friends' information sharing on Facebook. It is suggested that the users who are aware of that information are significantly more likely to share it and in less time compared to those who are not. The authors suggest that, although the stronger ties are individually more influential, the weak ties, which surpass them in numbers, are responsible for the propagation of information.

Kwak et al. (Kwak et al., 2010) examine the "retweet" functionality offered by Twitter and study the information diffusion as a result of this action, as it has the power to spread information broader. The authors state that the retweets counters are measurements of the popularity of the tweets and consequently of their authors. According to the study, once a message gets retweeted it will almost instantly be spread up to 4 hops away from the source, leading in fast diffusion after the 1st retweet. That popularity measurement has been utilized in our work (Section 3.1).

All the related studies have shown that the most active users or those with the most followers are not necessarily the most influential. This fact has also been spotted by our work. Our Influence Metric depends on a set of factors (Section 3.1) and the activity of a user is only one of them. Therefore it can affect the result but the overall value of the Metric is not dependent directly on it.

Contrary to the aforementioned studies, for the calculation of our Influence Metric we neither set a lower threshold on the number of the user-generated tweets, nor we utilize only a subset of these tweets which fulfill certain criteria (i.e. those which contain URL). All Twitter accounts can be used as seed for the calculation of the Influence Metric. Our work also differentiates in the fact that our Influence Metric does not deal with the mentions of the examined user in other users' tweets or the topical similarity in the "Follower-Following" relations. It is concentrated on the characteristics of the Twitter account. In the future we plan to utilize the way in which other users interact with the examined user's tweets (marked as favorites or being retweeted) as a quality measurement of their content. Most of the presented studies related with the information diffusion aim in identifying nodes of high influence as responsible for affecting neighboring ones to behave the same, in terms of spreading information of the same topic. As the results show (Section 4.2) our proposed

Influence Metric succeeds in indentifying the nodes as to maximize the information diffusion in an OSN.

3   Methodology

As already mentioned, the contribution of this paper is twofold. Firstly, a methodology is proposed for calculating the importance and the influence of an account in Twitter. Secondly, a framework is described regarding its evaluation.

3.1   Calculation of Account's Importance and Influence

Twitter accounts form a Social Network. If depicted in a graph, they would be represented by nodes. The edges that connect these nodes are the relations of "Follower-Following", introduced by Twitter. Obviously, some accounts are more influential than others. The methodology of calculating the importance and influence that an account has in an OSN is presented here. That measurement should not depend merely on the number of "Followers" of an account, even if that number is big enough and the account's tweets are received by a large number of other accounts (followers). In case that the number of "Following" is larger, then the user could be characterized as a "passive" one. That type of users are regarded as those who are keener on viewing or being informed through tweets rather than composing new ones. Therefore, a suitable factor is the ratio of "Followers to Following" (FtF ratio).But this ratio is also not sufficient. Another important factor is the tweets creation rate (TCR). For example, let us see the case where two accounts have nearly the same FtF ratio. Obviously the account with the higher TCR has more impact on the Network. In our methodology, in order to calculate that rate, we process the accounts' latest 100 tweets according to the Twitter API. That leads to the TCR, and consequently the Influence Metric, being dynamic as it depends on the most recent accounts' activity in Twitter. The proposed Influence Metric depends on all of the aforementioned characteristics of the examined account, as defined in Equation 1. The FtF ratio is placed inside a base-10 log for avoiding outlier values. Moreover, the ratio is added by 1 so as to avoid the metric being equal to 0 in cases that the value of "Followers" is equal to the "Following".

$$\text{Influence Metric} = \frac{\text{tweets}_k}{\text{Days}_{\text{since } k_{th}\text{tweet}}} * \text{OOM(Followers)} * \log_{10}\left(\frac{\text{Followers}}{\text{Following}} + 1\right),$$

where OOM: Order Of Magnitude   (1)

Each Tweet is associated with several other kinds of information. Two of them are the "Retweets" and "Favorites" counts which represent how many times a Tweet has been retweeted or marked as favorite by other users. In our methodology, we utilize these counts in order to calculate the h-index of the retweets and favorites counts over the last 100 tweets of an examined account. The aim of these measurements is to provide a quality overview of the tweets of a Twitter account in terms of likeability and impact. These indexes are based on the established h-index (Hirsch, 2005) measurement and are named "ReTweet h-index - Last 100 Tweets" and "Favorite h-index - Last 100 Tweets". The most important factor regarding them is that they reflect other users' assessment of the content of the tweets.

Consequently, a Twitter account has "ReTweet h-index - Last 100 Tweets" h, if h over the last Nt tweets have at least h retweets each, and the remaining (Nt - h) of these tweets have no more than h retweets each (max. Nt=100).

The "ReTweet h-index - Daily" and the "Favorite h-index - Daily" are two similar metrics which represent the estimated daily value of "ReTweet and Favorite h-index" during the lifespan of a Twitter account over the last Nt tweets.

These h-index values are separately calculated and presented in influencetracker.com web site. However, we are currently working towards incorporating them in Equation 1, and more specifically on the evaluation of their impact over the proposed Influence Metric.

3.2 Information Diffusion/Tweet Transmission

An important functionality offered by Twitter is the "Retweet". It allows users to repost a received tweet to their Followers. This results in viewing the tweets of accounts that are not being directly followed. That fact leads in the diffusion of information to users not targeted (to the followers of their followers). The same process can be repeatedly take place by the new viewers of the message and so on.

The most important factor which affects the transmission of the tweets is the followers' probability of retweeting. The higher this value is, the higher the probability of transmitting tweets to other users, initially not targeted by the source. Another dependency of the transmission of the tweets is the followers' TCR. The value of this rate includes both the accounts' generated tweets, as well as their retweets. The final dependency of that measurement is the "TCR of Follower to TCR of Account" ratio. Increased values of that ratio lead in bigger flow of tweets between these Twitter accounts. The Tweet Transmission measurement depends on all of the aforementioned characteristics of the directly related accounts and it is defined in Equation 2.

$$\text{Tweet Transmission} = \frac{TCR_{n+1}}{TCR_n} * RT_{n+1}, \text{where } n \geq 0, n \in Z, 0 \leq RT \leq 1 \qquad (2)$$

3.3 Proposed Evaluation Framework

In order to evaluate the above metrics we employ the evaluation framework illustrated at Figure 1.

The framework is split into seven Phases, presented in Figure 1. During the 1st Phase of the process, the Twitter account under examination is selected. In Phase 2, we fetch a large number of Followers (N_f) and their Twitter-related characteristics. These are necessary in order to calculate their Influence Metric measurement (Equation 1).

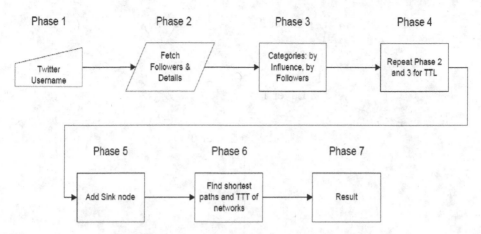

Fig. 1. The seven phases of the proposed framework

In Phase 3, the Followers are placed in two categories. The first one is classified by the value of our Influence Metric, while the second one by the absolute amount of followers each follower has. Both of these categories are sorted in descending format. After that, we select the top-k followers of these two categories.

Similarly, for these top-k accounts, Phases 2 and 3 are repeated. This process is continued until a specified distance threshold (layers) between Twitter accounts is reached (Phase 4). In computer networks, this distance is expressed by the Time-To-Live value (TTL) and corresponds to the amount of hops between different nodes a transmitted packet can perform before being rejected by the network. For the purposes of this work this threshold is set equal to 3.

The examined account, all the followers, as well as their relations and characteristics are modeled as a separate network. Nodes depict accounts, while edges depict their relations containing specific attributes. As a result of that process, two structures of the initial account and the followers of followers are created. The one depicts the top-k accounts by Influence, while the other one the top-k accounts by Followers. An example of such graph is presented in Figure 2.

In Figure 2 a 3-layered structure graph is displayed. The blue node represents the initial examined account. This account is connected to the yellow nodes, which stand for the top-3 followers either by the Influence Metric or by the amount of their followers (1st layer of distance). The process is iteratively continued with these nodes. The green and red nodes, 2nd and 3rd layer respectively, represent the followers of the previously examined followers and so on. We should note here, that a node can be connected with others, independent if they belong to the same layer or not.

During the fifth Phase an ending node (sink) is added to each of the two generated networks. This node is connected with all the accounts-followers of the last layer. These are the red nodes of Figure 2 which belong to the 3rd layer. That results in a fixed starting and ending point of the network. Figure 3 presents the network illustrated in Figure 2 including the sink node (black node in the center).

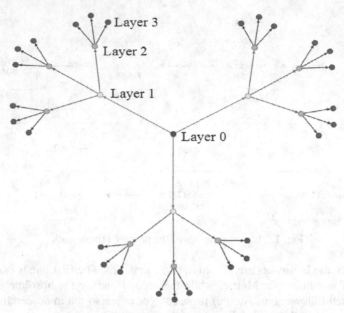

Fig. 2. A 3-layered structure graph of the initial account and the top-3 followers of followers

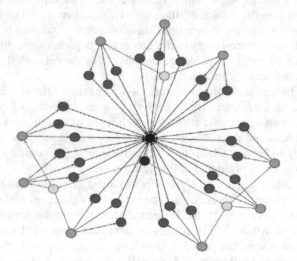

Fig. 3. A 3-layered structure graph with a sink node

When all previous phases are completed, the sixth and final Phase is initiated. Its purpose is to discover all the paths, starting from the initial examined account (blue node) and ending to the sink (black node) and consist of exactly 4 steps. The number of 4 steps is very important, since it is the necessary and sufficient condition in order to find all the shortest paths between the initial account and the sink, which contain exactly one node belonging to every layer.

Furthermore, that number of steps ensures us that any possible loops will be avoided during the traversal of the networks from the initial account to the sink. A possible case of loop is when the examined Twitter account appears as a follower of another account. In such a case the examined account could also appear at the first (as a yellow node) or the second layer (as a green node).

The Tweet Transmission (TT) value, presented in Equation 2, is calculated for each layer of every shortest path. Then the TT value of the shortest path for all layers is calculated. This process is repeated until the TT values of all shortest paths of the two networks are computed. The network with the higher Total Tweet Transmission (TTT) value is considered the one with the higher disseminated information.

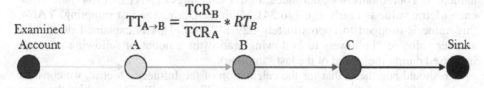

$$TT_{A \to B} = \frac{TCR_B}{TCR_A} * RT_B$$

Examined Account   A   B   C   Sink

Fig. 4. Calculation of a TT value

Figure 4 displays an example of a path derived from the network illustrated in Figure 3. This path consists of four edges (Examined Account→A, A→B, B→C, and C→Sink). For this path, three TT values are calculated, namely from the Examined Account to A ($TT_{ExaminedAccount \to A}$), from A to B ($TT_{A \to B}$), and finally from B to C ($TT_{B \to C}$). The total TT value of the whole path is the multiplication of these three calculated values and is assigned to the sink node. Figure 4 depicts the calculation of $TT_{A \to B}$.

4   Evaluation Results

In this section, we will present and analyze the results and the evaluation concerning the calculation of the importance and influence of a user in an OSN, and of the framework regarding the maximization of diffusion of information. As case study, we evaluate six real Twitter accounts. Three belong to political persons (@AdonisGeorgiadi, @IliasKasidiaris,and @PanosKammenos), one belongs to the Hellenic Fire Brigade (@Pyrosvestiki), and the rest belong to a Greek news media channel (@SkaiGr) and to the international information network of activists and hacktivists named Anonymous (@YourAnonNews).

The experiments took place between 14/12/2013 and 31/1/2014. For each account four separate samplings were made, during which the number of the followers and the top-k accounts were gradually increased. The distance threshold, which defines the number of layers as described in Section 3.3, was set equal to 3.

4.1   Accounts' Influence

In this section, we present the Influence Metric measurements in respect to the examined Twitter accounts. We also provide the sampling date, the value of the Influence Metric, as well as other metrics are depicted in Table 1.

As we can see, the Influence Metric is directly dependant of the accounts' activity, which is measured by the TCR value. Account's "@SkaiGr" Influence value during the first three samplings (SG1 to SG3) is approximately the same (nearly 35 Millions). However, during the fourth sampling (SG4) that value was almost the half. This was caused by the fact that the TCR value was dropped to half, despite that the "Followers to Following" ratio was slightly increased. In the case of the account named "@YourAnonNews", during the first 3 samplings (YAN1 to YAN3) the Influence Metric value is nearly equal to 341 Millions. During the last sampling, YAN4, this value is dropped to approximately 329 Millions. This is explained due to the smaller value of "Followers to Following" ratio (the amount of following accounts increased during the period of the last sampling).

We should note here, that for the calculation of the Influence Metric, we consider the latest 100 accounts' tweets directly from the Twitter API. This enables the measurement to be dynamic and in accordance to the latest trend activity of the examined Twitter account.

Table 1. The Influence Metric measurement and the Twitter related characteristics of the examined Twitter accounts

ID	Username	Date	Influence	TCR	Followers	Following
AG1	@AdonisGeorgiadi	14/12/2013	126,857.416	15.50	33,410	3,574
AG2	@AdonisGeorgiadi	18/12/2013	112,929.569	11.11	33,566	3,576
AG3	@AdonisGeorgiadi	29/12/2013	511,537.359	50.00	34,164	3,579
AG4	@AdonisGeorgiadi	16/01/2014	148,166.219	14.29	35,430	3,584
IK1	@IliasKasidiaris	16/12/2013	26,686.871	1.11	14,148	56
IK2	@IliasKasidiaris	19/12/2013	26,927.978	1.12	14,150	56
IK3	@IliasKasidiaris	26/12/2013	25,492.531	1.06	14,172	56
IK4	@IliasKasidiaris	26/01/2014	23,840.975	0.99	14,278	56
PK1	@PanosKammenos	17/12/2013	63,708.939	3.33	33,889	419
PK2	@PanosKammenos	21/12/2013	56,266.498	2.94	33,940	419
PK3	@PanosKammenos	29/12/2013	46,724.779	2.44	34,029	419
PK4	@PanosKammenos	12/01/2014	41,621.203	2.17	34,274	419
P1	@Pyrosvestiki	01/01/2014	23,516.011	0.62	18,619	3
P2	@Pyrosvestiki	30/01/2014	23,894.273	0.63	18,612	3
P3	@Pyrosvestiki	31/01/2014	23,516.156	0.62	18,620	3
P4	@Pyrosvestiki	31/01/2014	23,516.011	0.62	18,619	3
SG1	@SkaiGr	17/12/2013	35,356,300.107	100.00	178,446	52
SG2	@SkaiGr	21/12/2013	35,363,204.477	100.00	178,730	52
SG3	@SkaiGr	31/12/2013	35,380,441.726	100.00	179,441	52
SG4	@SkaiGr	01/01/2014	17,733,148.729	50.00	179,505	51
YAN1	@YourAnonNews	18/12/2013	341,594,730.673	100.00	1,185,201	455
YAN2	@YourAnonNews	23/12/2013	341,102,758.175	100.00	1,184,723	460
YAN3	@YourAnonNews	27/12/2013	340,808,348.148	100.00	1,184,390	463
YAN4	@YourAnonNews	24/01/2014	328,801,969.528	100.00	1,189,204	613

4.2   Information Diffusion/Tweet Transmission

In this section, the TTT values of the created networks in respect of the examined accounts are presented. Table 2 is divided in four parts. Each part refers to the separate samplings mentioned above, while the related information per part is:

- the number of followers that is iteratively fetched,
- the number of top-k followers of the two generated categories, these are "by Influence Metric" and "by Followers", which are used for the creation of the respective layered networks,
- the distance threshold value that reflects the layers of the examined account's networks (TTL),
- the account which is the root of the two resulting networks (the six examined accounts),
- the TTT values of the two networks, according to the "by Influence Metric" and "by Followers", and finally
- the difference of the above TTT values for both generated networks.

Table 2. The tables containing the details of each sampling set

Followers = 50, top-k users = 3, TTL = 3				Followers = 180, top-k users = 7, TTL = 3		
Username	By Influence	By Followers	Difference	Username	By Influence	By Followers	Difference
@AdonisGeorgiadi	2,174	12,933	-10,759	@AdonisGeorgiadi	45,831	20,038	25,794
@IliasKasidiaris	57,833	42,027	15,806	@IliasKasidiaris	682,280	592,961	89,319
@PanosKammenos	22,527	30,074	-7,547	@PanosKammenos	723,534	373,959	349,575
@SkaiGr	1,016	0,465	0,551	@SkaiGr	4,773	3,172	1,600
@YourAnonNews	0,038	0,018	0,020	@YourAnonNews	2,234	0,980	1,255
@Pyrosvestiki	0,496	0,864	-0,368	@Pyrosvestiki	909,388	263,730	645,658

Followers = 100, top-k users = 5, TTL = 3				Followers = 360, top-k users = 7, TTL = 3		
Username	By Influence	By Followers	Difference	Username	By Influence	By Followers	Difference
@AdonisGeorgiadi	12,733	8,632	4,102	@AdonisGeorgiadi	50,686	15,503	35,183
@IliasKasidiaris	116,048	241,823	-125,775	@IliasKasidiaris	124,871	265,954	-141,083
@PanosKammenos	134,417	30,997	103,420	@PanosKammenos	549,347	108,909	440,438
@SkaiGr	3,462	0,302	3,160	@SkaiGr	3,768	2,628	1,141
@YourAnonNews	1,762	0,446	1,316	@YourAnonNews	3,844	2,866	0,978
@Pyrosvestiki	210,442	85,437	125,005	@Pyrosvestiki	917,656	533,136	384,520

In addition, the green-highlighted values in column "Difference" correspond to the cases where the TTT value is larger in the "By Influence" category, thus indicating that our approach manages to create a network of followers who are more influential in comparison to the network of category "By Followers". Red-highlighted values reflect to the opposite cases. As we can see the wider the examined networks are in terms of the top-k accounts and their followers up to the third layer, the more influential network of accounts we have.

As can be observed from Table 2, the TTT values of the two networks are escalated as both the numbers of the followers and of the top-k accounts are also increased. The results of the use cases used for the evaluation of the influence metric calculation, show that the number of followers an account has, is not solely sufficient to guarantee the maximum diffusion of information in Twitter (and practically to any similar OSN). This is because, these followers should not only be active Twitter accounts, but also have impact on the network. The latter is calculated by the Influence Metric value.

5   Conclusions and Future Work

In this paper, we proposed a methodology for calculating the importance and the influence of a Twitter account, as well as a methodology regarding the maximization of diffusion of information. For evaluation purposes a framework was applied. As a case study, we evaluated six real Twitter accounts. The experiments took place between 14/12/2013 and 31/1/2014 and for each account four separate samplings were made.

The results of the use cases show that the number of followers an account has, is not sufficient to guarantee the maximum diffusion of information in Twitter (and practically to any similar OSN). This is because, these followers should not only be active Twitter accounts, but also have impact on the network. The latter is calculated by our Influence Metric.

Ongoing research is performed on how the proposed Influence Metric can be improved. We are currently working towards incorporating the "ReTweet h-index - Daily" and the "Favorite h-index - Daily" in Equation 1 and more specifically on the evaluation of their impact over the proposed Influence Metric.

Additionally effort is conducted over the evaluation of how conversational a Twitter account is ("Reply Percentage" in InfluenceTracker, which indicates the percentage of the account's latest tweets which are replies to other accounts' tweets). The work of (Leavitt et al., 2009) concluded that news media are better at spreading content, while celebrities are better at making conversation. So, for example we can use the aforementioned metric in order to categorize the Twitter accounts in news/blogs media (with automated content) and user ones (human generated content). Also if combined with the "Daily h-indexes" measurements, representing the quality of the content, spam accounts can also be discovered.

References

[Bakshy et al., 2012] Bakshy, E., Rosenn, I., Marlow, C., Adamic, L.A.: The role of social networks in information diffusion. In: WWW, Lyon, France, pp. 519–528 (2012)

[Cha et al., 2010] Cha, M., Haddadi, H., Benevenuto, F., Krishna Gummadi, P.: Measuring User Influence in Twitter: The Million Follower Fallacy. In: ICWSM, Washington, DC, USA (2010)

[Goyal et al.] Goyal, A., Bonchi, F., Lakshmanan, L.V.S.: Learning influence probabilities in social networks. In: WSDM, pp. 241–250 (2010)

[Hirsch, 2005] Hirsch, J.: An index to quantify an individual's scientific research output. PNAS USA 7(46), 16569–16572 (2005)

[Kimura and Saito, 2006] Kimura, M., Saito, K.: Tractable models for information diffusion in social networks. In: Fürnkranz, J., Scheffer, T., Spiliopoulou, M. (eds.) PKDD 2006. LNCS (LNAI), vol. 4213, pp. 259–271. Springer, Heidelberg (2006)

[Kwak et al., 2010] Kwak, H., Lee, C., Park, H., Moon, S.B.: What is Twitter, a social network or a news media? In: WWW, Raleigh, North Carolina, USA, pp. 591–600 (2010)

[Leavitt et al.] Leavitt, A., Burchard, E., Fisher, D., Gilber, S.: The Influentials: New Approaches for Analyzing Influence on Twitter. Web Ecology Project 92, 1–18 (2009)

[Lerman and Ghosh, 2010] Lerman, K., Ghosh, R.: Information Contagion: An Empirical Study of the Spread of News on Digg and Twitter Social Networks. In: ICWSM, Washington, DC, USA (2010)

[Romero et al., 2011] Romero, D.M., Asur, S., Galuba, W., Huberman, B.A.: Influence and Passivity in Social Media. In: WWW (Companion Volume), pp. 113–114 (2011)

[Weng at al., 2010] Weng, J., Lim, E.-P., Jiang, J., He, Q.: TwitterRank: finding topic-sensitive influential twitterers. In: WSDM, pp. 261–270 (2010)

[Yang and Counts, 2010] Yang, J., Counts, S.: Predicting the Speed, Scale, and Range of Information Diffusion in Twitter. In: ICWSM, Washington, DC, USA (2010)

[Yang and Leskovec, 2010] Yang, J., Leskovec, J.: Modeling Information Diffusion in Implicit Networks. In: ICDM, pp. 599–608 (2010)

A New Framework for Bridging the Gap from Protein-Protein Interactions to Biological Process Interactions

Christos Dimitrakopoulos[1], Andreas Dimitris Vlantis[1], Konstantinos Theofilatos[1], Spiros Likothanassis[1], and Seferina Mavroudi[1,2]

[1] Department of Computer Engineering and Informatics, University of Patras, Greece
[2] Department of Social Work, School of Sciences of Health and Care, Technological Educational Institute of Western Greece, Greece
{dimitrakop,vlantis,theofilk,likothan,mavroudi}@ceid.upatras.gr

Abstract. Proteins and their interactions have been proven to play a central role in many cellular processes and have been extensively studied so far. However of great importance, little work has been conducted for the identification of biological process interactions in the higher cellular level which could provide knowledge about the high level cellular functionalities and maybe enable researchers to explain mechanisms that lead to diseases. Existing computational approaches for predicting Biological Process interactions used PPI graphs of low quality and coverage but failed to utilize weighted PPI graphs to quantify the quality of the interactions. In the present paper, we propose a unified two-step framework to reach the goal of predicting biological process interactions. After conducting a comparative study we selected as a first step the EVOKALMA model as a very promising algorithm for robust PPI prediction and scoring. Then, in order to be able to handle weights, we combined it with a novel variation of an existing algorithm for predicting biological processes interactions. The overall methodology was applied for predicting biological processes interactions for Saccharomyces Cerevisiae and Homo Sapiens organisms, uncovering thousands of interactions for both organisms. Most of the linked processes come in agreement with the existing knowledge but many of them should be further studied.

Keywords: Protein-Protein Interactions, protein-protein interaction networks, Biological Process Interactions, EvoKalma Model, protein function, statistical enrichment.

1 Introduction

Among the numerous participants in molecular interactions, proteins are considered most important ones. In specific, proteins transmit regulatory signals throughout the cell, catalyze a huge number of chemical reactions, and are critical for the stability of numerous cellular structures. The total number of possible interactions within the cell is extremely large and the full identification of all true PPIs is a very challenging task. However, the identification of all true PPIs may contribute in understanding cellular functionality, designing more efficient medicines and uncovering the mechanisms that

L. Iliadis et al. (Eds.): AIAI 2014 Workshops, IFIP AICT 437, pp. 196–204, 2014.

lead to diseases. For these reasons, this problem has been extensively studied in the last decades and many experimental and computational techniques have been combined to solve it [1].

A Biological Process Network is a model designed to offer an insight at the interactions occurring among biological processes. The traditional approach for studying complex biological networks is based on the identification of interactions between genes and proteins. As a result, little is known about interactions of the higher order biological systems, such as biological processes. The knowledge derived by process interactions can be effectively used in protein function prediction, increasing both coverage and accuracy of predictions. Moreover, uncovering the interactions of biological processes can be a step towards understanding the cellular functions in a high level.

Existing methods for predicting interactions between biological processes take as inputs the PPI networks and functional annotations. In [2], processes are considered as interacting when more proteins annotated by them interact than expected by chance. Other methods also require gene expression information [3] to examine how interactions may change in different experimental conditions. In addition, this method incorporates weights at the protein interactions. In [4], the aim is to include as few inter-process links that successfully present as many gene interactions, to reduce complexity and allow further exploration in a greater detail. Despite the promising results of these approaches, their basic limitation is the utilization of PPI graphs of low quality and coverage as inputs. Moreover, most of them do not utilize confidence scores - weights for each PPI and others use confidence scores which only refer to the reference rate in the bibliography of a certain interaction.

In the present paper, we propose a computational framework for the prediction of PPIs, the prediction of a confidence score for each interaction which will reflect function, structural and sequential information and the prediction of interactions between biological processes. For the prediction and scoring of PPIs the method EVOKALMA was utilized [5]. The selection was only performed after the conduction of extended experiments in order to ensure the superior performance of the EVOKALMA model when compared with existing state of the art classification techniques. The experimental results confirmed that EVOKALMA method outperformed the other methodologies in all the examined metrics. For the prediction of biological process interactions we enhanced the methodology proposed in [2] in order to be able to handle weighted PPIs. In this way, the method accounted for the high false positive and false negative rates encountered in PPI datasets ensuring that only high-confidence interactions contribute significantly to the method's output while low-confidence interactions have only a low effect.

The proposed algorithmic framework was applied for Saccharomyces Cerevisiae and Homo Sapiens organisms and useful conclusions were made about the interactions of biological processes in these organisms.

The rest of the paper is organized as follows: In section 2 the datasets and the proposed algorithmic framework are described in detail. In section 3 the experimental results are presented and analyzed while in section 3 conclusions and proposals for future work are presented.

2 Datasets and Methods

2.1 Datasets

For the training and testing process of the methods for predicting PPIs 1000 positive interactions referred in HPRD [6] and 1000 negative protein interactions were selected. HPRD database is assumed to be highly reliable as it contains protein interactions that are supported by low and high throughput experimental evidence. The negative samples were created randomly from the unique identities of the whole set of proteins leaving out protein pairs which have been reported in iRefindex [7] as protein-protein interactions. For every protein pair in the deployed dataset 22 informative features were calculated including several co-expression features, function similarity features, sequence similarity, a homology based feature, domain-domain interaction feature and co-localization features. More details about the utilized features is available at [5]. All feature values are normalized in the range [0, 1] and missing are estimated using the kNN-impute methodology [8]. The trained EVOKALMA model (see section 2.2.1) was applied on an extended set of over 600000 protein pairs to predict most human PPIs and predicted 211367 PPIs. Moreover for every interaction a confidence score has been calculated. This confidence score indicates the strength of the interaction, its frequency and possibility to exist. These interactions alongside their weights (confidence scores) are stored in the HINT-KB.

The input graph datasets for predicting interactions among biological processes were: a) the protein interaction network for Saccharomyces Cerevisiae, obtained from Uniprot Database [9] (22969 interactions), b) the protein interaction network for Homo Sapiens, created from the interactions reported from iRefIndex database (115404 interactions) and, c) the weighted protein interaction network for H. Sapiens obtained from HINT-KB database (211367 interactions).

The input genetic interaction networks for S. Cerevisiae and H. Sapiens were created from BioGRID 3.2.98 database [10] and comprised 1606 genetic interactions for H. Sapiens and 145265 genetic interactions for S. Cerevisiae.

2.2 Methods

The overall methodology is consisted of a two step procedure. First, for each specific organism PPIs are predicted with computational prediction methods or retrieved through public available databases. Then, functional, structural and sequential information about the proteins are combined to predict a confidence score for these interactions. Finally, the constructed weighted PPI graphs are used as input for the biological process network prediction method to predict interactions between biological process networks. All these methodologies are described in sections 2.2.1 -2.2.2.

2.2.1 EVOKALMA Model

The main idea of the proposed classification methodology, called (Evolutionary Kalman Mathematical Modelling - EvoKalMaModel) is to find a simple mathematical equation that governs the best classifier and enables the extraction of biological knowledge. This method was based on previous methodologies proposed by our authoring group [11, 12] and it was initially published in [5]. It combines a state-of-the-art adaptive filtering technique named Kalman Filtering [13] with an adaptive genetic algorithm. The adaptive genetic algorithm is used to detect the optimal subset of terms in order to build the mathematical model for our predictor and then Extended Kalman Filters to compute its optimal parameters. The final model is in the form of a mathematical equation including a subset of the available mathematical terms and inputs. The evolutionary process is guided with a problem specific fitness function and utilizes an adaptive mutation rate to improve its convergence behavior.

2.2.2   Proposed Biological Process Network Method

Protein-protein interaction (PPI) networks are defined as graphs whose nodes represent proteins and the edges interactions between nodes/proteins. The weights of the edges (if they exist) represent the confidence scores of these interactions. Genetic interaction (GI) network is a graph whose nodes represent genes and the edges represent interactions between nodes/genes.

The definition of linked biological processes is based on an input undirected graph whose nodes represent proteins and edges represent an interaction. Two types of networks were used: physical level PPI networks and functional level GI networks. Only proteins annotated by at least one function (Gene Ontology function [14]) were considered. A link between process i and process j suggests that the proteins annotated by i tend to interact with proteins annotated by j more than expected by chance. The statistical enrichment of process j is calculated based on the set of proteins connected with proteins annotated by i, except those annotated by both i and j.

The formal definition is as follows:

Suppose N_i be the set of nodes annotated with process i (and possibly with other processes as well) and NB_i be the set of nodes not annotated with process i and connected with at least one node annotated with process i then:

$$NBi = \{v: <u,v> \in E, v \notin Ni, u \in Ni\} \tag{1}$$

Based on the above definitions, a process i is connected with a process j, if j is statistically enriched in NB_i set, when $P(i,j) < threshold$ where:

$$P(i,j) = \sum_{x=|NBi \cap Nj|}^{\min\{|Nj|,|NBi|\}} \frac{\binom{|Nj|}{x}\binom{N-|Nj|}{|NBi|-x}}{\binom{N}{|NBi|}} \tag{2}$$

For the aforementioned threshold a very strict value (0.001) was assigned (as proposed in [2]) in order to filter false positive results. N is defined as the sum of all proteins annotated with processes i and j, so P is the probability a link to connect a protein annotated with process i and a protein annotated with process j, among all possible links between proteins annotated with i or j.

The methodology so far, is similar to the methodology initially proposed in by [2]. The drawback of this approach is the assumption that interactions have a dual nature. However, recent approaches for constructing PPI graphs achieved to predict a confidence score for each interaction. Thus, in our approach for the quantity $|NB_i|$ in equation 2, we utilized the sum of the weights (confidence scores) that connects the nodes not annotated with process i and the nodes annotated with processes i. Thus, connections of low confidence score will not be significant for the algorithm's output, while connections of high confidence score would be extremely significant.

To calculate semantic similarity of two biological processes the following formula was used:

$$\text{similarity}(t_i, t_j) = \frac{2*\log P(t_{ancestor})}{\log P(t_i) + \log P(t_j)} \tag{3}$$

where $P(t_i)$ is the probability of a gene to be annotated with t_i, that is the number of genes annotated with t_i divided by the total number of genes and $t_{ancestor}$ is the most specific common ancestor of t_i and t_j in Gene Ontology.

The pseudocode of the method is provided in Table 1:

Table 1. Pseudocode of BPN methodology

FOR processes i, j
Find proteins annotated by these processes
Exclude common proteins
Find interactions among the remaining proteins
Calculate P(i,j)
IFP(i,j)<0.001 create link between i and j
REPEAT FOR EVERY i, j

3   Results and Discussion

For the problem of predicting and scoring PPIs many methods were applied and comparative results are provided in Table 2. The methodologies used for comparative reasons with the proposed EVOKALMA model, include the Naive Bayesian Classifier which is the algorithm utilized by most PPI databases that include computationally predicted PPIs [15, 16]. Moreover, two methodologies which have already provided encouraging results in predicting PPIs were used (Random Forests [17] and jGEPModel2.0 [18]) alongside with the hybrid combinations of Genetic Algorithms [19], Particle Swarm Optimization [20] and Differential Evolution [21] with Support Vector Machines (SVM) which have several applications in many fields.

Classifiers' performance is theoretically supposed to be more confident when more testing datasets are pooled. Therefore, our experimental setup was extended from one testing dataset to multiple testing datasets and performed double loop cross validation [22]. During the training 10-fold external and 9-fold internal cross validation were used. In particular, the external folds alter the subset of data used for testing the trained models (10 non-overlapping different subsets) after internal algorithms iterations (either heuristic iterations either parameter optimization steps), whereas the internal folds vary the subset of data used for validating the under training models (9 non-overlapping different subsets) during the algorithms internal iterations. The optimal model of each implementation of the 9 internal folds was finally kept for every algorithm. Then, the average of the metrics in the external fold was calculated for the 10 optimal models for every algorithm and Table 2 presents these results.

Table 2. Comparative Results of PPI classification methods

ALGORITHM	ACCURACY	GEOMETRIC MEAN	SENSITIVITY	SPECIFICITY
Naive Bayesian Classifier	73.59%	74.84%	63.64%	88.00%
Random Forests	81.83%	81.82%	81.45%	82.20%
GA-SVM	79.19%	79.04%	74.34%	84.04%
PSO-SVM	81.64%	81.99%	82.08%	81.20%
DE-SVM	81.84%	81.84%	82.28%	81.40%
jGEPModel 2.0	82.67%	82.66%	83.28%	82.06%
EVOKALMA	**87.43%**	**87.40%**	**85.33%**	**89.51%**

From Table 2 it is clearly observed that EVOKALMA model outperformed significantly all the other deployed methodologies in all the utilized metrics. This strengthens our belief that EVOKALMA algorithm is the most appropriate solution for predicting and scoring PPIs.

The main concepts of the methodology are Interacting Processes and the method's output, the Biological Processes Network (BPN) which nodes are biological processes terms, as described by Gene Ontology database. This method, combined with experimental or computational data of genetic or protein interactions, led to the discovery of a variety of process interactions. Two types of networks were created: a) networks based on protein interactions and, b) networks based on genetic interactions. Some of the discovered connections are consistent with our biological knowledge, while others require further research. For example, the process "protein ubiquitination" (GO:0016567) was found to be PPI-linked with protein catabolism-related proteins and GI-linked with processes related with cell cycle.

A process a was defined as PPI-linked with a process b if the number of proteins which are annotated with b and interact with those annotated with a is greater than expected by chance. Based on the process interactions the BPNs were created. The network is directed, its nodes represent processes and a directed edge from process a to b dictates interaction from a to b. The network created for S. Cerevisiae contains 5285 edges and the network created for H. Sapiens contains 24758 edges.

Gene Ontology database uses various metrics for the semantic similarity of processes which are calculated by the distance of the processes on the hierarchical tree of Gene Ontology. These metrics provide values mostly between 0 and 1, where 0 means that the only common ancestor of two processes is the term GO:0008150 ("biological process"). Many of the PPI-linked processes share a high semantic similarity. While some others have low semantic similarity, it could be useful to be further studied.

As an example, the method discovered a link between "response to DNA damage stimulus" (GO:0006974) and "chromatin modification" (GO:0016568). Despite the fact that they have semantic similarity 0, chromatin and histone modification are utilized in the DNA damage response pathway [23].

Using the genetic interaction datasets, two more networks were created. Two genes are considered as interacting when their mutations show a combined effect which does not appear by either mutation alone. The network created for S. Cerevisiae contains 32967 edges and the network created for H. Sapiens contains 245 edges.

The resulting linked processes for the two organisms show great overlap, which is consistent with the fact that they share a large number of protein and genetic interactions. Unique linked processes also appear, such as the link between "hyperosmotic response" (GO:0006972) and "regulation of MAP kinase activity" (GO:0043405), which exists only in the S. Cerevisiae's network. The use of MAP kinase activity for the regulation of the hyperosmotic shock response, is a known mechanism of S. Cerevisiae.

Many of the proteins that genetically interact, also participated in PPI-linked processes. For S. Cerevisiae, the probability two proteins to participate in PPI-linked processes is 11% (only those processes with at least 50 genes participating were considered). If it is considered as known that these proteins interact genetically, the probability reaches 36%. Therefore, PPI-linked processes could be used by genetic interaction prediction algorithms to increase their performance. Also, two proteins are more likely to interact if they both belong to PPI-linked processes.

4   Conclusions and Future Work

In the present paper, we proposed a new holistic algorithmic framework to construct accurate biological process network. This framework is consisted of a PPI prediction and scoring algorithm which outperforms other existing methodologies and a variation of an existing algorithms for prediction biological process interactions which enables it to handle weighted PPI networks.

Utilizing the proposed methodology, various interactions between biological processes were predicted for both examined organisms, S. Cerevisiae and H. Sapiens. Some were consistent with previous knowledge, while others require further research. The method presented, focuses on the computational detection of links between processes, rather than their confirmation or biological interpretation. A challenge derived by this work is the need to review the structure of biological information, since many of the computed interactions show little semantic similarity based on the current structure of Gene Ontology database.

It was proven that PPI-linked processes could be used to enhance the performance of genetic interaction prediction algorithms. Identifying GI-linked processes cal lead to a new direction in generic interaction research. As a next step, the method could incorporate gene expression information, to study processes interactions at a treatment-control basis, as proposed in [3, 4].

Another interesting idea for future work, is to investigate which processes show different behavior in various conditions and how the interactions between them may change. A way to achieve that, is by incorporating gene expression information from different experimental conditions, such as infected and healthy tissue samples. Using microarray experiments, genes whose expression changes greatly (often determined by using simple statistical tests) can be defined as perturbed. A process is considered as perturbed when the genes annotated with this process are perturbed. A link between processes is perturbed when both incident processes are perturbed. A major drawback of this analysis is the report of all possible links between processes, which includes unperturbed links and possibly identical links. To overcome this constraint the utilized process links could be minimized by rewarding the inclusion of links representing many gene interactions.

References

1. Theofilatos, K., Dimitrakopoulos, C., Tsakalidis, A., Likothanassis, S., Papadimitriou, S., Mavroudi, S.: Computational Approaches for the Prediction of Protein-Protein Interactions: A Survey. Current Bioinformatics 6(4), 398–414 (2011)
2. Dotan-Cohen, D., Letovsky, S., Melkman, A.A., Kasif, S.: Biological Process Linkage-Networks. PLoS ONE 4(4), e5313 (2009), doi:10.1371/journal.pone.0005313
3. Lasher, C.D., Rajagopalan, P., Murali, T.M.: Discovering networks of perturbed biological processes in hepatocyte cultures. PLoS ONE 6(1), e15247 (2011)
4. Lasher, C., Rajagopalan, P., Murali, T.M.: Summarizing cellular responses as biological process networks. BMC Systems Biology (2013),
   http://dx.doi.org/10.1186/1752-0509-7-68
5. Theofilatos, K., Dimitrakopoulos, C., Likothanassis, S., Kleftogiannis, D., Moschopoulos, C., Alexakos, C., Papadimitriou, S., Mavroudi, S.: The Human Interactome Knowledge Base (HINT-KB): An integrative Human protein interaction database enriched with predicted protein protein interaction scores using a novel hybrid technique (Evolutionary Kalman Mathematical Modelling - EvoKalMaModel). Artificial Intelligence Review, 1–17 (2013), , doi: 10.1007/s10462-013-9409-8
6. Keshava Prasad, T.S., Goel, R., Kandasamy, K., et al.: Human Protein Reference Database-2009 update. Nucleic Acids Res. 37, D767–D772 (2009)
7. Razick, S., Magklaras, G., Donaldson, I.M.: iRefIndex: A consolidated protein interaction database with provenance. BMC Bioinformatics 9(1), 405 (2008)
8. Troyanskaya, O., Cantor, M., Sherlock, G., et al.: Missing value estimation methods for DNA microarrays. Bioinformatics 17(6), 520–525 (2001)
9. The UniProt Consortium: Reorganizing the protein space at the Universal Protein Resource (UniProt). Nucleic Acids Res. 40, D71-D75 (2012)
10. Stark, C., Breitkreutz, B., Reguly, T., et al.: BioGRID: a general repository for interaction datasets. Nucleic Acids Res. 34, D535-D539 (2006)
11. Theofilatos, K.A., Dimitrakopoulos, C.M., Tsakalidis, A.K., et al.: A new hybrid method for predicting protein interactions using Genetic Algorithms and Extended Kalman Filters. In: Proceedings of the IEEE/EMBS Region 8 International Conference on Information Technology Applications in Biomedicine (ITAB). art. no. 5687765 (2010), doi : 10.1109/ITAB.2010.5687765

12. Dimitrakopoulos, C.M., Theofilatos, K.A., Georgopoulos, E.F., et al.: Efficient Computational Construction of Weighted Protein-Protein Interaction Networks Using Adaptive Filtering Techniques Combined with Natural-Selection Based Heuristic Algorithms. International Journal of Systems Biology and Biomedical Technologies (IJSBBT) 1(2), 20–34 (2011)

13. Welch, G., Bishop, G.: An Introduction to the Kalman Filter. University of North Carolina at Chapel Hill (1995)

14. Ashburner, M., Ball, C.A., Blake, J.A., et al.: Gene ontology: tool for the unification of biology. The Gene Ontology Consortium. Nat. Genet. 25, 25–29 (2000)

15. Scott, M., Barton, G.: Probabilistic prediction and ranking of human protein-protein interactions. BMC Bioinformatics 8, 239 (2007)

16. Zhang, Q., Petrey, D., Garzon, J., et al.: PrePPI: a structure-informed database of protein-protein interactions. Nucl. Acids Res (2012), doi:10.1093/nar/gks1231

17. Liu, Y., Kim, I., Zhao, H.: Protein interaction predictions from diverse sources. Drug Discov. Today 13, 409–416 (2008)

18. Theofilatos, K., Dimitrakopoulos, C., Antoniou, M., Georgopoulos, E., Papadimitriou, S., Likothanassis, S., Mavroudi, S.: Efficient Computational Prediction and Scoring of Human Protein-Protein Interactions Using a Novel Gene Expression Programming Methodology. In: Jayne, C., Yue, S., Iliadis, L. (eds.) EANN 2012. CCIS, vol. 311, pp. 472–481. Springer, Heidelberg (2012)

19. Holland, J.: Adaptation in natural and artificial systems: an introductory analysis with applications to biology, control, and artificial intelligence. MIT Press, Cambridge (1995)

20. Kennedy, J., Eberhart, R.: Particle swarm optimization. In: Proceedings of IEEE International Conference on Neural Networks, Piscataway, NJ, pp. 1942–1948 (1995)

21. Storn, R., Price, K.: Differential evolution - a simple and efficient heuristic for global optimization over continuous spaces. Journal of Global Optimization 11, 341–359 (1997)

22. Veenman, C.J., Tax, D.M.: LESS: a model-based classifier for sparse subspaces. IEEE Trans. Pattern Anal. Mach. Intell. 27(9), 1496–1500 (2005)

23. Unal, E., Arbel-Eden, A., Sattler, U., Shroff, R., et al.: DNA damage response pathway uses histone modification to assemble a double-strand break-specific cohesin domain. Mol. Cell. 16, 991–1002 (2003)

Adjusting the Tests According to the Perception of Greek Students Who Are Taught Russian Motion Verbs via Distance Learning

Oksana Kalita[1] and Georgios Pavlidis[2]

[1] Peoples' Friendship University, Moscow, Russia
kalitaxenia@gmail.com
[2] University of Patras, Patra, Greece
pvlds@ceid.upatras.gr

Abstract. Nowadays, the quantity of digital data is so large that its analysis and evaluation can only be performed through (semi-) automatic methods. In a distance learning context, such problems arise for the teacher who needs to personalize the educational material for specific students. The present study focuses on the personalization of the educational material for Greek students learning the Russian language in a distance learning environment. We discovered that it is important for the Intelligent Tutoring System and more specifically for the Intelligent Agents (IA) to have a set of key-characteristics for a proper representation of the states. By having more features an agent has more accurate results whereas useless features are ignored.

Keywords: intelligent tutoring system, trained agents, intelligent test agent, Russian motion verbs, Greek students' preferences.

1 Introduction

Distance learning platforms and more specifically, the Intelligent Tutoring Systems (ITS) play an increasing and important role in society. Towards an efficient and effective implementation of such systems, the technology of Intelligent Agents (IA) is gradually integrated. Processes are assigned to IA that must be complete under specific limitation [1]. In this way, acting on behalf of the users, they implement an additional type of interaction which is often referred as indirect administration [2].

At the same time, IA can be reeducated and formed in such a way that they will perform an optimal gathering and processing of information. In other words, they will gradually become more efficient by learning the interests, the habits and the preferences of the users and their community [3]. Moreover, they help the members of the educational community to effectively cooperate, to schedule their workload and to take part in common events.

The fact that the IA act without human intervention or intervention from other systems, gives them the desirable autonomy meaning that they have their own control practice and can take their own decisions which are presented to the users [4,5]. For

L. Iliadis et al. (Eds.): AIAI 2014 Workshops, IFIP AICT 437, pp. 205–210, 2014.
© IFIP International Federation for Information Processing 2014

this reason, they must be reliable. By taking decisions on behalf of the users, they should guarantee the quality of the services that are provided. Finally, for the efficient communication among the IA, it is essential that they act simultaneously and without interruption so that every process is handled directly [1].

2   Competence and Confidence

The purpose of IA is to help the educational community; however this presupposes the resolving of two basic problems. The first one is the competence, i.e. how an IA receives the desired information (knowledge) so as to decide when and how it can help the educational community. Under this scope, the efficiency and the effectiveness will depend on the number of key characteristics as well as on the quality of the information that are accessible from the IA. The second problem concerns confidence, i.e. how the educational community can feel secure when assign tasks to an IA.

An implementation, which is in accordance with the necessary requirements and provides solutions to the above-mentioned problems, is the trained IA, presented by Pattie Maes [6]. One trained IA can develop its competence through four different ways. First, it is trained by monitoring the activities of the user; it records the user's actions and seeks possible repetitions of behavioral patterns which can be automated. Secondly, it is adjusted based on direct or indirect user feedback. Indirect feedback is received when the user ignores the suggestion made by the IA or when negative feedback is given to an automatic IA action. Third, it is trained through explicit examples by the user, who is the one to show what action should be taken at various hypothetical situations. Fourth, it seeks advice coming from others either more experienced or supplementary IA, which support users in similar processes (situations).

This approach boosts confidence by allowing the IA to provide explanations both for the reasoning and for the behavior in a familiar to the user form. For example, it should explain to the students that, we propose you to proceed to this action:

- due to the fact that there is similarity among past preferences for something
- or there is similarity among past actions
- or it was applied by another student with similar preferences.

3   Filtering the Information

Filtering the information that enters and circulates in a ITS was and still remains an open issue theoretically and practically. Its goal is to introduce to users only the information that is relevant to them. Generally, the existing Information Filtering Systems represent a type of an information retrieval system designed for the management of large volume unstructured or semi structured data [7].

There are three basic methods of information filtering: simple, based on the content and collaborative [8,9]. Simple Filtering is based on the categorization of the users in groups according to their personal features and their profiles. Thus, an IA will be in the position to send the appropriate information to the members of each category.

In Content-Based Filtering, the content of the existing information is analyzed and a representation of the interests (views) of the educational community members is formed. For this purpose, the teacher has to establish a set of key characteristics for every learning item and then the student has to assign the appropriate values to each characteristic. Content-based filtering uses the Euclidean distance in order to analyze the evaluations of key-characteristics and thus to establish the closer value that could be proposed to the "new" user.

In Collaborative Filtering, the viewpoints of the users about the educational objects are collected, either directly or indirectly, in order to develop clusters of likeminded persons. This system, in contrast with the content-based filtering, proposes information that has been evaluated by students of similar abilities and preferences with the user that receives the new educational material. In this case, the Euclidean distance is also used for the analysis of the values and the designation of information that will be proposed to the user.

4   Intelligent Tutoring Systems

The Intelligent Tutoring Systems (ITS) represent applications with major knowledge on a specific subject and aim to transfer this knowledge to the students through an interactive individualized process. They are trying to simulate the educational process so that a virtual teacher will guide the student to the learning process. The purpose of each ITS is to effectively convey its incorporated knowledge.

The main problem of ITS lies on their adaptation to the students' needs. In addition, students continuously interact with the system. The adaptation of the user is provided through pedagogical strategies which determine the sequence, the type of given assistance during the teaching and learning procedure, the time frame and the way of presenting the teaching material (problems, definitions, examples etc).

Generally, the architecture of an ITS consists of four basic components: the Student Model, the Field Model, the Pedagogical Unit and the Educational Unit [10]. Moreover, its function is supported by the following intelligent agents: the Preference agent, the Accounting agent, the Activity agent and the Test agent [11]. These IA must cooperate, help and complement each other in a dynamic way.

5   Application Example of a Test Agent

By extending the functionality of an ITS, we will analyze a way of calculating the suggestions produced by the Test agent, on the basis of the preferences of Greek students who learn Russian motion verbs with prefixes. These verbs cannot be understood by the students since the application of a prefix entails multiple meanings and multiple ways of use. In addition, the same prefixes have similar meanings indicating either proximity or withdrawal at different degree of achievement. There are no equivalent verbs in Greek language, which makes the situation more complex and enforces the development of many exercises on behalf of the teachers.

Having successfully finished with the exercises in the test phase, our research indicates confusion among students when all prefixes are used in combination with all
pairs of motion verbs. More specifically, some prefixes in Greek language do not
exist or have a similar meaning. The testbed of 10 classes of 10 students per class
faced a particular difficultly as far as the following three groups of prefixes are con-
cerned: (i) |При-| До-| Под-|, (ii) |Пере-| Про-|, and (iii) |Вы-| У-| (Table 1).

Table 1. Prefixes with similar meaning and usage

Directional prefixes		Meaning in Greek (English)
with opposites	**При-**	Άφιξη: μπαίνω μέσα (Arrival: to come to see a person / to come to a place)
	У-	Αναχώρηση (Departure: to leave a person / to leave a place)
	Вы-	Έξοδος (Exit: to go out of a place)
	Под-	Προσέγγιση (Approaching: to go up to / to approach)
without opposites	**Про-**	Κίνηση διαμέσου, διέλευση δίπλα από κάποιο αντικείμενο (Motion through or pass: to go through / to pass)
	До-	Φτάνω μέχρι (Reach a destination: to go as far as)
	Пере-	Κίνηση από τη μία άκρη στην άλλη, από ένα σημείο σε άλλο (Motion across: to cross / to go across)

IA suggests some tests to the students from the same category according to collaborative filtering technique the tests are evaluated later. Then, these grades are compared with those of other users, their similarity is estimated and the current user is
provided with highly-evaluated activities by users of the same preferences as him/her.

As mentioned above, collaborative filtering technique makes use of the Euclidean
distance in order to calculate the degree of evaluation similarity among users. It is
assumed that each student can evaluate the activities with the use of a scale ranging
from 1 (he/she does not like the activity at all) to 5 (he/she likes it very much), while
3 denotes a neutral opinion.

Let's assume the students A, B and C have evaluated three tests each as Table 2
shows.

We assume that the current user is C. At first, the Test agent will calculate the Euclidean distance of C from the other students. For example, the distance between C and
B for i activities is calculated as follows:

$$d(X,Y) = \sqrt{\sum_{i=1}^{2}(x_i - y_i)^2} = \sqrt{(x_1 - y_1)^2 + (x_2 - y_2)^2} = \sqrt{5} \approx 2,24$$

It is noted that only common ratings, in our case that are those of tests 1 and 2,
are taken into account. The corresponded Euclidean distance of student C from student A is 3,61. It is observed that the student who has the nearest distance from C

Table 2. Ratings of the tests by the students plus a possible execution route

Test		Evaluations			Possible route of execution
		Student A	Student B	Student C	
Group 1. |При-|До-| |Под-|	1	5☆	2☆	4☆	1
	...	...	...	...	
	n	...	...	...	1.6
Group 2. |Пере-| |Про-|	1	3☆	-	5☆	
	...	...	...	...	2.3
	m	...	...	...	2.4
Group 3. |Вы-| У-|	1	-	4☆	1☆	3.1
	...	...	...	...	3.7
	p	...	...	...	↓

is A. Finally, the Test agent will suggest to student C the most highly-rated activities of student A which haven't been already executed and rated by him/her. In a similar way we implement the proposals from the Activity agent.

6   Conclusions

Traditionally, when a teacher and his/her team, prepares the educational material and determines the outlines of the teaching process, he does not take into account the personality and the special characteristics of the students in the audience. The teacher assumes that the audience is a team which is logical because he/she has never a permanent, face to face, contact with all the students. However, this is not the case when the educational process takes place in a virtual, digital world. In such a world, personal contact is restricted or non-existent; nevertheless there is an increasing amount of old and new, primary or secondary external information regarding each of the students.

The ITS analyze and collect all necessary data from the interaction of students with the system, thus the rate of presenting the contents of the educational material, the learning ability are adjusted to the preferences and the learning styles of each student. They offer invaluable assistance to the teachers, because the student is in a controlled but at the same time in a friendly environment. Moreover, individual proposals allow the avoidance of the student's information overload and improve the interaction with the system.

Our research will be continued with the study of more complex situations of interconnection. Specifically, our goal is the smooth and effective transition from the exercises to the tests, through the collaboration between the Test and the Exercise agent.

References

1. Jennings, N.R., Wooldridge, M.J.: Agent Technology. Springer, Berlin (1998)
2. Kay, A.: User Interface: A Personal View. In: The Art of Human-Computer Interface Design, pp. 121–131 (1989)
3. Bentley, T.J.: Managing information: Avoiding Overload. CIMA, London (1998)
4. Leng, J., Lim, C.P., Li, J., Li, D., Jain, L.: A Role-Based Cognitive Architecture for Multi-Agent Teaming. In: Håkansson, A., et al. (eds.) Agent and Multi-agent Technology for Internet and Enterprise Systems. SCI, vol. 289, pp. 229–255. Springer, Heidelberg (2010)
5. Franklin, S., Graesser, A.: Is it an Agent, or Just a Program? A Taxonomy for Autonomous. In: Jennings, N.R., Wooldridge, M.J., Müller, J.P. (eds.) ECAI-WS 1996 and ATAL 1996. LNCS, vol. 1193, pp. 21–35. Springer, Heidelberg (1997)
6. Maes, P.: Agents that Reduce Work and Information Overload. Communications of the ACM 34(7), 30–40 (1994)
7. Hanani, U., Shapira, B., Shoval, P.: Information Filtering: Overview of Issues, Research and Systems. In: User Modeling and User-Adapted Interaction, pp. 203–259 (2001)
8. Palme, J.: Information Filtering, Stockholm (1998)
9. Juszczyszyn, K., Kazienko, P., Musiał, K.: Personalized Ontology-Based Recommender Systems for Multimedia Objects. In: Håkansson, A., et al. (eds.) Agent and Multi-agent Technology for Internet and Enterprise Systems. SCI, vol. 289, pp. 275–292. Springer, Heidelberg (2010)
10. Gascueña, G.M., Fernández-Caballero, A.: An Agent-based Intelligent Tutoring System for Enhancing E-Learning/E-Teaching. International Journal of Instructional Technology and Distance Learning 2(11), 11–24 (2005)
11. Håkansson, A., Hartung, R., Nguyen, N.T.: Agent and Multi-Agent Technology for Internet and Enterprise Systems. Springer, Berlin (2010)

CSMR: A Scalable Algorithm for Text Clustering with Cosine Similarity and MapReduce

Giannakouris-Salalidis Victor, Plerou Antonia, and Sioutas Spyros

Ionian University, Department of Informatics, Greece
{p12gian1,tplerou,sioutas}@ionio.gr

Abstract. As Internet develops rapidly huge amounts of texts need to be processed in a short time. This entails the necessity of fast, scalable methods for text processing. In this paper a method for pairwise text similarity on massive data-sets, using the Cosine Similarity metric and the tf-idf (Term Frequency-Inverse Document Frequency) normalization method is proposed. The research approach is mainly focused on the MapReduce paradigm, a model for processing large data-sets in parallel manner, with a distributed algorithm on computer clusters. Through MapReduce model application on each step of the proposed method, text processing speed and scalability is enhanced in reference to other traditional methods. The CSMR (Cosine Similarity with MapReduce) method's implementation is currently at the implementation stage. Precise and analytical conclusions concerning the efficiency of the proposed method are to be reached upon completion and review of the overall project phases.

Keywords: MapReduce, Hadoop, TF-IDF, Text Mining, Cosine Similarity.

1 Introduction

Nowadays, as the data amount grows rapidly, challenge of big data need to be faced [1] in various domains such as Business Intelligence [2] or Bioinformatics [3, 4]. With ever increasing volume of text documents, the abundant texts flowing over the Internet, huge collections of documents in digital libraries and digitized personal information are collected quickly every day [5]. In this paper, an innovative method for text similarity measuring with the use of common techniques and metrics is proposed. In particular, a prospective of applying tf-idf [6] and Cosine Similarity [7] measurements on distributed text processing is further analyzed. The CSMR (Cosine Similarity with MapReduce) method includes the component of document pairwise similarity calculation. Especially, CSMR method performs pairwise text similarity with the use of a parallel and distributed algorithm which scales up, regardless the massive input size. This is utilized with the use of MapReduce component of the Hadoop Framework. The authors' proposed method consists by two main components: tf-idf and Cosine Similarity. In this study, these components are designed by following the concept of the MapReduce programming model. Initially, the terms of each document are counted. Secondly, texts are normalized with the use of tf-idf. Finally, Cosine Similarity of each document pair is calculated and results are given as an output. The

L. Iliadis et al. (Eds.): AIAI 2014 Workshops, IFIP AICT 437, pp. 211–220, 2014.

CSMR method is proposed as a faster and more efficient method comparing to the traditional methods. This is due to MapReduce model implementation in each algorithmic step tends to enhance method's efficiency as well as to the aforementioned techniques innovative blend.

2   Related Work

There are quite many cases where several methods have been used for measuring similarity among texts.

Tamer Elsayed et.al [8] method focuses on a MapReduce algorithm for computing pairwise document similarity in large document collections. The algorithm proposed exhibits linear growth in running time and space, in terms of the number of documents. This algorithm is suggested as an example of a programming paradigm that could be useful for a broad range of text analysis problem. Another approach has been proposed by Bin Li et.al [9], i.e. a tf-idf algorithm based on the Hadoop framework. This method is using the MapReduce model provided by Hadoop in order to improve the efficiency of traditional tf-idf algorithm. This case study showed that in the case of massive data computing, Hadoop framework implementation is more efficient comparing to the traditional method.

Jacob Bank et.al [10] use a different approach in order to analyze the vast amounts of data associated with large-scale social networks on the web with the use of the MapReduce program. The Jaccard similarity coefficient between users of Wikipedia based on co-occurrence of page edits is proposed. After several separate linear time computations it was con-firmed that this approach was superior to quadratic computations on long lists of data. Calculating the Jaccard Similarity Coefficient with Map Reduce for Entity Pairs in Wikipedia.

Furthermore, Jian Wan et.al [11] proposed an approach about how document clustering for large collection could be efficiently implemented with MapReduce. Additionally tf-idf and K-Means algorithm on MapReduce design and implementation is described in order to improve algorithm efficiency and effectiveness. Experimentation confirmed the scalability of processing mass data proposed method.

Ping Zhou et.al [12] supplementary research in reference to large-scale data sets clustering amplification a parallel K-Means algorithm based on MapReduce framework is proposed. Model's implementation results illustrated that the proposed clustering algorithm running on Hadoop cluster preserve a higher performance while handling large-scale document automatic classification. In the above mentioned methods dealing with text clustering, there is none or only a slight and indirect approach via Cosine Similarity in order to improve processing speed and scalability.

Finally, according to Rada Mihalcea et.al [13] approach a method for measuring the semantic similarity of short texts, using corpus-based and knowledge-based measures of similarity is presented. Through experiments per-formed on a paraphrase data set, semantic similarity method outer-forms methods based on simple lexical matching, resulting in up to 13% error rate reduction with respect to the traditional vector-based similarity metric. On the contrary, they focus to the aspect that Cosine Similarity, tf-idf as well as other methods can be used for text similarity measuring.

There are also some approaches using MapReduce but, according to authors' knowledge, none of them proposes a model with tf-idf and Cosine function.

Authors' proposed method combines overall of these 3 powerful techniques, i.e. tf-idf, Cosine Similarity and MapReduce and provides a powerful and scalable algorithm suitable for various purposes on Data Mining, especially on Text Processing on big, massive data-sets.

3   Basic Background

According to the project needs, three techniques had been chosen: The Vector Space Model, tf-idf and Cosine Similarity. Each of these techniques is being described in detail below.

3.1   Vector Space Model

Vector Space Model is an algebraic model for representing text documents as vectors. [14] With the use of this model, each term of a document and each number of occurrences in the document could be represented [15]. For instance, the document $d1$ = "This is a vector, this is algebra" based on a vocabulary $V(t)$ could be represented as follows:

$$V(t) = \begin{cases} 1, t = "this" \\ 2, t = "is" \\ 3, t = "a" \\ 4, t = "vector" \\ 5, t = "algebra" \end{cases}$$

$$d1 = \left(tf(1,d1), tf(2,d1), tf(3,d1), tf(4,d1), tf(5,d1) \right)$$

$$= (2,2,1,1,1)$$

Where $d1$ is the document and $tf(t, d_i)$ is the term frequency of the t-term in the ith document.

3.2   Tf-Idf

In Text Mining, tf-idf (Term Frequency-Inverse Document Frequency) [6] is a numerical statistic that reflects the significance of a term in a document in a corpus. The importance increases proportionally to the number of times a word appears in the document but is offset by the frequency of the word in the corpus. Tf-idf algorithm is usually used in search engine, web data mining, text similarity computation and other applications [16]. These applications are often faced with massive data processing.

According to Bin Li [9]approach the tf-idf of a term is calculated with the use of the following formula:

$$TF \times IDF = \frac{n_{i,j}}{|t \in d_j|} \times \log \frac{|D|}{|d \in D : t \in d|}$$

3.3   Cosine Similarity

Cosine Similarity is a measure of similarity between two vectors of an inner product space that measures the cosine of the angle between them [17]. For document clustering, there are different similarity measures available. The Cosine function is proposed as the most commonly used. For two documents A and B, the similarity between them is calculated with the use of the following formula:

$$\cos(A, B) = \frac{A \cdot B}{\| A \| \| B \|} = \sum_{i=1}^{n} \frac{A_i \times B_i}{\sqrt{\sum_{i=1}^{1}(A_i)^2} \times \sqrt{\sum_{i=1}^{n}(B_i)^2}}$$

When the cosine value is computed to be 1, that indicates that the two documents are identical and while it is computed to be 0 if there is nothing in common between them (i.e., their document vectors are orthogonal to each other). The attribute vectors A and B are usually the term frequency vectors of the documents.

3.4   Hadoop and MapReduce

Hadoop software library [18], is a framework developed by Apache, suitable for scalable, distributed computing. It allows storage and large-scale data processing across clusters of commodity servers [19]. The innovative aspect of Hadoop is that there is no absolute necessity of expensive, high-end hardware. Instead, it enables distributed parallel processing of massive amounts of data [20] on industry-standard servers with high scalability for both data storing and processing. Therefore it is considered to be one of the most popular frameworks for Big Data Analytics. Especially, Hadoop has two main subprojects: HDFS (Hadoop Distributes File System) & MapReduce.

MapReduce [21] is the main component of Hadoop. It's a programming model that allows massive data processing across thousands of servers in a Hadoop cluster. The MapReduce paradigm is derived from the Map and Reduce functions of the Functional Programming model [22]. A MapReduce program constitutes from the Mappers and the Reducers. In the Map phase, the master node the master node divides the input into smaller partitions and distributes them to the worker nodes. Then a worker node may repeat the same step recursively. As soon as this procedure is completed, the master node collects the key-value pairs resulted from the Mappers and distributes them to the Combiners to combine the pairs with the same key. This phase is known as the Shuffle & Sort phase. Finally, the key-value pairs are distributed to the Reducers that produce the final output. This step is called the Reduce phase. MapReduce program procedure is visualized as follows:

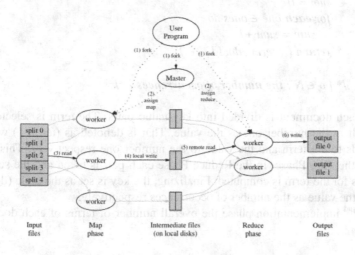

Fig. 1. MapReduce Procedure Visualization

4   Method

4.1   Description

Authors' proposed method for measuring text similarity applying MapReduce consists of 4 stages. At the first stage, occurrences of each term in our documents are counted. Then, the term frequency of every one term in each document is measured. Thereafter the tf-idf of each term is measured and finally the cosines of the pairs are calculated in order to estimate the similarity among them. MapReduce model was used in order to design each one of the above mentioned steps. The algorithm paradigm in pseudocode and further analysis of each step is disposed in details in the next section.

4.2   MapReduce Stages

In the 1st implementation stage the occurrences of each term in every document are counted. The algorithm applied is as follows:

Algorithm 1: Word Count
```
1:   class Mapper
2:      method Map( document )
3:         for each term ∈ document
4:            write ( ( term , docId ) , 1 )
```

```
5:
6:    class Reducer
7:        method Reduce( ( term , docId ) , ones[ 1 , 1 , … , n ] )
8:            sum = 0
9:            for each one ∈ ones do
10:               sum = sum +1
11:           return ( ( term , docId ) , o )
12:
13:    /* { o ∈ N : the number of occurrences } */
```

Initially, each document is divided into key-value pairs. The term is selected as the key as well as the number one as the value. That is denoted as (term, 1) where key corresponds to the term and the value to the number one respectively. This phase is known as the Map Phase. In the Reduce Phase each pair is taken and the sum of the list of ones for the term is computed. Finalizing, the key is set as the tuple (document, term) and the value as the number of occurrences respectively.

In the 2nd implementation phase the overall number of terms of each document is computed.

Algorithm 2: Term Frequency
```
1:    class Mapper
2:        method Map( ( term , docId ) , o )
3:            for each element ∈ ( term , docId )
4:                write ( docId, ( term, o ) )
5:
6:    class Reducer
7:        method Reduce( docId, (term, o) )
8:            N = 0
9:            for each tuple ∈ ( term, o ) do
10:               N = N + o
11:           return ( (docId, N), (term, o) )
```

By this algorithm implementation, concerning the Map Phase, the input is divided into key-value pairs while the *docId* is set as the key in addition to the tuple *(term, o)* as the value. In the reduce phase the total of terms in each document is counted and the key-value pairs are returned with the *(DocId, N)* as the key as well as the tuples *(term, o)* as the value (*N* is the total of terms in the document). The key-value pairs are returned with the tuples (docId, N) as the key and the tuples (term, o) as the value, where N is the total of terms in the document.

In the 3rd implementation stage the tf-idf of each term in a document is computed with the use of the following formula:

$$tfidf = \frac{n}{N} \frac{|D|}{|\{d \in D : t \in d\}|}$$

Where |D| is the number of the documents in corpus and $|\{d \in D : t \in d\}|$ number of documents where t-term appears.

Algorithm 3: Tf-Idf

```
 1:   class Mapper
 2:      method Map( ( docId , N ), ( term , o  ) )
 3:         for each element ∈ ( term , o  )
 4:            write ( term, ( docId, o, N ) )
 5:
 6:   class Reducer
 7:      method Reduce( term, ( docId , o , N ) )
 8:         n = 0
 9:         for each element ∈ ( docId , o , N ) do
10:            n = n + 1
11:         tf = o / N
12:         idf =  log(|D|/(1+n))
13:            return ( docId, ( term , tf×idf  ) )
14:
15:      /* Where |D| is the number of documents in the corpus */
```

Applying the aforementioned algorithm, during the Map Phase the term is set as the key as well as the tuple *(docId, o, N)* as the value. In that case, the number of documents is calculated by the reducer, where the term appears and the result to the n variable is set. The term frequency is subsequently calculated plus the inverse document frequency of each term as well. Finally, key-value pairs with the *docId* as the key and the tuple *(term, tf×idf)* as the value are taken as a result.

In the 4th and final implementation phase all the possible combinations of two documents pairs are provided and cosine for each of them is computed. Assuming that there are n documents in the corpus, a similarity matrix of size is generated as follows:

$$\binom{n}{2} = \frac{n!}{2!(n-2)!}$$

Algorithm 4: Cosine Similarity

```
16: class Mapper
17:    method Map( docs )
18:       n = docs.length
19:
20:    for i = 0 to docs.length
21:       for j = i+1 to docs.length
```

22:            *write* ((*docs[i].id, docs[j].id*),(*docs[i].tfidf, docs[j].tfidf*))

23:

24: *class* Reducer

25:    *method* Reduce((*docId_A, docId_B*),(*docA.tfidf, docB.tfidf*))

26:      A = *docA.tfidf*

27:      B = *docB.tfidf*

28:    $cosine = sum(\ A \times B\ )/ \left(sqrt(\ sum(A^2)\ ) \times sqrt(\ sum(B^2)\ ) \right)$

29:    *return* ((*docId_A, docId_B*), *cosine*)

In the Map phase, implementing the abovementioned algorithm, every potential combination of the input documents is generated and the document IDs for the key as well as the tf-idf vectors for the value is set. Within the Reduce phase, cosine for each document pair is calculated and the similarity matrix is also provided. Algorithm 4 is visualized as follows:

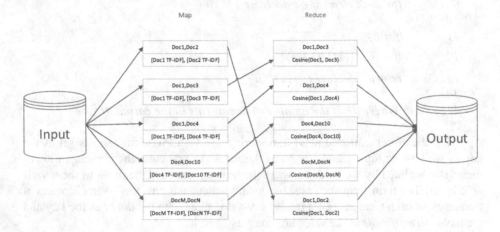

Fig. 2. Algorithm 4 Visualization

5   Discussion

In this case study, popular methods for measure the similarity of texts had been used. In particular, tf-idf and Cosine Similarity were adjusted with the MapReduce model in order to propose an innovative and scalable method. Authors' approach enhances the innovative aspect of the MapReduce programming paradigm in the field of text processing. The key contribution is that the proposed method enhances the procedure of measuring the text similarity with the Cosine metric and increase algorithm scalability. The implementation of the aforementioned techniques on computer clusters running the Hadoop Distributed File System (HDFS) blended with MapReduce also ensure algorithms effectiveness. The proposed method is currently at the design and implementation stage. Therefore, more clear and specific conclusions for its efficiency, as well as proposals for revisions and improvement, are to be provided after the projects' overall implementation.

Authors' future work concerns the finalized proposed method version as well as statistically analyzed results of the data collected during piloting implementation procedure presentation. In addition the software's design and development for the CSMR algorithm implementation on real text files is ongoing. An additional research approach is the implementation of the abovementioned algorithm with the use of tools like Apache Spark and Scala [23] as well as an Open Source project implemented in Java.

Acknowledgements. This research has been co-financed by the European Union (European Social Fund – ESF) and Greek national funds through the Operational Program "Education and Lifelong Learning" of the National Strategic Reference Framework (NSRF) - Research Funding Program: Thales. Investing in knowledge society through the European Social Fund.

References

1. Wu, X., Zhu, X., Member, S.: Data Mining with Big Data 26, 97–107 (2014)
2. Chen, H., Storey, V.C.: Business Intelligence and Analytics: From Big Data to Big Impact 36, 1165–1188 (2012)
3. Taylor, R.C.: An overview of the Hadoop/MapReduce/HBase framework and its current applications in bioinformatics. BMC Bioinformatics 11(suppl. 1), S1 (2010), doi: 10.1186/1471-2105-11-S12-S1
4. Matsunaga, A., Tsugawa, M., Fortes, J.: CloudBLAST: Combining MapReduce and Virtualization on Distributed Resources for Bioinformatics Applications. In: 2008 IEEE Fourth Int. Conf. eScience, pp. 222–229 (2008), doi:10.1109/eScience.2008.62
5. Huang, A.: Similarity Measures for Text Document Clustering (2008)
6. Ramos, J., Eden, J.: Edu R Using TF-IDF to Determine Word Relevance in Document Queries
7. Tata, S., Patel, J.M., Science, C., Arbor, A.: Estimating the Selectivity of tf-idf based Cosine Similarity Predicates 36, 7–12 (2007)
8. Elsayed, T., Lin, J., Oard, D.W.: Pairwise Document Similarity in Large Collections with MapReduce, 265–268 (2008)
9. Bin, L., Yuan, G.: Improvement of TF-IDF Algorithm Based on Hadoop Framework. In: Proc 2nd Int. Conf. Comput. Appl. Syst. Model., pp. 391–393 (2012), doi:10.2991/iccasm.2012.98
10. Bank, J., Cole, B.: Calculating the Jaccard Similarity Coefficient with Map Reduce for Entity Pairs in Wikipedia (2008)
11. Wan, J., Yu, W., Xu, X.: Design and Implement of Distributed Document Clustering Based on MapReduce 7, 278–280 (2009)

12. Zhou, P., Lei, J., Ye, W.: Large-Scale Data Sets Clustering Based on MapReduce and Hadoop 16, 5956–5963 (2011)
13. Mihalcea, R., Corley, C., Strapparava, C.: Corpus-based and Knowledge-based Measures of Text Semantic Similarity (2005)
14. Turney, P.D.: From Frequency to Meaning: Vector Space Models of Semantics 37, 141–188 (2010)
15. Raghavan, V.V., Wong, S.K.M.: A critical analysis of vector space model for information retrieval. J. Am. Soc. Inf. Sci. 37, 279–287 (1986), doi:10.1002/asi.4630370502
16. Terms RT NRC Publications Archive Archives des publications du CNRC Coherent Keyphrase Extraction via Web Mining Coherent Keyphrase Extraction via Web Mining
17. Kalaivendhan, K., Sumathi, P.: An Efficient Clustering Method To Find Similarity Between The Documents 2, 2532–2535 (2014)
18. Shvachko, K., Kuang, H., Radia, S., Chansler, R.: The Hadoop Distributed File System. In: 2010 IEEE 26th Symp. Mass. Storage Syst. Technol., pp. 1–10 (2010), doi:10.1109/MSST.2010.5496972
19. Lin, X., Meng, Z., Xu, C., Wang, M.: A Practical Performance Model for Hadoop MapReduce. In: 2012 IEEE Int. Conf. Clust. Comput. Work, pp. 231–239 (2012), doi:10.1109/ClusterW.2012.24
20. Ekanayake, J., Pallickara, S., Fox, G.: MapReduce for Data Intensive Scientific Analyses. In: 2008 IEEE Fourth Int Conf eScience, pp. 277–284 (2008), doi:10.1109/eScience.2008.59
21. Dean, J., Ghemawat, S.: MapReduce: Simplified Data Processing on Large Clusters, 1–13
22. Lämmel, R.: Google's MapReduce programming model — Revisited. Sci. Comput. Program 70, 1–30 (2008), doi:10.1016/j.scico.2007.07.001
23. Zaharia, M., Chowdhury, M., Franklin, M.J., et al.: Spark: Cluster Computing with Working Sets

From Conformal to Probabilistic Prediction

Vladimir Vovk, Ivan Petej, and Valentina Fedorova

Computer Learning Research Centre,
Department of Computer Science,
Royal Holloway, University of London,
Egham, Surrey, UK
{volodya.vovk,ivan.petej,alushaf}@gmail.com

Abstract. This paper proposes a new method of probabilistic prediction, which is based on conformal prediction. The method is applied to the standard USPS data set and gives encouraging results.

1 Introduction

In essence, conformal predictors output systems of p-values: to each potential label of a test object a conformal predictor assigns the corresponding p-value, and a low p-value is interpreted as the label being unlikely. It has been argued, especially by Bayesian statisticians, that p-values are more difficult to interpret than probabilities; besides, in decision problems probabilities can be easily combined with utilities to obtain decisions that are optimal from the point of view of Bayesian decision theory. In this paper we will apply the idea of transforming p-values into probabilities (used in a completely different context in, e.g., [10], Sect. 9, and [7]) to conformal prediction: the p-values produced by conformal predictors will be transformed into probabilities.

The approach of this paper is as follows. It was observed in [12] that some criteria of efficiency for conformal prediction (called "probabilistic criteria") encourage using the conditional probability $Q(y \mid x)$ as the conformity score for an observation (x, y), Q being the data-generating distribution. In this paper we extend this observation to label-conditional predictors (Sect. 2).

Next we imagine that we are given a conformal predictor Γ that is nearly optimal with respect to a probabilistic criterion (such a conformal predictor might be an outcome of a thorough empirical study of various conformal predictors using a probabilistic criterion of efficiency). Essentially, this means that in the limit of a very large training set the p-value that Γ outputs for an observation (x, y) is a monotonic transformation of the conditional probability $Q(y \mid x)$ (Theorem 1 in Sect. 3).

Finally, we transform the p-values back into conditional probabilities using the distribution of p-values in the test set (Sect. 5). Following [10] and [7], we will say that at this step we *calibrate* the p-values into probabilities,

In Sect. 6 we give an example of a realistic situation where use of the techniques developed in this paper improves on a standard approach. The performance of the probabilistic predictors considered in that section is measured using standard loss functions, logarithmic and Brier (Sect. 4).

L. Iliadis et al. (Eds.): AIAI 2014 Workshops, IFIP AICT 437, pp. 221–230, 2014.
© IFIP International Federation for Information Processing 2014

Comparisons with Related Work

It should be noted that in the process of transforming p-values into probabilities suggested in this paper we lose a valuable feature of conformal prediction, its automatic validity. Our hope, however, is that the advantages of conformal prediction will translate into accurate probabilistic predictions.

There is another method of probabilistic prediction that is related to conformal prediction, Venn prediction (see, e.g., [13], Chap. 6, or [14]). This method does have a guaranteed property of validity (perhaps the simplest being Theorem 1 in [14]); however, the price to pay is that it outputs multiprobabilistic predictions rather than sharp probabilistic predictions. There are natural ways of transforming multiprobabilistic predictions into sharp probabilistic predictions (see, e.g., [14], Sect. 4), but such transformations, again, lead to the loss of the formal property of validity.

As preparation, we study label-conditional conformal prediction. For a general discussion of conditionality in conformal prediction, see [11]. Object-conditional conformal prediction has been studied in [5] (in the case of regression).

2   Criteria of Efficiency for Label-Conditional Conformal Predictors and Transducers

Let $\mathbf{X}$ be a measurable space (the *object space*) and $\mathbf{Y}$ be a finite set equipped with the discrete σ-algebra (the *label space*); the *observation space* is defined to be $\mathbf{Z} := \mathbf{X} \times \mathbf{Y}$. A *conformity measure* is a measurable function A that assigns to every sequence $(z_1, \ldots, z_n) \in \mathbf{Z}^*$ of observations a same-length sequence $(\alpha_1, \ldots, \alpha_n)$ of real numbers and that is equivariant with respect to permutations: for any n and any permutation π of $\{1, \ldots, n\}$,

$$(\alpha_1, \ldots, \alpha_n) = A(z_1, \ldots, z_n) \Longrightarrow (\alpha_{\pi(1)}, \ldots, \alpha_{\pi(n)}) = A(z_{\pi(1)}, \ldots, z_{\pi(n)}).$$

The *label-conditional conformal predictor* determined by A is defined by

$$\Gamma^\epsilon(z_1, \ldots, z_l, x) := \{y \mid p^y > \epsilon\}, \tag{1}$$

where $(z_1, \ldots, z_l) \in \mathbf{Z}^*$ is a training sequence, x is a test object, $\epsilon \in (0, 1)$ is a given *significance level*, and for each $y \in \mathbf{Y}$ the corresponding *label-conditional p-value* p^y is defined by

$$p^y := \frac{|\{i = 1, \ldots, l+1 \mid y_i = y \ \& \ \alpha_i^y < \alpha_{l+1}^y\}|}{|\{i = 1, \ldots, l+1 \mid y_i = y\}|}$$
$$+ \tau \frac{|\{i = 1, \ldots, l+1 \mid y_i = y \ \& \ \alpha_i^y = \alpha_{l+1}^y\}|}{|\{i = 1, \ldots, l+1 \mid y_i = y\}|}, \tag{2}$$

where τ is a random number distributed uniformly on the interval $[0, 1]$ and the corresponding sequence of *conformity scores* is defined by

$$(\alpha_1^y, \ldots, \alpha_l^y, \alpha_{l+1}^y) := A(z_1, \ldots, z_l, (x, y)).$$

It is clear that the system of *prediction sets* (1) output by a conformal predictor is nested, namely decreasing in ϵ.

The *label-conditional conformal transducer* determined by A outputs the system of p-values $(p^y \mid y \in \mathbf{Y})$ defined by (2) for each training sequence $(z_1, \ldots, z_l)$ of observations and each test object x.

Four Criteria of Efficiency

Suppose that, besides the training sequence, we are also given a test sequence, and would like to measure on it the performance of a label-conditional conformal predictor or transducer. As usual, let us define the performance on the test set to be the average performance (or, equivalently, the sum of performances) on the individual test observations. Following [12], we will discuss the following four criteria of efficiency for individual test observations; all the criteria will work in the same direction: the smaller the better.

- The sum $\sum_{y \in \mathbf{Y}} p^y$ of the p-values; referred to as the *S criterion*. This is applicable to conformal transducers (i.e., the criterion is ϵ-independent).
- The size $|\Gamma^\epsilon|$ of the prediction set at a significance level ϵ; this is the *N criterion*. It is applicable to conformal predictors (ϵ-dependent).
- The sum of the p-values apart from that for the true label: the *OF* ("observed fuzziness") *criterion*.
- The number of false labels included in the prediction set Γ^ϵ at a significance level ϵ; this is the *OE* ("observed excess") *criterion*.

The last two criteria are simple modifications of the first two (leading to smoother and more expressive pictures). Equivalently, the S criterion can be defined as the arithmetic mean $\frac{1}{|\mathbf{Y}|} \sum_{y \in \mathbf{Y}} p^y$ of the p-values; the proof of Theorem 1 below will show that, in fact, we can replace arithmetic mean by any mean ([3], Sect. 3.1), including geometric, harmonic, etc.

3 Optimal Idealized Conformity Measures for a Known Probability Distribution

In this section we consider the idealized case where the probability distribution Q generating independent observations $z_1, z_2, \ldots$ is known (as in [12]). The main result of this section, Theorem 1, is the label-conditional counterpart of Theorem 1 in [12]; the proof of our Theorem 1 is also modelled on the proof of Theorem 1 in [12]. In this section we assume, for simplicity, that the set $\mathbf{Z}$ is finite and that $Q(\{z\}) > 0$ for all $z \in \mathbf{Z}$.

An *idealized conformity measure* is a function $A(z, Q)$ of $z \in \mathbf{Z}$ and $Q \in \mathcal{P}(\mathbf{Z})$ (where $\mathcal{P}(\mathbf{Z})$ is the set of all probability measures on $\mathbf{Z}$). We will sometimes write the corresponding conformity scores as $A(z)$, as Q will be clear from the context. The *idealized smoothed label-conditional conformal predictor* corresponding to A outputs the following prediction set $\Gamma^\epsilon(x)$ for each object $x \in \mathbf{X}$ and each significance level $\epsilon \in (0, 1)$. For each potential label $y \in \mathbf{Y}$ for x define the corresponding *label-conditional p-value* as

$$
\begin{aligned}
p^y = p(x, y) := {} & \frac{Q(\{(x', y) \mid x' \in \mathbf{X} \ \& \ A((x', y), Q) < A((x, y), Q)\})}{Q_{\mathbf{Y}}(\{y\})} \\
& + \tau \frac{Q(\{(x', y) \mid x' \in \mathbf{X} \ \& \ A((x', y), Q) = A((x, y), Q)\})}{Q_{\mathbf{Y}}(\{y\})}
\end{aligned}
\tag{3}
$$

(this is the idealized analogue of (2)), where $Q_{\mathbf{Y}}$ is the marginal distribution of Q on $\mathbf{Y}$ and τ is a random number distributed uniformly on $[0, 1]$. The prediction set is

$$
\Gamma^\epsilon(x) := \{y \in \mathbf{Y} \mid p(x, y) > \epsilon\} .
\tag{4}
$$

The *idealized smoothed label-conditional conformal transducer* corresponding to A outputs for each object $x \in \mathbf{X}$ the system of p-values $(p^y \mid y \in \mathbf{Y})$ defined by (3); in the idealized case we will usually use the alternative notation $p(x, y)$ for p^y.

Four Idealized Criteria of Efficiency

In this subsection we will apply the four criteria of efficiency that we discussed in the previous section to the idealized case of infinite training and test sequences; since the sequences are infinite, they carry all information about the data-generating distribution Q. We will write $\Gamma_A^\epsilon(x)$ for the $\Gamma^\epsilon(x)$ in (4) and $p_A(x, y)$ for the $p(x, y)$ in (3) to indicate the dependence on the choice of the conformity measure A. Let U be the uniform probability measure on the interval $[0, 1]$.

An idealized conformity measure A is:

- *S-optimal* if $\mathbb{E}_{(x,\tau) \sim Q_{\mathbf{X}} \times U} \sum_y p_A(x, y) \leq \mathbb{E}_{(x,\tau) \sim Q_{\mathbf{X}} \times U} \sum_y p_B(x, y)$ for any idealized conformity measure B, where $Q_{\mathbf{X}}$ is the marginal distribution of Q on $\mathbf{X}$;
- *N-optimal* if $\mathbb{E}_{(x,\tau) \sim Q_{\mathbf{X}} \times U} |\Gamma_A^\epsilon(x)| \leq \mathbb{E}_{(x,\tau) \sim Q_{\mathbf{X}} \times U} |\Gamma_B^\epsilon(x)|$ for any idealized conformity measure B and any significance level ϵ;
- *OF-optimal* if

$$
\mathbb{E}_{((x,y),\tau) \sim Q \times U} \sum_{y' \neq y} p_A(x, y') \leq \mathbb{E}_{((x,y),\tau) \sim Q \times U} \sum_{y' \neq y} p_A(x, y')
$$

for any idealized conformity measure B;
- *OE-optimal* if

$$
\mathbb{E}_{((x,y),\tau) \sim Q \times U} |\Gamma_A^\epsilon(x) \setminus \{y\}| \leq \mathbb{E}_{((x,y),\tau) \sim Q \times U} |\Gamma_B^\epsilon(x) \setminus \{y\}|
$$

for any idealized conformity measure B and any significance level ϵ.

The *conditional probability (CP) idealized conformity measure* is

$$A((x,y),Q) := Q(y \mid x).$$

An idealized conformity measure A is a (label-conditional) *refinement* of an idealized conformity measure B if

$$B((x_1,y)) < B((x_2,y)) \Longrightarrow A((x_1,y)) < A((x_2,y)) \tag{5}$$

for all $x_1, x_2 \in \mathbf{Z}$ and all $y \in \mathbf{Y}$. (Notice that this definition, being label-conditional, is different from the one given in [12].) Let $\mathcal{R}(\text{CP})$ be the set of all refinements of the CP idealized conformity measure. If C is a criterion of efficiency (one of the four discussed above), we let $\mathcal{O}(C)$ stand for the set of all C-optimal idealized conformity measures.

Theorem 1. $\mathcal{O}(S) = \mathcal{O}(OF) = \mathcal{O}(N) = \mathcal{O}(OE) = \mathcal{R}(\text{CP})$.

Proof. We start from proving $\mathcal{R}(\text{CP}) = \mathcal{O}(N)$. Fix a significance level ϵ. A smoothed confidence predictor at level ϵ is defined as a random set of observations $(x,y) \in \mathbf{Z}$; in other words, to each observation (x,y) is assigned the probability $P(x,y)$ that the observation will be outside the prediction set. Under the restriction that the sum of the probabilities $Q(x,y)$ of observations (x,y) outside the prediction set (defined as $\sum_x Q(x,y)P(x,y)$ in the smoothed case) is bounded by $\epsilon Q_{\mathbf{Y}}(y)$ for a fixed y, the N criterion requires us to make the sum of $Q_{\mathbf{X}}(x)$ for (x,y) outside the prediction set (defined as $\sum_x Q_{\mathbf{X}}P(x,y)$ in the smoothed case) as large as possible. It is clear that the set should consist of the observations with the smallest $Q(y \mid x)$ (by the usual Neyman–Pearson argument: cf. [4], Sect. 3.2).

Next we show that $\mathcal{O}(N) \subseteq \mathcal{O}(S)$. Let an idealized conformity measure A be N-optimal. By definition,

$$\mathbb{E}_{x,\tau} |\Gamma_A^\epsilon(x)| \leq \mathbb{E}_{x,\tau} |\Gamma_B^\epsilon(x)|$$

for any idealized conformity measure B and any significance level ϵ. Integrating over $\epsilon \in (0,1)$ and swapping the order of integrals and expectations,

$$\mathbb{E}_{x,\tau} \int_0^1 |\Gamma_A^\epsilon(x)| \, d\epsilon \leq \mathbb{E}_{x,\tau} \int_0^1 |\Gamma_B^\epsilon(x)| \, d\epsilon. \tag{6}$$

Since

$$|\Gamma^\epsilon(x)| = \sum_{y \in \mathbf{Y}} 1_{\{p(x,y) > \epsilon\}},$$

we can rewrite (6), after swapping the order of summation and integration, as

$$\mathbb{E}_{x,\tau} \sum_{y \in \mathbf{Y}} \left(\int_0^1 1_{\{p_A(x,y) > \epsilon\}} \, d\epsilon \right) \leq \mathbb{E}_{x,\tau} \sum_{y \in \mathbf{Y}} \left(\int_0^1 1_{\{p_B(x,y) > \epsilon\}} \, d\epsilon \right).$$

Since

$$\int_0^1 1_{\{p(x,y)>\epsilon\}} \, d\epsilon = p(x,y),$$

we finally obtain

$$\mathbb{E}_{x,\tau} \sum_{y \in \mathbf{Y}} p_A(x,y) \le \mathbb{E}_{x,\tau} \sum_{y \in \mathbf{Y}} p_B(x,y).$$

Since this holds for any idealized conformity measure B, A is S-optimal.

The argument in the previous paragraph in fact shows that $\mathcal{O}(S) = \mathcal{O}(N) = \mathcal{R}(CP)$.

The equality $\mathcal{O}(S) = \mathcal{O}(OF)$ follows from

$$\mathbb{E}_{x,\tau} \sum_y p(x,y) = \mathbb{E}_{(x,y),\tau} \sum_{y' \ne y} p(x,y') + \frac{1}{2},$$

where we have used the fact that $p(x,y)$ is distributed uniformly on $[0,1]$ when $((x,y),\tau) \sim Q \times U$ (see [13] and [12]).

Finally, we notice that $\mathcal{O}(N) = \mathcal{O}(OE)$. Indeed, for any significance level ϵ,

$$\mathbb{E}_{x,\tau} |\Gamma^\epsilon(x)| = \mathbb{E}_{(x,y),\tau} |\Gamma^\epsilon(x) \setminus \{y\}| + (1 - \epsilon),$$

again using the fact that $p(x,y)$ is distributed uniformly on $[0,1]$ and so $\mathbb{P}_{(x,y),\tau}(y \in \Gamma^\epsilon(x)) = 1 - \epsilon$. $\qquad\square$

4   Criteria of Efficiency for Probabilistic Predictors

Given a training set $(z_1, \ldots, z_l)$ and a test object x, a probabilistic predictor outputs a probability measure $P \in \mathcal{P}(\mathbf{Y})$, which is interpreted as its probabilistic prediction for the label y of x; we let $\mathcal{P}(\mathbf{Y})$ stand for the set of all probability measures on $\mathbf{Y}$. The two standard way of measuring the performance of P on the actual label y are the *logarithmic* (or *log*) *loss* $-\ln P(\{y\})$ and the *Brier loss*

$$\sum_{y' \in \mathbf{Y}} \left(1_{\{y'=y\}} - P(\{y'\})\right)^2,$$

where 1_E stands for the indicator of an event E: $1_E = 0$ if E happens and $1_E = 0$ otherwise. The efficiency of probabilistic predictors will be measured by these two loss functions.

Suppose we have a test sequence $(z_{l+1}, \ldots, z_{l+k})$, where $z_i = (x_i, y_i)$ for $i = l+1, \ldots, l+k$, and we want to evaluate the performance of a probabilistic predictor (trained on a training sequence $z_1, \ldots, z_l$) on it. In the next section we will use the *average log loss*

$$-\frac{1}{k} \sum_{i=l+1}^{l+k} \ln P_i(\{y_i\})$$

Algorithm 1. Conformal-type probabilistic predictor

Input: training sequence $(z_1, \ldots, z_l) \in \mathbf{Z}^l$
Input: calibration sequence $(x_{l+1}, \ldots, x_{l+k}) \in \mathbf{X}^k$
Input: test object x_0
Output: probabilistic prediction $P \in \mathcal{P}(\mathbf{Y})$ for the label of x_0
  for $y \in \mathbf{Y}$ **do**
    for each x_i in the calibration sequence find the p-value p_i^y by (2)
      (with $l + i$ in place of $l + 1$)
    let g_y be the antitonic density on $[0, 1]$ fitted to $p_{l+1}^y, \ldots, p_{l+k}^y$
    find the p-value p_0^y by (2) (with 0 in place of $l + 1$)
    for each $y \in \mathbf{Y}$, set $P'(\{y\}) := g_y(1)/g_y(p_0^y)$
  end for
  set $P(\{y\}) := P'(\{y\})/\sum_{y'} P'(\{y'\})$ for each $y \in \mathbf{Y}$

---

and the *standardized Brier loss*

$$\sqrt{\frac{1}{k\,|\mathbf{Y}|} \sum_{i=l+1}^{l+k} \sum_{y' \in \mathbf{Y}} \left(1_{\{y'=y_i\}} - P_i(\{y'\})\right)^2},$$

where $P_i \in \mathcal{P}(\mathbf{Y})$ is the probabilistic prediction for x_i. Notice that in the binary case, $|\mathbf{Y}| = 2$, the average log loss coincides with the mean log error (used in, e.g., [14], (12)) and the standardized Brier loss coincides with the root mean square error (used in, e.g., [14], (13)).

5   Calibration of p-Values into Conditional Probabilities

The argument of this section will be somewhat heuristic, and we will not try to formalize it in this paper. Fix $y \in \mathbf{Y}$. Suppose that $q := P(y \mid x)$ has an absolutely continuous distribution with density f when $x \sim Q_\mathbf{X}$. (In other words, f is the density of the image of $Q_\mathbf{X}$ under the mapping $x \mapsto P(y \mid x)$.) For the CP idealized conformity measure, we can rewrite (3) as

$$p(q) := \int_0^q q' f(q') dq' \bigg/ D, \tag{7}$$

where $D := Q_\mathbf{Y}(\{y\})$; alternatively, we can set $D := \int_0^1 q' f(q') dq'$ to the normalizing constant ensuring that $p(1) = 1$. To see how (7) is a special case of (3) for the CP idealized conformity measure, notice that the probability that $Y = y$ and $P(Y \mid X) \in (q', q' + dq')$, where $(X, Y) \sim f$, is $q' f(q') dq'$. In (7) we write $p(q)$ rather than p^y since p^y depends on y only via q.

    We are more interested in the inverse function $q(p)$, which is defined by the condition

$$p = \int_0^{q(p)} q' f(q') dq' \bigg/ D.$$

When $q \sim f$, we have

$$\mathbb{P}(p(q) \leq a) = \mathbb{P}(q \leq q(a)) = \int_0^{q(a)} f(q')dq'.$$

Therefore, when $q \sim f$, we have

$$\mathbb{P}(a \leq p(q) \leq a + da) = \int_{q(a)}^{q(a+da)} f(q')dq' \approx \frac{1}{q(a)} \int_{q(a)}^{q(a+da)} q'f(q')dq' = \frac{Dda}{q(a)},$$

and so

$$q(c) \approx D \left/ \frac{\mathbb{P}(c \leq p(q) \leq c + dc)}{dc} \right..$$

This gives rise to the algorithm given as Algorithm 1, which uses real p-values (2) instead of the ideal p-values (3). The algorithm is transductive in that it uses a training sequence of labelled observations and a calibration sequence of unlabelled objects (in the next section we use the test sequence as the calibration sequence); the latter used for calibrating p-values into conditional probabilities. Given all the p-values for the calibration sequence with postulated label y, find the corresponding antitonic density $g(p)$ (remember that the function $q(p)$ is known to be monotonic, namely isotonic) using Grenander's estimator (see [2] or, e.g., [1], Chap. 8). Use $D/g(p)$ as the calibration function, where $D := g(1)$ is chosen in such a way that a p-value of 1 is calibrated into a conditional probability of 1. (Alternatively, we could set D to the fraction of observations labelled as y in the training sequence; this approximates setting $D := Q_{\mathbf{Y}}(\{y\})$.) The probabilities produced by this procedure are not guaranteed to lead to a probability measure: the sum over y can be different from 1 (and this phenomenon has been observed in our experiments). Therefore, in the last line of Algorithm 1 we normalize the calibrated p-values to obtain genuine probabilities.

6   Experiments

In our experiments we use the standard USPS data set of hand-written digits. The size of the training set is 7291, and the size of the test set is 2007; however, instead of using the original split of the data into the two parts, we randomly split all available data (the union of the original training and test sets) into a training set of size 7291 and test set of size 2007. (Therefore, our results somewhat depend on the seed used by the random number generator, but the dependence is minor and does not affect our conclusions at all; we always report results for seed 0.)

A powerful algorithm for the USPS data set is the 1-Nearest Neighbour (1-NN) algorithm using tangent distance [8]. However, it is not obvious how this algorithm could be transformed into a probabilistic predictor. On the other hand, there is a very natural and standard way of extracting probabilities from support vector machines, which we will refer to it as *Platt's algorithm* in this paper: it is the combination of the method proposed by Platt [6] with pairwise coupling [15] (unlike our algorithm, which is applicable to multi-class problems directly, Platt's

Table 1. The performance of the two algorithms, Platt's (with the optimal values of parameters) and the conformal-type probabilistic predictor based on 1-Nearest Neighbour with tangent distance

algorithm	average log loss	standardized Brier loss
optimized Platt	0.06431	0.05089
conformal-type 1-NN	0.04958	0.04359

Table 2. The performance of Platt's algorithm with the polynomial kernels of various degrees for the cost parameter $C = 10$

degree	average log loss	standardized Brier loss
1	0.12681	0.07342
2	0.09967	0.06109
3	0.06855	0.05237
4	0.11041	0.06227
5	0.09794	0.06040

method is directly applicable only to binary problems). In this section we will apply our method to the 1-NN algorithm with tangent distance and compare the results to Platt's algorithm as implemented in the function svm from the e1071 R package (for our multi-class problem this function calculates probabilities using the combination of Platt's binary method and pairwise coupling).

There is a standard way of turning a distance into a conformal predictor ([13], Sect. 3.1): namely, the conformity score α_i of the ith observation in a sequence of observations can be defined as

$$\frac{\min_{j:y_j \neq y_i} d(x_i, x_j)}{\min_{j \neq i:y_j = y_i} d(x_i, x_j)}, \tag{8}$$

where d is the distance; the intuition is that an object is considered conforming if it is close to an object labelled in the same way and far from any object labelled in a different way.

Table 1 compares the performance of the conformal-type probabilistic predictor based on the 1-NN conformity measure (8), where d is tangent distance, with the performance of Platt's algorithm with the optimal values of its parameters. The conformal predictor is parameter-free but Platt's algorithm depends on the choice of the kernel. We chose the polynomial kernel of degree 3 (since it is known to produce the best results: see [9], Sect. 12.2) and the cost parameter $C := 2.9$ in the case of the average log loss and $C := 3.4$ in the case of the standardized Brier loss (the optimal values in our experiments). (Reporting the performance of Platt's algorithm with optimal parameter values may look like data snooping, but it is fine in this context since we are helping our competitor.) Table 2 reports the performance of Platt's algorithm as function of the degree of the polynomial kernel with the cost parameter set at $C := 10$ (the dependence on C is relatively mild, and $C = 10$ gives good performance for all degrees that we consider).

Acknowledgments. We thank the reviewer for useful comments. In our experiments we used the R package e1071 (by David Meyer, Evgenia Dimitriadou, Kurt Hornik, Andreas Weingessel, Friedrich Leisch, Chih-Chung Chang, and Chih-Chen Lin) and the implementation of tangent distance by Daniel Keysers. This work was partially supported by EPSRC (grant EP/K033344/1, first author) and Royal Holloway, University of London (third author).

References

1. Devroye, L.: A Course in Density Estimation. Birkhäuser, New York (1987)
2. Grenander, U.: On the theory of mortality measurement. Part II. Skandinavisk Aktuarietidskrift 39, 125–153 (1956)
3. Hardy, G.H., Littlewood, J.E., Pólya, G.: Inequalities, 2nd edn. Cambridge University Press, Cambridge (1952)
4. Lehmann, E.L.: Testing Statistical Hypotheses, 2nd edn. Springer, New York (1986)
5. Lei, J., Wasserman, L.: Distribution free prediction bands for nonparametric regression. Journal of the Royal Statistical Society B 76, 71–96 (2014)
6. Platt, J.C.: Probabilities for SV machines. In: Smola, A.J., Bartlett, P.L., Schölkopf, B., Schuurmans, D. (eds.) Advances in Large Margin Classifiers, pp. 61–74. MIT Press (2000)
7. Sellke, T., Bayarri, M.J., Berger, J.: Calibration of p-values for testing precise null hypotheses. American Statistician 55, 62–71 (2001)
8. Simard, P., LeCun, Y., Denker, J.: Efficient pattern recognition using a new transformation distance. In: Hanson, S., Cowan, J., Giles, C. (eds.) Advances in Neural Information Processing Systems, vol. 5, pp. 50–58. Morgan Kaufmann, San Mateo (1993)
9. Vapnik, V.N.: Statistical Learning Theory. Wiley, New York (1998)
10. Vovk, V.: A logic of probability, with application to the foundations of statistics (with discussion). Journal of the Royal Statistical Society B 55, 317–351 (1993)
11. Vovk, V.: Conditional validity of inductive conformal predictors. In: Hoi, S.C.H., Buntine, W. (eds.) JMLR: Workshop and Conference Proceedings, vol. 25, pp. 475–490 (2012); Asian Conference on Machine Learning. Full version: Technical report arXiv:1209.2673 [cs.LG], arXiv.org e-Print archive (September 2012), The journal version: Machine Learning (ACML 2012 Special Issue) 92, 349–376 (2013).
12. Vovk, V., Fedorova, V., Gammerman, A., Nouretdinov, I.: Criteria of efficiency for conformal prediction, On-line Compression Modelling project (New Series), Working Paper 11 (April 2014) http://alrw.net
13. Vovk, V., Gammerman, A., Shafer, G.: Algorithmic Learning in a Random World. Springer, New York (2005)
14. Vovk, V., Petej, I.: Venn–Abers predictors. In: Zhang, N.L., Tian, J. (eds.) Proceedings of the Thirtieth Conference on Uncertainty in Artificial Intelligence, pp. 829–838. AUAI Press, Corvallis (2014), http://auai.org/uai2014/proceedings/uai-2014-proceedings.pdf
15. Wu, T.F., Lin, C.J., Weng, R.C.: Probability estimates for multi-class classification by pairwise coupling. Journal of Machine Learning Research 5, 975–1005 (2004)

Aggregated Conformal Prediction

Lars Carlsson[1], Martin Eklund[2], and Ulf Norinder[3]

[1] AstraZeneca Research and Development, SE-431 83 Mölndal, Sweden
lars.a.carlsson@astrazeneca.com

[2] Department of Surgery, University of California San Francisco (UCSF), 1600
Divisadero St, San Francisco CA 94143, USA
martin.eklund@farmbio.uu.se

[3] H. Lundbeck A/S, Ottiliavej 9, 2500 Valby, Denmark
ulfn@lundbeck.com

Abstract. We present the aggregated conformal predictor (ACP), an extension to the traditional inductive conformal prediction (ICP) where several inductive conformal predictors are applied on the same training set and their individual predictions are aggregated to form a single prediction on an example. The results from applying ACP on two pharmaceutical data sets (CDK5 and GNRHR) indicate that the ACP has advantages over traditional ICP. ACP reduces the variance of the prediction region estimates and improves efficiency. Still, it is more conservative in terms of validity than ICP, indicating that there is room for further improvement of efficiency without compromising validity.

1 Introduction

Quantitative Structure-Activity Relationship (QSAR) modeling for predicting properties, e.g. solubility or toxicity, of chemical compounds using statistical learning techniques is a widespread approach within the pharmaceutical industry to prioritize compounds for experimental testing or to alert for potential toxicity. In making informed decisions based on predictions from QSAR models, the confidence in such predictions is of vital importance and conformal predictors have been successfully applied to the drug discovery setting [1]. In particular, we have shown that a Mondrian inductive conformal predictor is efficient (i.e. informative) and almost valid when applied to binary categorical data; exact validity was not achieved due to deviations from the exchangeability assumption [2]. Furthermore, we have demonstrated that an inductive conformal predictor (ICP) applied to regression data from the pharmaceutical industry was valid. However, for regression the prediction regions were in many cases as wide or wider than the possible ranges of the true responses (i.e. the range of the experimental assay that generates the true label). An important problem is thus to improve the efficiency of the ICPs when applied in the QSAR domain.

There are many ways to improve efficiency of conformal predictions, for example by using improved nonconformity scores, choosing different machine-learning methods, or using a transductive approach [3]. The transductive approach is the

L. Iliadis et al. (Eds.): AIAI 2014 Workshops, IFIP AICT 437, pp. 231–240, 2014.

most appealing in terms of validity, but is often computationally costly and the nonconformity scores can be difficult to compute.

Another interesting approach to improve efficiency is the *cross conformal predictor* (CCP) [4]. Here the training data is divided into separate non-overlapping folds and each fold is used as a calibration set and the remainder of the data is used as a proper training set. This division allows for more data to be used for calibration and p-values are averaged over all folds. Similarly, the *bootstrap conformal predictor* (BCP) bootstraps datasets and uses the out-of-bag examples as a calibration set. p-values are then averages across all bootstrap replications.

In this paper we attempt to generalize the BCP and the CCP in what we term the *aggregated inductive conformal predictor* (ACP). We will empirically assess the ACP using data from the pharmaceutical domain and we show through a theoretical argument and experiments that ACP seems to have advantages over the standard ICP.

2   Aggregated Conformal Predictor

Consider the standard prerequisite for a description of a conformal predictor (CP), a bag of examples $\{z_1, \ldots, z_i, \ldots, z_l\}$ drawn from an exchangeable distribution Q. Each example $z_i = (x_i, y_i)$ can be described by its object $x_i \in \mathbf{X}$ and its label $y_i \in \mathbf{Y}$. The labels can be either categorical or continuous. For an inductive conformal predictor (ICP) the bag $\{z_i\}$ is partitioned into two different bags, one holding the proper training examples $\{z_1, \ldots, z_m\}$ and the other holding the calibration examples $\{z_{m+1}, \ldots, z_l\}$. The ICP p-value is then computed as

$$p = \frac{|\{j = m+1, ..., l : \alpha_j \geq \alpha_{l+1}\}|}{l - m + 1}$$

The prediction region of an ICP is determined by the "borderline" p-value, p_t, i.e. the smallest value p can obtain and still satisfy $p > \epsilon$. We can thus view p_t as the ϵth sample quantile (estimated from above)

$$\hat{p}_t = U_{l-m}^{-1}(\epsilon),$$

where U_{l-m} is the empirical cumulative probability distribution of the p-values defined by

$$U_{l-m}(p) = \frac{1}{l - m + 1} \sum_{j=m+1}^{l} I(p_j < p),$$

where $I(\cdot) = 1$ if $p_j < p$ and $I(\cdot) = 0$ otherwise. We now introduce definitions of *Exchangeable resampling* and *Consistent resampling*, after which we define what we mean by an aggregated conformal predictor (ACP).

Definition 1 (Exchangeable resampling). *Let $\{z_1^*, \ldots, z_n^*\}$ be a bag of examples resampled from the empirical distribution Q_l. We call this resampling exchangeable if*

$$P\{(z_1^*, \ldots, z_n^*)\} = P\{(z_{\pi(1)}^*, \ldots, z_{\pi(n)}^*)\},$$

where π is any permutation of $\{1, \ldots, n\}$.

Definition 2 (Consistent resampling). Let $T = T(z_1, \ldots, z_l, Q)$ be a statistic and $T^* = T(z_1^*, \ldots, z_n^*, Q_l)$ be an exchangeably resampled version of T. Further, let G_l and G_l^* be the probability distributions of T and T^*, respectively. We call the sampling process consistent (with respect to T) if

$$\sup_z | G_l - G_l^* | \to 0 \text{ as } l \to \infty \text{ and } n \to \infty.$$

Definition 3 (ACP: Aggregated Conformal Predictor). The following procedure is repeated B times, for $b = 1, \ldots, B$: Resample a bag $\{z_1^*, \ldots, z_{n_b}^*\}$ of examples from $\{z_1, \ldots, z_l\}$ using a consistent resampling procedure with respect to α_t. Compute the ICP p-value using the resampled bag,

$$p_b^* = \frac{| \{j = m_b + 1, \ldots, n_b : \alpha_j^* \geq \alpha_{n_b+1}^*\} |}{n_b - m_b + 1}, \tag{1}$$

where α_j^* are the nonconformity scores computed using $\{z_1^*, \ldots, z_{n_b}^*\}$ (m_b and n_b are indexed with b to make explicit that they may differ for different values of b). We define the ACP p-value as

$$p_B = \frac{1}{B} \sum_{b=1}^{B} p_b^* \tag{2}$$

and the corresponding prediction region as

$$\Gamma^\epsilon(z_1, \ldots, z_l, x_{l+1}) := \{y | p_B > \epsilon\}. \tag{3}$$

A smoothed ACP can be defined analogously.

We note that the cross-conformal predictor and the bootstrapped conformal predictor suggested by Vovk [5] are two examples of ACPs.

Proposition 1. The aggregated conformal predictor is conservatively valid.

Proof. Since we use an exchangeable resampling procedure to construct the ACP and since an ICP is conservatively valid (Proposition 4.1, [3]), each resampled ICP in the ACP is conservatively valid (by symmetry). From this follows that the ACP also is conservatively valid.

Remark 1. Proposition 1 only holds unconditionally. The situation is different conditional on the particular dataset we have observed.

2.1   How Does the ACP Improve on the ICP?

For a p in an ICP, p_t is a hard threshold in the sense that the label y corresponding to p is either inside or outside the prediction region. Heuristically, the ACP averages over thresholds varying around p_t (since p_t^* based on a resampled bag $\{z_1^*, ..., z_l^*\}$ fluctuates around p_t), resulting in a smoothed threshold estimate with decreased variance compared to the estimate p_t.

We can use the method used by Bühlmann and Yu [6] to analyze this in a bit more detail. Consider the function

$$\delta(p) = I(p_t < p),$$

which indicates whether a p is smaller or larger than the borderline value p_t in an ICP (and thus if its corresponding label y either is inside or outside the prediction region). The sample quantile estimate p_t follows a normal distribution with mean $\epsilon = U^{-1}(\epsilon)$ and variance

$$\sigma^2 = \frac{(1-\epsilon)\epsilon}{(l-m+1)\left[f(q)\right]^2} = \frac{1-\epsilon}{(l-m+1)\epsilon},$$

where F is the population cumulative distribution function with density function f [7]. For a p in the neighborhood of ϵ

$$p = p(c) = \epsilon + c\sigma\sqrt{l-m+1} \tag{4}$$

we have the approximation

$$\delta(p(c)) \approx I(W < c), \quad W \sim N(0,1), \tag{5}$$

where W is the limiting random quantity from the asymptotic distribution of p_t (because of the construction of $p(c)$ in Equation (4)). For a fixed c, this is a hard threshold function of W. It follows that

$$\mathrm{E}\left[\delta(p(c))\right] \to P(W < c) = \Phi(c) \text{ as } l \to \infty \text{ and } l - m \to \infty$$
$$\mathrm{Var}\left[\delta(p(c))\right] \to \Phi(c)(1 - \Phi(c)) \text{ as } l \to \infty \text{ and } l - m \to \infty, \tag{6}$$

where $\Phi(\cdot)$ is the cumulative distribution function of the standard normal distribution. Note that the variance does not converge to zero; $\delta(p(c))$ assumes the values 0 and 1 with a positive probability even as l tends to infinity. However, for the ACP the situation looks different,

$$\delta_B(p(c)) = \mathrm{E}^* \left[I(p_t^* \leq p(c))\right]$$
$$= \mathrm{E}^* \left[I\left(\sqrt{l-m+1}(p_t^* - p_t)/\sigma \leq \sqrt{l-m+1}(p(c) - p_t)/\sigma\right)\right]$$
$$= \Phi(\sqrt{l-m+1}(p(c) - p_t)) + o_p(1) \approx \Phi(c - W), \quad W \sim N(0,1).$$

where the first approximation over the second equal sign follows because we by definition of the ACP have a consistent resampling process. Comparing with

Equation (5) for an ICP, the ACP produces a smoothed decision function of Z and therefore reduces variance. Again, following Bühlmann and Yu [6], we can study the case $p = p(0) = \epsilon$, i.e. when we are right at the population threshold and therefore has maximum variance. Then

$$\delta_B(p(0)) \to \Phi(-W) \sim U[0,1]$$

and, therefore,

$$\mathrm{E}\left[\delta_B(p(0))\right] \to \mathrm{E}\left[U\right] = 1/2 \text{ as } l \to \infty$$
$$\mathrm{Var}\left[\delta_B(p(0))\right] \to \mathrm{Var}(U) = 1/12 \text{ as } l \to \infty. \qquad (7)$$

Comparing Equation (6) to Equation (7), we see that the variance is reduced to one third for ACP compared to ICP.

3   Empirical Results of ACP

We used two different machine learning methods; the support vector machine (SVM) [8] implemented in the Java library version of libsvm [9] with a Gaussian radial basis kernel function

$$K(x, x') = \exp\left(-\gamma \|x - x'\|^2\right),$$

with $\gamma = 0.002$ and $C = 50$, and Random Forest (RF) [10], for which the default settings were used and each ensemble contained 100 trees.

Following Equation (16) in [11], we defined the nonconformity measure used in combination with SVM according to

$$\alpha_i = \frac{|y_i - \hat{y}_i|}{\exp\left(\hat{\mu}_i\right)}, \qquad (8)$$

where $\hat{\mu}_i$ is the prediction of the value $\ln\left(|y_i - \hat{y}_i|\right)$ produced by a support vector regression machine trained on the proper training sets. After training the underlying SVM of the ICP, we calculate the residuals $|y_j - \hat{y}_j|$ for all proper training examples $j = 1, ..., m$ and train an SVM on the pairs $(x_i, \ln(|y_i - \hat{y}_i|))$. Measure (8) normalizes the absolute prediction error with the predicted accuracy of the SVM on a given example. The nonconformity measure used with RF was

$$\alpha_i = \frac{|y_i - \hat{y}_i|}{\hat{\nu}_i},$$

where $\hat{\nu}_i$ is the RF prediction of the value $|y_i - \hat{y}_i|$ with the same settings as for the other RF model.

Two public dataset (CDK5 and GNRHR) from the pharmaceutical domain was used [12]. The datasets consist of 230 and 198 examples, respectively. Each example (chemical compound) was described (characterised) by so-called signature descriptors [13] in the same way as described in [2].

The data was randomly split into two parts: A training set (80% of the original data) and a working set (20%). This procedure was repeated 50 times as to generate 50 training and working set, respectively. Furthermore, each training set was then, subsequently, randomly split into a proper training set (70% of the training set) and a calibration set (30%), similar to the 2:1 recommendation in [5]. This random selection of proper training and calibration examples was, in turn, repeated 100 times enabling the construction of 100 inductive conformal predictors for each working set. This sampling procedure is often called the $m - n$ sampling or non-replacement subsampling in the bootstrap literature and consistent with Definition 2 [14].

The results are presented in Tables 1- 8 and in Figures 1 and 2.

Table 1. The fraction of true labels within the prediction ranges for an ACP and a, for each run, randomly selected ICP at different significance levels for the 50 runs on the CDK data using SVM. Here, Q_1 and Q_3 are the lower and upper quartile, respectively.

Method	ϵ	min	Q_1	median	mean	Q_3	max
ACP	0.4	0.6522	0.7391	0.7717	0.7717	0.8043	0.8913
ICP	0.4	0.5100	0.5622	0.5997	0.5990	0.6303	0.6826
ACP	0.3	0.7174	0.8261	0.8478	0.8474	0.8696	0.9348
ICP	0.3	0.6191	0.6535	0.6879	0.6896	0.7202	0.7876
ACP	0.2	0.8043	0.8913	0.9130	0.9170	0.9565	0.9783
ICP	0.2	0.7296	0.7712	0.7923	0.7989	0.8282	0.8848
ACP	0.1	0.8913	0.9565	0.9565	0.9643	0.9783	1.0000
ICP	0.1	0.8378	0.8626	0.8933	0.8890	0.9101	0.9537

Table 2. The fraction of true labels within the prediction ranges for an ACP and a, for each run, randomly selected ICP at different significance levels for the 50 runs on the GNRHR data using SVM. Here, Q_1 and Q_3 are the lower and upper quartile, respectively.

Method	ϵ	min	Q_1	median	mean	Q_3	max
ACP	0.4	0.3590	0.5962	0.6667	0.6505	0.7179	0.8205
ICP	0.4	0.4297	0.5560	0.5960	0.5947	0.6569	0.7321
ACP	0.3	0.5128	0.6731	0.7179	0.7227	0.7628	0.8718
ICP	0.3	0.5215	0.6428	0.6736	0.6750	0.7208	0.7926
ACP	0.2	0.7179	0.7949	0.8205	0.8286	0.8462	0.9744
ICP	0.2	0.6782	0.7466	0.7749	0.7791	0.8126	0.8797
ACP	0.1	0.8205	0.8974	0.9231	0.9258	0.9487	1.0000
ICP	0.1	0.8072	0.8633	0.8745	0.8835	0.9104	0.9605

Table 3. The fraction of true labels within the prediction ranges for an ACP and a, for each run, randomly selected ICP at one significance level for the 50 runs on the CDK data using RF. Here, Q_1 and Q_3 are the lower and upper quartile, respectively.

Method	ϵ	min	Q_1	median	mean	Q_3	max
ACP	0.2	0.7174	0.8478	0.8696	0.8696	0.9130	0.9565
ICP	0.2	0.6926	0.7786	0.8071	0.7982	0.8249	0.8930

Table 4. The fraction of true labels within the prediction ranges for an ACP and a, for each run, randomly selected ICP at one significance level for the 50 runs on the GNRHR data using RF. Here, Q_1 and Q_3 are the lower and upper quartile, respectively.

Method	ϵ	min	Q_1	median	mean	Q_3	max
ACP	0.2	0.7179	0.8205	0.8718	0.8564	0.8974	1.0000
ICP	0.2	0.6854	0.7706	0.7919	0.8013	0.8271	0.9162

Table 5. The efficiency for an ACP and an ICP at one significance level for the 50 runs on the CDK5 data using SVM. Here, Q_1 and Q_3 are the lower and upper quartile, respectively.

Method	ϵ	min	Q_1	median	mean	Q_3	max
ACP	0.2	0.5874	1.2190	1.6880	3.2000	3.3220	67.6700
ICP	0.2	0.1550	0.9128	1.2620	3.5760	1.8970	398.3000

Table 6. The efficiency for an ACP and an ICP at one significance level for the 50 runs on the CDK5 data using RF. Here, Q_1 and Q_3 are the lower and upper quartile, respectively.

Method	ϵ	min	Q_1	median	mean	Q_3	max
ACP	0.2	0.4752	1.0060	1.2640	1.3750	1.5870	3.9780
ICP	0.2	0.0653	0.8659	1.1830	1.3680	1.6380	8.5950

4   Discussion

The results presented in Tables 1- 8 and in Figures 1 and 2 indicate that the ACP methodology has advantages over traditional ICP. The former adds stability and robustness to the predictions, which is particularly clear in Figures 1- 2 where the variance in the prediction ranges is smaller than from an individual ICP. This is of considerable importance within the pharmaceutical domain where precision as well as robustness in predictions are key elements for successful application in ongoing discovery projects. Although the traditional ICP is valid on average,

Table 7. The efficiency for an ACP and an ICP at one significance level for the 50 runs on the GNRHR data using SVM. Here, Q_1 and Q_3 are the lower and upper quartile, respectively.

Method	ϵ	min	Q_1	median	mean	Q_3	max
ACP	0.2	1.2770	1.625	1.733	1.907	1.903	25.830
ICP	0.2	0.5486	1.457	1.704	1.957	2.000	62.400

Table 8. The efficency for an ACP and an ICP at one significance level for the 50 runs on the GNRHR data using RF. Here, Q_1 and Q_3 are the lower and upper quartile, respectively.

Method	ϵ	min	Q_1	median	mean	Q_3	max
ACP	0.2	0.7481	1.6050	1.9620	2.1290	2.5490	6.3040
ICP	0.2	0.07138	1.3940	1.8230	2.0980	2.4840	10.3100

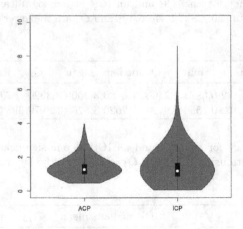

Fig. 1. The distribution of prediction interval size for an ACP and a randomly sampled ICP for all 50 runs on the CDK5 data. The results are shown at $\epsilon = 0.2$.

the variance is very large. This means that there is a relatively large proportion of very tight predictions regions that in fact are far from valid (for example, for the CDK5 data, an ICP with a confidence of 80% produces 46% error in the quartile with tightest prediction regions; the corresponding figure for the ACP is 21%. These tight prediction regions are offset by some prediction regions that are very wide (as wide or wider than the possible range of the response value) that are *always* valid, which produces a conformal predictor that is valid on average. Neither the too tight prediction regions that cannot be trusted nor non-informative regions are helpful for the researcher using the model.

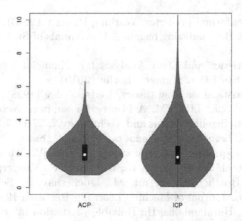

Fig. 2. The distribution of prediction interval size for an ACP and a randomly sampled ICP for all 50 runs on the GNRHR data. The results are shown at $\epsilon = 0.2$.

The results in Tables 1- 8 show that the ACP (as used in this paper) is too conservative on datasets with a relatively small number of examples, which is clear from Equations (1) and (2). This indicates that improved efficiency can be achieved without compromising the validity of the ACP, e.g. by more clever resampling (a large body of literature exists that address this problem, see e.g. [15]) or smoothing of p-values on either side of ϵ.

To conclude: We have introduced the ACP, a generalization of the BCP and CCP introduced by Vovk [5] and we have shown that it improves on classical ICP by reducing the variance in the estimated prediction regions (analogously to how a bagged predictor improves the prediction by reducing variance). ACP seems to represent a pragmatic and useful way forward for obtaining models and predictions with good precision and small prediction ranges.

Future ways to develop the ACP include to (i) study ACP for categorical labels (e.g. aggregation through voting); (ii) other resampling schemas.

References

1. Eklund, M., Norinder, U., Boyer, S., Carlsson, L.: Application of Conformal Prediction in QSAR. In: Iliadis, L., Maglogiannis, I., Papadopoulos, H., Karatzas, K., Sioutas, S. (eds.) AIAI 2012, Part II. IFIP AICT, vol. 382, pp. 166–175. Springer, Heidelberg (2012)
2. Eklund, M., Norinder, U., Boyer, S., Carlsson, L.: The application of conformal prediction to the drug discovery process. Annals of Mathematics and Artificial Intelligence (2013)
3. Vovk, V., Gammerman, A., Shafer, G.: Algorithmic Learning in a Random World, 1st edn. Springer (2005)

4. Vovk, V.: Cross-conformal predictors. Annals of Mathematics and Artificial Intelligence, 1–20 (2013)
5. Vovk, V.: Cross-conformal predictor. Working Paper 6 (2013), http://alrw.net
6. Bühlmann, P., Yu, B.: Analyzing bagging. The Annals of Statistics 30(4), 927–961 (2002)
7. Ruppert, D.: Statistics and Data Analysis for Financial Engineering, 1st edn. Springer Texts in Statistics. Springer, Berlin (2010)
8. Vapnik, V.N.: Statistical learning theory, 1 edn. Wiley (1998)
9. Chang, C.C., Lin, C.J.: LIBSVM: A library for support vector machines. ACM Transactions on Intelligent Systems and Technology 2, 27:1–27:27 (2011), Software available at http://www.csie.ntu.edu.tw/~cjlin/libsvm
10. Pedregosa, F., Varoquaux, G., Gramfort, A., Michel, V., Thirion, B., Grisel, O., Blondel, M., Prettenhofer, P., Weiss, R., Dubourg, V., Vanderplas, J., Passos, A., Cournapeau, D., Brucher, M., Perrot, M., Duchesnay, E.: Scikit-learn: Machine learning in Python. Journal of Machine Learning Research 12, 2825–2830 (2011)
11. Papadopoulos, H., Haralambous, H.: Reliable prediction intervals with regression neural networks. Neural Networks 24(8), 842–851 (2011)
12. Chen, H., Carlsson, L., Eriksson, M., Varkonyi, P., Norinder, U., Nilsson, I.: Beyond the scope of free-wilson analysis: Building interpretable qsar models with machine learning algorithms. Journal of Chemical Information and Modeling 53, 1324–1336 (2013)
13. Faulon, J.-L., Collins, M.J., Carr, R.D.: The signature molecular descriptor. 4. Canonizing molecules using extended valence sequences. J. Chem. Inf. Comput. Sci. 44(2), 427–436 (2004)
14. Politis, D.N., Romano, J.P., Wolf, M.: Subsampling. Springer, New York (1999)
15. Egloff, D., Leippold, M.: Quantile estimation with adaptive importance sampling. The Annals of Statistics 38(2), 1244–1278 (2010)

A Cross-Conformal Predictor
for Multi-label Classification

Harris Papadopoulos

Computer Science and Engineering Department, Frederick University,
7 Y. Frederickou St., Palouriotisa, Nicosia 1036, Cyprus
h.papadopoulos@frederick.ac.cy

Abstract. Unlike the typical classification setting where each instance
is associated with a single class, in multi-label learning each instance
is associated with multiple classes simultaneously. Therefore the learning task in this setting is to predict the subset of classes to which each
instance belongs. This work examines the application of a recently developed framework called Conformal Prediction (CP) to the multi-label
learning setting. CP complements the predictions of machine learning algorithms with reliable measures of confidence. As a result the proposed
approach instead of just predicting the most likely subset of classes for a
new unseen instance, also indicates the likelihood of each predicted subset being correct. This additional information is especially valuable in
the multi-label setting where the overall uncertainty is extremely high.

1 Introduction

Most machine learning research on classification deals with problems in which
each instance is associated with a single class y from a set of classes $\{Y_1, \ldots, Y_c\}$.
As opposed to this standard setting, in *multi-label classification* each instance can
belong to multiple classes, so that each instance is associated with a set of classes
$\psi \subseteq \{Y_1, \ldots, Y_c\}$, called a *labelset*. There are many real-world problems in which
such a setting is natural. For instance in the categorization of news articles an
article discussing the positions of political parties on the educational system of a
country can be classified as both politics and education. Although until recently
the main multi-label classification application was the categorization of textual
data, in the last few years an increasing number of new applications started to
attract the attention of more researchers to this setting. Such applications include
the semantic annotation of images and videos, the categorization of music into
emotions, functional genomics, proteomics and directed marketing.

As a result of the increasing attention to the multi-label classification setting,
many new machine learning techniques have been recently developed to deal
with problems of this type, see e.g. [2, 8–11]. However, like most machine learning methods, these techniques do not produce any indication about the likelihood
of each of their predicted labelsets being correct. Such an indication though can
be very helpful in deciding how much to rely on each prediction, especially since

L. Iliadis et al. (Eds.): AIAI 2014 Workshops, IFIP AICT 437, pp. 241–250, 2014.
© IFIP International Federation for Information Processing 2014

the certainty of predictions may vary to a big degree between instances. To address this problem this paper examines the application of a recently developed framework for providing reliable confidence measures to predictions, called *Conformal Prediction* (CP) [7], to the multi-label classification setting. Specifically it follows the newly proposed *Cross-Conformal Prediction* (CCP) [6] version of the framework, which allows it to overcome the prohibitively large computational overhead of the original CP. The proposed approach computes a p-value for each of the possible labelsets, which can be used either for accompanying each prediction with confidence measures that indicate its likelihood of being correct, or for producing sets of labelsets that are guaranteed to contain the true labelset with a frequency equal to or higher than a required level of confidence.

In the remaining paper, a description of the general idea behind CP and of the CCP version of the framework, is first provided in Section 2. Section 3 defines the proposed approach while Section 4 presents the experiments performed and the obtained results. Finally Section 5 gives the conclusions of this work.

2   Conformal and Cross-Conformal Prediction

Typically in classification we are given a set of training examples $\{z_1, \ldots, z_l\}$, where each $z_i \in \mathcal{Z}$ is a pair (x_i, y_i) consisting of a vector of attributes $x_i \in \mathbb{R}^d$ and the classification $y_i \in \{Y_1, \ldots, Y_c\}$. We are also given a new unclassified example x_{l+1} and our task is to state something about our confidence in each possible classification of x_{l+1} without assuming anything more than that all (x_i, y_i), $i = 1, 2, \ldots$, are independent and identically distributed.

The idea behind CP is to assume every possible classification Y_j of the example x_{l+1} and check how likely it is that the extended set of examples

$$\{(x_1, y_1), \ldots, (x_l, y_l), (x_{l+1}, Y_j)\} \tag{1}$$

is i.i.d. This in effect will correspond to the likelihood of Y_j being the true label of the example x_{l+1} since this is the only unknown value in (1).

First a function A called *nonconformity measure* is used to map each pair (x_i, y_i) in (1) to a numerical score

$$\alpha_i = A(\{(x_1, y_1), \ldots, (x_l, y_l), (x_{l+1}, Y_j)\}, (x_i, y_i)), \quad i = 1, \ldots, l, \tag{2a}$$

$$\alpha_{l+1}^{Y_j} = A(\{(x_1, y_1), \ldots, (x_l, y_l), (x_{l+1}, Y_j)\}, (x_{l+1}, Y_j)), \tag{2b}$$

called the *nonconformity score* of instance i. This score indicates how nonconforming, or strange, it is for i to belong in (1). In effect the nonconformity measure is based on a conventional machine learning algorithm, called the *underlying algorithm* of the corresponding CP and measures the degree of disagreement between the actual label y_i and the prediction $\hat{y}_i$ of the underlying algorithm, after being trained on (1). The nonconformity measure for multi-label learning used in this work is defined in Section 3.

The nonconformity score $\alpha_{l+1}^{Y_j}$ is then compared to the nonconformity scores of all other examples to find out how unusual (x_{l+1}, Y_j) is according to the nonconformity measure used. This comparison is performed with the function

$$p((x_1, y_1), \ldots, (x_l, y_l), (x_{l+1}, Y_j)) = \frac{|\{i = 1, \ldots, l, l+1 : \alpha_i^{Y_j} \geq \alpha_{l+1}^{Y_j}\}|}{l+1}, \quad (3)$$

the output of which is called the p-value of Y_j, also denoted as $p(Y_j)$. An important property of (3) is that $\forall \delta \in [0, 1]$ and for all probability distributions P on $\mathcal{Z}$,

$$P^{l+1}\{((x_1, y_1), \ldots, (x_l, y_l), (x_{l+1}, y_{l+1})) : p(y_{l+1}) \leq \delta\} \leq \delta; \quad (4)$$

a proof can be found in [7]. According to this property, if $p(Y_j)$ is under some very low threshold, say 0.05, this means that Y_j is highly unlikely as the probability of such an event is at most 5% if (1) is i.i.d.

There are two standard ways to use the p-values of all possible classifications for producing the output of a CP:

- Predict the classification with the highest p-value and output one minus the second highest p-value as confidence to this prediction and the p-value of the predicted classification (i.e. the highest p-value) as credibility.
- Given a confidence level $1 - \delta$, output the prediction set $\{Y_j : p(Y_j) > \delta\}$.

In the first case, confidence is an indication of how likely the prediction is of being correct compared to all other possible classifications, whereas credibility indicates how suitable the training set is for the particular instance; specifically a very low credibility value indicates that the particular instance does not seem to belong to any of the possible classifications. In the second case, the prediction set will not contain the true label of the instance with at most δ probability.

The important drawback of the above process is that since the last example in (1) changes for every possible classification, the underlying algorithm needs to be trained c times. Moreover the whole process needs to be repeated for every test example. This makes it extremely computationally inefficient in many cases and especially in a multi-label setting where there are 2^n possible labelsets for n classes, or $2^n - 1$ if we exclude the empty labelset.

To overcome this computational inefficiency problem an inductive version of the framework was proposed in [3] and [4] called *Inductive Conformal Prediction* (ICP). ICP is based on the same theoretical foundations described above, but follows a modified version of the approach, which allows it to train the underlying algorithm only once. This is achieved by dividing the training set into two smaller sets, the *proper training set* and the *calibration set*. The proper training set is then used for training the underlying algorithm of the ICP and only the examples in the calibration set are used for calculating the p-value of each possible classification for every test example.

Although ICP is much more computationally efficient than the original CP approach, the fact that it does not use the whole training set for training the underlying algorithm and for calculating its p-values results in lower informational efficiency. That is the resulting prediction sets might be larger than the ones produced by the original CP approach. Cross-Conformal Prediction, which was recently proposed in [6], tries to overcome this problem by combining ICP with cross-validation. Specifically, CCP partitions the training set in K subsets (folds) $S_1, \ldots, S_K$ and calculates the nonconformity scores of the examples in each subset S_k and of (x_{l+1}, Y_j) for each possible classification Y_j as

$$\alpha_i = A(\cup_{m \neq k} S_m, (x_i, y_i)), \quad i \in S_k, \quad m = 1, \ldots, K, \tag{5a}$$

$$\alpha_{l+1}^{Y_j, k} = A(\cup_{m \neq k} S_m, (x_{l+1}, Y_j)), \quad m = 1, \ldots, K, \tag{5b}$$

where A is the given nonconformity measure. Note that for (x_{l+1}, Y_j) K nonconformity scores $\alpha_{l+1}^{Y_j, k}$, $k = 1, \ldots, K$ are calculated, one with each of the K folds. Now the p-value for each possible classification Y_j is computed as

$$p(Y_j) = \frac{\sum_{k=1}^{K} |\{(x_i, y_i) \in S_k : \alpha_i \geq \alpha_{l+1}^{Y_j, k}\}| + 1}{l + 1}. \tag{6}$$

The CCP version of the framework was chosen to be followed here due to its big advantage in computational efficiency over CP, since it needs to train the underlying algorithm only K times, and its advantage over ICP since it utilizes the whole training set for producing its p-values. As opposed to CP and ICP, the validity of which has been proven theoretically, at the moment there are no theoretical results about the validity of CCP. However in [6] its outputs have been shown to be empirically valid. The same is shown in the experimental results of this work, presented in Section 4.

3   ML-RBF Cross-Conformal Predictor

This section describes the proposed approach, which in effect comes down to the definition of a suitable nonconformity measure for multi-label classification and its use with the CCP framework. In order to use the CCP framework in the multi-label setting, the set of possible classifications $\{Y_1, \ldots, Y_c\}$ is replaced by the powerset $\mathcal{P}(\{Y_1, \ldots, Y_n\})$, where n is the number of the original classes of the problem. In the experiments that follow RBF Neural Networks for multi-label learning (ML-RBF) [8] is used as underlying algorithm as it seems to be one of the best performing algorithms designed specifically for multi-label problems. However the proposed approach is general and can be used with any other method which gives scores for each class.

After being trained on a given training set, for every test example ML-RBF produces a score for each possible class of the task at hand. It then outputs as its prediction the labelset containing all classes with score higher than zero; higher score indicates higher chance of the class to be in the labelset. In order to use the scores produced by ML-RBF for computing the nonconformity scores of the proposed Cross-Conformal Predictor, the former were transformed to the range $[0, 1]$ with the logistic sigmoid function

$$f(x) = \frac{1}{1 + e^{-x}}.$$

The nonconformity measure (5) for the multi-label CCP can now be defined based on the transformed outputs of ML-RBF for (x_i, ψ_i) after being trained on $\cup_{m \neq k} S_m$, $m = 1, \ldots, K$ as

$$\alpha_i = \sum_{j=1}^{n} |t_i^j - o_i^j|^d, \tag{7}$$

where n is the number of possible classes, t_i^j is 1 if $Y_j \in \psi_i$ and 0 otherwise, and o_i^j is the transformed output of the ML-RBF corresponding to class Y_j. Finally d is a parameter of the algorithm which controls the sensitivity of the nonconformity measure to small differences in o_i^j when $|t_i^j - o_i^j|$ is small as opposed to when it is large. This nonconformity measure takes into account the distance of all outputs from the true values. Note that in the case of test examples t_i^j is the corresponding value for each assumed labelset.

The nonconformity measure (7) takes into account only the outputs of the ML-RBF for each instance. However, it is very strange to have a pair of labels in a labelset that have never appeared together in the training set. So (7) was extended to take into account the occurrence of each pair of labels in the training set $\cup_{m \neq k} S_m$, $m = 1, \ldots, K$. This extended nonconformity measure is defined as:

$$\alpha_i = \sum_{j=1}^{n} |t_i^j - o_i^j|^d + \lambda \sum_{1 \leq j < r \leq n} t_i^j t_i^r \mu_{j,r}, \tag{8}$$

where $\mu_{j,r}$ is 0 if the labels Y_j and Y_r have been observed together in the labelset of at least one instance of the training set and 1 otherwise. In effect the additional part of this nonconformity measure adds λ to the nonconformity score of an example for each pair of labels in the labelset of the example ($t_i^j t_i^r = 1$) which have never been observed together in any instance of the training set ($\mu_{j,r} = 1$). The parameter λ adjusts the sensitivity to this part of the measure. A high value of λ makes this part of the measure dominate when a pair of labels that has never been observed before exists in the labelset of the example.

The complete proposed approach is derived by plugging in (8) as A in (5) and computing the p-value of each possible labelset with (6). The resulting p-values can be used in either of the two ways described in Section 2.

4   Experiments and Results

4.1   Data Sets

To evaluate the performance of the proposed approach two data sets from different application domains were used, one from the semantic scene analysis domain and one from the bioinformatics domain. The first data set, *scene* [1], is concerned with the semantic classification of pictures into one or more of the classes: beach, sunset, foliage, field, mountain and urban. It consists of 1211 training and 1196 test examples, each described by 294 features. The second data set, *yeast*

[2], is concerned with predicting the functional classes of genes in the Yeast Saccharomyces cerevisiae. Each gene is described by the concatenation of microarray expression data and a phylogenetic profile, and is associated with a set of 14 functional classes. The data set contains 1500 genes as training set and 917 genes as test set, each described by 103 features. Both data sets were obtained from the website of the Mulan library [5].

4.2   Single Prediction Evaluation

The first set of experiments evaluates the quality of the single predictions produced by the ML-RBF CCP and compares it to that of its underlying method and those of three other popular multi-label techniques, namely BP-MLL [10], ML-kNN [11] and ML-NB [9]. Four evaluation measures for multi-label classification were used. The first is Hamming Loss (HL), which is the most popular measure for multi-label problems, defined as:

$$HL = \frac{1}{g} \sum_{i=l+1}^{l+g} \frac{|\psi_i \, \triangle \, \hat{\psi}_i|}{n},$$ (9)

where $\{(x_{l+1}, \psi_{l+1}), \ldots, (x_{l+g}, \psi_{l+g})\}$ are the test examples, $\hat{\psi}_i$ is the predicted labelset for example i and $\triangle$ is the symmetric difference between two sets. The second measure is Classification Accuracy (CA) defined as:

$$CA = \frac{1}{g} \sum_{i=l+1}^{l+g} I(\psi_i = \hat{\psi}_i),$$ (10)

where $I(true) = 1$ and $I(false) = 0$. This measure is rather strict as it requires the predicted and true labelsets to be identical. The third and fourth measures are the *macro averaged* and *micro averaged* F-measure, which is the harmonic mean of precision and recall. The F-measure for a single label is defined as:

$$F(tp, tn, fp, fn) = \frac{2tp}{2tp + fp + fn},$$ (11)

where tp is the number of true positives, tn the number of true negatives, fp the number of false positives and fn the number of false negatives. In the multi-label case, if tp_j, tn_j, fp_j and fn_j are the same values for each label Y_j, then its macro averaged version is defined as:

$$F_{macro} = \frac{1}{n} \sum_{j=1}^{n} F(tp_j, tn_j, fp_j, fn_j).$$ (12)

The micro averaged version of the F-measure is defined as:

$$F_{micro} = F \left(\sum_{j=1}^{n} tp_j, \sum_{j=1}^{n} tn_j, \sum_{j=1}^{n} fp_j, \sum_{j=1}^{n} fn_j \right).$$ (13)

Table 1. Performance of the proposed approach and comparison to that of other multi-label algorithms on the scene data set

Algorithm	Evaluation Measure			
	HL	CA	F_{macro}	F_{micro}
ML-RBF CCP with $\lambda = 0$	0.0928	0.6798	**0.7417**	**0.7363**
ML-RBF CCP with $\lambda = 1$	**0.0927**	**0.6831**	0.7410	0.7358
Orginial ML-RBF	0.0959	0.5468	0.6922	0.6890
BP-MLL	0.2903	0.1630	0.0509	0.1665
ML-kNN	0.0953	0.6012	0.7189	0.7183
ML-NB	0.1309	0.4105	0.6230	0.6221

Table 2. Performance of the proposed approach and comparison to that of other multi-label algorithms on the yeast data set

Algorithm	Evaluation Measure			
	HL	CA	F_{macro}	F_{micro}
ML-RBF CCP with $\lambda = 0$	**0.1954**	0.1821	**0.3896**	**0.6432**
ML-RBF CCP with $\lambda = 1$	**0.1954**	0.1821	**0.3896**	**0.6432**
Orginial ML-RBF	0.1970	**0.1865**	0.3891	0.6407
BP-MLL	0.2272	0.0960	0.3047	0.6212
ML-kNN	0.1980	0.1658	0.3567	0.6360
ML-NB	0.2115	0.1254	0.3428	0.6152

Tables 1 and 2 present the performance of the proposed approach together with that of its underlying algorithm and that of BP-MLL, ML-kNN and ML-NB on the scene and yeast data sets respectively. The best values for each measure are highlighted in bold. For ML-RBF the fraction parameter was set to 0.01 and the scaling factor to 1 as in [8]. In the case of the CCP the number of folds for each data set was chosen so that each fold contained approximately 100 training instances, therefore 12 folds were used for the scene data set and 15 folds for the yeast dataset. The parameter d of the nonconformity measure (8) was set to 4, which seems to be a good choice based on the performed experiments, while for λ the two extreme values of 0 and 1 were used. Setting λ to 0 in effect corresponds to nonconformity measure (7) and setting it to 1 makes the nonconformity score of any labelset containing a pair of classes that has never been observed in the training set always higher than others. For BP-MLL the number of hidden neurons was set to 20% of the input dimensionality, the learning rate to 0.05 and the training epochs to 100 as in [10]. For ML-kNN the number of nearest neighbours was set to 10 and the smoothing parameter to 1 as in [11]. For ML-NB the percentage of remaining features after PCA was set to 0.3 as in [9].

The results in these tables show that not only the proposed approach provides important additional information about the likelihood of each of its predictions being correct, but it also outperforms its underlying algorithm and the three other popular multi-label techniques. The only exception is the classification

Table 3. Prediction set sizes and error rates at the 95%, 90% and 80% confidence levels for the scene data set

# of labelsets	With $\lambda = 0$ Confidence Level			With $\lambda = 1$ Confidence Level		
	95%	90%	80%	95%	90%	80%
1	0.00%	2.34%	54.93%	6.77%	18.23%	62.63%
2	0.08%	25.50%	34.20%	9.62%	32.69%	29.43%
3 to 2^2	25.50%	65.64%	10.87%	43.31%	45.07%	7.94%
$(2^2 + 1)$ to 2^3	72.74%	6.52%	0.00%	40.05%	4.01%	0.00%
$(2^3 + 1)$ to 2^4	1.67%	0.00%	0.00%	0.25%	0.00%	0.00%
Errors	3.76%	9.28%	21.07%	3.60%	9.28%	20.99%

accuracy of the ML-RBF in the case of the yeast data set, but even in this case the accuracy of the proposed approach with both values of λ is very close. Comparing the performance of the ML-RBF CCP with the two different λ values one can see that in the case of the scene data there is only a negligible difference while in the case of the yeast data the performance remains the same for all evaluation measures. Therefore the value of λ does not have any important effect on the performance of the single predictions produced by the ML-RBF CCP. It does however affect the quality of the resulting p-values as will be shown in the next subsection.

4.3   Prediction Region Evaluation

The main advantage of the proposed approach over other multi-label techniques is the production of a p-value for each possible labelset of a new unseen instance, which can be translated either to confidence and credibility measures for its prediction or to prediction sets that are guaranteed to contain the true labelset at a required confidence level. This subsection examines the informativeness and reliability of the resulting prediction sets and consequently of the computed p-values and confidence measures. More specifically given a required level of confidence $1 - \delta$, the ML-RBF CCP produces a set of labelsets that has at most δ chance of not containing the true labelset of the unseen instance. The informativeness of this set of labelsets can be assessed in terms of its size, while its reliability can be assessed by the percentage of cases for which it does not contain the true labelset, this percentage should be less than or very near δ.

Tables 3 and 4 present the results of the proposed approach in this setting for the scene and yeast data sets respectively with the nonconformity measure parameter λ set to 0 and 1. The same parameters reported in Subsection 4.2 were used. The two tables report the sizes of the prediction sets produced for the 95%, 90% and 80% confidence levels together with the observed error percentages, i.e. the percentages of prediction sets that did not contain the true labelset.

Table 3 reports the results for the scene data set in terms of the percentage of prediction sets containing only 1, 2, 3 to 4, 5 to 8 and 9 to 16 labelsets for each

Table 4. Prediction set sizes and error rates at the 95%, 90% and 80% confidence levels for the yeast data set

# of labelsets	With $\lambda = 0$ Confidence Level			With $\lambda = 1$ Confidence Level		
	95%	90%	80%	95%	90%	80%
$(2^6 + 1)$ to 2^7	0.00%	0.00%	1.42%	0.00%	0.00%	1.42%
$2^7 < l \leq 2^8$	0.00%	0.11%	3.71%	0.00%	0.22%	4.58%
$2^8 < l \leq 2^9$	0.55%	3.82%	14.07%	0.76%	4.36%	18.65%
$2^9 < l \leq 2^{10}$	3.05%	7.09%	60.96%	3.93%	10.14%	60.32%
$2^{10} < l \leq 2^{11}$	8.94%	34.46%	19.85%	13.85%	45.58%	15.05%
$2^{11} < l \leq 2^{12}$	38.71%	52.89%	0.00%	57.58%	39.48%	0.00%
$2^{12} < l \leq 2^{13}$	48.64%	1.64%	0.00%	23.88%	0.22%	0.00%
$2^{13} < l \leq 2^{14}$	0.11%	0.00%	0.00%	0.00%	0.00%	0.00%
Errors	4.69%	9.38%	19.85%	4.80%	9.60%	19.96%

confidence level; there was no prediction set containing more than 16 labelsets out of the possible 63. The last row of the table reports the percentage of errors observed for each confidence level. Comparing the results obtained with the nonconformity measure parameter λ set to 0 with those obtained with $\lambda = 1$, one can see that the latter produces much more informative prediction sets. Taking into account the classification accuracy of this data set (68.31%) and the large number of possible labelsets, the resulting prediction sets are quite tight. One can be 95% confident in about 60% of the test instances by considering less than 4 out of the possible 63 labelsets. By reducing the required confidence to 90% one can be certain in a single labelset for about 18% of the test instances and between one or two labelsets for about half the test instances. Finally with a confidence level of 80% a single labelset is given for more than 60% of the test instances. In terms of empirical reliability, only the percentages of errors for the 80% confidence level are slightly higher than the required significance level, which can be attributed to statistical fluctuations.

Table 4 presents the same results for the yeast data set. In this case the prediction sets contained a much higher number of labelsets so the table reports the percentage of prediction sets containing between $2^i + 1$ and 2^{i+1} labelsets with $i = 6, \ldots, 13$ for each confidence level; there were no prediction sets containing less than 2^6 labelsets. The rather big size of the resulting prediction sets is not strange baring in mind the very low classification accuracy of this data set, which is only 18.65%. Comparing the results obtained with the nonconformity measure parameter λ set to 0 with those obtained with $\lambda = 1$, again shows the superiority of nonconformity measure (8), as it produces smaller prediction sets. Considering the high difficulty of the particular task one can say that the resulting prediction sets are quite informative. The number of labelsets needed to satisfy the 80% confidence level is 1/16th or less ($\leq 2^{10}$) of all the possible labelsets for 85% of the test instances. Finally in terms of empirical reliability, the percentage of errors observed is in all cases below the required significance level.

5   Conclusions

This work examined the application of the conformal prediction framework to the multi-label setting. Unlike the other techniques developed for multi-label problems, the proposed approach accompanies each of its predictions with reliable measures of confidence. Experimental results on two popular multi-label data sets have shown that not only the proposed approach provides important additional information for each prediction, but it also outperforms other popular multi-label techniques. Furthermore its confidence measures have been shown to be informative and reliable. The provision of confidence measures can be very helpful in practical applications, considering the high uncertainty that exists in this setting.

Future work includes the development of additional nonconformity measures and the experimentation with more multi-label data. In addition generating separate p-values for each class and combining them for obtaining the p-value of each labelset could also be examined as an alternative. Finally the possibility of generating a ranking of the possible classes for each instance would also be a good addition to multi-label CP.

References

1. Boutell, M., Luo, J., Shen, X., Brown, C.: Learning multi-label scene classification. Pattern Recognition 37, 1757–1771 (2004)
2. Elisseeff, A., Weston, J.: A kernel method for multi-labelled classification. In: Advances in Neural Information Processing Systems 14, pp. 681–687. MIT Press (2002)
3. Papadopoulos, H., Proedrou, K., Vovk, V., Gammerman, A.: Inductive confidence machines for regression. In: Elomaa, T., Mannila, H., Toivonen, H. (eds.) ECML 2002. LNCS (LNAI), vol. 2430, pp. 345–356. Springer, Heidelberg (2002)
4. Papadopoulos, H., Vovk, V., Gammerman, A.: Qualified predictions for large data sets in the case of pattern recognition. In: Proceedings of the 2002 International Conference on Machine Learning and Applications (ICMLA 2002), pp. 159–163. CSREA Press (2002)
5. Tsoumakas, G., Katakis, I., Vlahavas, I.: Mining multi-label data. In: Data Mining and Knowledge Discovery Handbook, pp. 667–685 (2010)
6. Vovk, V.: Cross-conformal predictors. Annals of Mathematics and Artificial Intelligence (2013), http://dx.doi.org/10.1007/s10472-013-9368-4
7. Vovk, V., Gammerman, A., Shafer, G.: Algorithmic Learning in a Random World. Springer, New York (2005)
8. Zhang, M.L.: ML-RBF: RBF neural networks for multi-label learning. Neural Processing Letters 29(2), 61–74 (2009)
9. Zhang, M.L., Peña, J.M., Robles, V.: Feature selection for multi-label naive bayes classification. Information Sciences 179(19), 3218–3229 (2009)
10. Zhang, M.L., Zhou, Z.H.: Multi-label neural networks with applications to functional genomics and text categorization. IEEE Transactions on Knowledge and Data Engineering 18(10), 1338–1351 (2006)
11. Zhang, M.L., Zhou, Z.H.: ML-kNN: A lazy learning approach to multi-label learning. Pattern Recognition 40(7), 2038–2048 (2007)

SVM Venn Machine with k-Means Clustering

Chenzhe Zhou, Ilia Nouretdinov, Zhiyuan Luo, and Alex Gammerman

Computer Learning Research Centre,
Royal Holloway, University of London
Egham, Surrey, TW20 0EX, UK

Abstract. In this paper, we introduce a new method of designing Venn Machine taxonomy based on Support Vector Machines and k-means clustering for both binary and multi-class problems. We compare this algorithm to some other multi-probabilistic predictors including SVM Venn Machine with homogeneous intervals and a recently developed algorithm called Venn-ABERS predictor. These algorithms were tested on a range of real-world data sets. Experimental results are presented and discussed.

Keywords: Venn Machine, Support Vector Machine, k-means clustering.

1 Introduction

Classification is one of the major tasks in machine learning. It gives predictions for the new objects based on known properties learned from the training data set. However, most algorithms could only give single prediction (i.e. label). Demand of probabilistic prediction has arisen in view of the fact that sometimes we appreciate probabilities more than single predictions. A simple example is the probabilistic weather forecasting.

But in some area, single probabilistic prediction has not yet been enough. The term *multi-probabilities* is then brought to mind, namely, we announce several probability distributions for the new label rather than a solitary one. Venn predictor (or Venn Machine) is one of the multi-probabilistic classification systems [8]. There are many Venn predictors, each taxonomy used in the algorithm defines a Venn predictor even if the underlying algorithms are the same.

In our previous paper [10], we introduced a Venn predictor with Support Vector Machines (SVM) as its underlying algorithm, which converts numerical predictions of SVM into a taxonomy. That approach was applicable to any method that initially supplied predictions with prediction scores such as the distance to the hyperplane in SVM. Nonetheless, the process is very simple: all available scores are firstly sorted and then divided into several groups by equal-length intervals according to which interval the score lies. Each of these groups is a category. However, that approach could only be applied in binary cases. In this paper, we propose a method to generalize binary Venn Machine with SVM to a method capable for multi-class cases. Then we consider two alternative methods that may be more accurate: SVM Venn Machine with k-means clustering and Venn-ABERS predictor. These two algorithms are also applicable to any machine learning algorithms with prediction scores.

L. Iliadis et al. (Eds.): AIAI 2014 Workshops, IFIP AICT 437, pp. 251–260, 2014.
© IFIP International Federation for Information Processing 2014

2   Methodology

In this section, two kinds of Venn predictors that use SVM as their underlying algorithm will be introduced together with our alternative methods. They are SVM Venn Machine with homogeneous intervals (VM-SVM-HI) generalized from the binary-only version of [10] together with our alternative method SVM Venn Machine with k-means clustering (VM-SVM-KM) and the Venn-ABERS predictor based on SVM (VA-SVM) proposed by Vladimir Vovk [7]. The former two algorithms could be implemented in both multi-class cases and binary cases, while VA-SVM could only deal with binary data sets.

2.1   Venn Machine

Venn Machine is a multi-probabilistic predictor described in [8]. The basic idea of Venn Machine is to divide every example into its corresponding category based on certain rules and then the frequencies of labels in the chosen category are used as probabilities for the new object's label. Taxonomy is the way how the examples are divided into categories. The underlying algorithm is the algorithm used in the taxonomy.

Assuming a standard machine learning classification problem: given a training set of examples $z_1, z_2, \ldots, z_{n-1}$. Each z_i consists of a pair of object x_i and label y_i. The possible labels y_i ($y_i \in \mathbf{Y}$) are finite. And we are also given a test object x_n. Our task is to predict the label y_n for the new object x_n and give the estimation of the likelihood that our prediction is correct.

Supposing we have a taxonomy A_n, consider a label $y \in \mathbf{Y}$ for the new object x_n. A_n assigns a category τ_i to an example z_i

$$\tau_i = A_n(\langle z_1, \ldots, z_{i-1}, z_{i+1}, \ldots, z_n \rangle, z_i) \tag{1}$$

where n is the number of objects in the bag, $\tau_i \in \mathbf{T}$ is one of the finite categories and z_i is the pair (x_i, y_i), z_n is the pair (x_n, y).

Moreover, we assign z_i and z_j to the same category if and only if

$$A_n(\langle z_1, \ldots, z_{i-1}, z_{i+1}, \ldots, z_n \rangle, z_i) = A_n(\langle z_1, \ldots, z_{j-1}, z_{j+1}, \ldots, z_n \rangle, z_j) \tag{2}$$

The category τ_n contains $z_n = (x_n, y)$. Let p_y be the empirical probability distribution of the labels in category τ_n.

$$p_y(y') := \frac{|\{(x^*, y^*) \in \tau_n : y^* = y'\}|}{|\tau_n|} \tag{3}$$

p_y is a probability distribution on $\mathbf{Y}$.

Having tried every possible label for x_n, we get a Venn predictor. The predictor $P_n := \{p_y : y \in \mathbf{Y}\}$ is a multi-probabilistic predictor consists of K distributions, where $K = |\mathbf{Y}|$. Then we could calculate a $K \times K$ frequency matrix P. The *quality* of a column is the minimum entry of the column. Let the *best* column which has the highest *quality* be j_{best}. Then our predicted label is j_{best} and the interval of possibility that our prediction is correct is

$$[\min_{i=1,\ldots,K} P_{i,j_{best}} \, , \, \max_{i=1,\ldots,K} P_{i,j_{best}}] \tag{4}$$

Underlying Algorithm for Taxonomy. Any algorithm that generates or predicts a numeric score for the example could be implemented in our taxonomy. However, we mainly focus on Venn predictors with SVM as the underlying algorithm in this paper. The decision function in SVM is a kind of scoring functions. Therefore, we use the values derived from the decision function of SVM (i.e. the values prior to applying a sign function) as part of our design.

Homogenous Intervals. One of the simplest ways to design taxonomies is stated as follows. Firstly we use the training set to train an SVM and calculate the decision values $(d(x) = \langle w, x_i \rangle + d)$ of all examples in the training set and the new object. Secondly the whole range of decision values obtained will be divided into several intervals of equal length. Each interval is a category and objects of which the decision values fall into the same interval are of the same category. This design was introduced in [10] and could only used in binary case. Now we will discuss the generalization and alternative to it.

Combined Decision Function. In multi-class cases, we will have several binary SVM classifiers regardless of whether One-vs-One or One-vs-All approach is used. A scheme for multi-class SVM using One-vs-All approach was developed by Lambrou et al. in [5], which uses the largest decision value as the score. Generally, One-vs-One SVM is more efficient in accuracy than One-vs-All SVM. Therefore, we need to develop a new function to combine the outputs of all One-vs-One SVM classifiers and transform them into a single prediction score which could be used by Venn Machine. We call such function a *Combined Decision Function*.

For a data set with k possible labels: $\{0, 1, \ldots, k-1\}$, there are $k(k-1)/2$ binary SVM if we use One-vs-One approach. For each possible label, there are $k-1$ related SVM decision functions. Then we use (5) to calculate the combined decision function $D(x)$ for the new example x,

$$D(x) = \hat{y} + \frac{1}{k-1} \sum_{i=0, i \neq \hat{y}}^{k-1} N(f_{\hat{y}i}(x)) \tag{5}$$

where $\hat{y}$ is the overall predicted label done by max-wins voting strategy in One-vs-One SVM, $f_{\hat{y}i}(x)$ is the decision function of SVM classifier on $\hat{y}$-vs-i, N is a function that does the normalized transformation to $[0, 1]$. Another point we need to declare here is that in $f_{\hat{y}i}(x)$ we always put $\hat{y}$ before i which means we need to apply an opposite operation when $\hat{y}$ is greater than i. Since the examples of label $\hat{y}$ are treated as negative examples in i-vs-$\hat{y}$ classifier of a binary SVM.

This function firstly selects all $k-1$ related SVM and applies an opposite operation if $\hat{y}$ is not treated as the positive class in the binary SVM classifier. Then it does the normalisation to transform the values into $[0, 1]$. Finally, we output the arithmetic mean of them added with $\hat{y}$ as the combined decision value of new example x. The reason that adding $\hat{y}$ to the arithmetic mean is that it could prevent the decision values of different classes stack at the same area.

Dividing Intervals by k-Means Clustering. Instead of dividing the intervals homogeneously, we came up with a new dividing scheme, which uses k-means clustering [4,6] to divide all decision values.

k-means clustering is a cluster analysis method which aims to divide n objects into k clusters in which each object belongs to the cluster with the nearest mean. Given a set of objects $(x_1, x_2, \ldots, x_n)$, where each object $x_i \in \mathbf{R}^d$ is a d-dimensional real vector, k-means clustering aims to partition the n objects into k sets $(k \leq n)$ $\mathbf{S} = \{S_1, S_2, \ldots, S_k\}$ so as to minimise the within-cluster sum of squares (WCSS):

$$\arg \min_{\mathbf{S}} \sum_{i=1}^{k} \sum_{x_j \in S_i} \|x_j - \mu_i\|^2 \tag{6}$$

where μ_i is the mean value of points in S_i.

In our design, dimension d is fixed to "1", while the number of clusters is equal to the number of possible labels. So the heuristic algorithm we used could be described as below.

1. k initial means values are randomly generated within the data domain.
2. k clusters are created (or reassigned) by associating every object with the nearest mean value.
3. The centroid of each of the k clusters becomes the new mean value.
4. Steps 2 and 3 are repeated until the change of WCSS (6) between two states declines to be less than $\epsilon = 10^{-4}$.

Having applied k-means clustering, we divided the decision values into categories which could be used to calculate the matrix for new examples and make the probabilistic predictions as the standard Venn Machine does.

2.2   Venn-ABERS Predictor

Venn-ABERS predictor is a recently developed algorithm for multi-probabilistic prediction. It is modified from Zadrozny and Elkan's procedure of probability forecasting [9], which cannot be well calibrated. The modification introduced Venn predictors into the procedure to overcome the problem of potentially weak calibration as a result of the fact that Venn predictors are always well calibrated and guaranteed to be well calibrated under the exchangeability assumption. The basic idea of pre-trained Venn-ABERS predictor is that the training set is split into two parts: the proper training set and the calibration set. The proper training set is used to train the learning machine and predict the label for new examples, while the calibration set is used to calculate the probabilistic outputs for the predicted labels. The calibration set will be turned into a monotonically increasing set in this algorithm according to [1].

Before we discuss Venn-ABERS predictor, there are some notions to be introduced yet. First notion is the term "*scoring algorithm*". Scoring algorithm is an algorithm that trains a classifier on the training set and uses the classifier to output a prediction score $s(x)$ for the new example x and predicts the label of x to be "1" if and only if $s(x) \geq c$ (c is a fixed threshold). So s is hereby called the *scoring function*. Many machine learning algorithms for classification are scoring algorithms. In our case, as what SVM defines, the decision function of SVM is a scoring function, since we assign a new example the positive label "+1" if and only if its decision value is greater than zero and vice versa for the negative label. The second notion is *"isotonic calibrator"*, which is a monotonically increasing function on the set $\{s(x_1), \ldots, s(x_l)\}$ that maximizes the likelihood

$$\prod_{i=1}^{l} p_i, \text{ where } p_i := \begin{cases} g(s(x_i)) & \text{if } y_i = 1 \\ 1 - g(s(x_i)) & \text{if } y_i = 0 \end{cases} \tag{7}$$

this function g is unique and can be found by using the "pool-adjacent violators algorithm" (PAVA) introduced in [1].

The workflow of Venn-ABERS predictor is as follows. Assuming a standard binary machine learning problem: a training set of examples $z_1, z_2, \ldots, z_l$. Each z_i consists of a pair of object x_i, and label y_i. The possible labels are binary, that is, $y \in \{0, 1\}$. And we are also given a test object x. Our task is to predict the label y for the new object x and give the estimation of the likelihood that our prediction is correct.

Let us split the training set $\{z_1, z_2, \ldots, z_l\}$ into two parts: the proper training set $\{z_1, z_2, \ldots, z_m\}$ of size m ($m < l$) and the calibration set $\{z_{m+1}, z_{m+2}, \ldots, z_l\}$. And $s : \mathbf{X} \to \mathbb{R}$ is the scoring function of training set $\{z_1, z_2, \ldots, z_m\}$. Given a new example x, we have two calibrators. Let g_0 be the isotonic calibrator for $\{(s(x_{m+1}), y_{m+1}), (s(x_{m+2}), y_{m+2}), \ldots, (s(x_l), y_l), (s(x), 0)\}$, g_1 be the calibrator for $\{(s(x_{m+1}), y_{m+1}), (s(x_{m+2}), y_{m+2}), \ldots, (s(x_l), y_l), (s(x), 1)\}$.

To achieve the isotonic calibrator, we do the followings according to the definition of PAVA. First we arrange the pairs $(s(x_i), y_i)$ in the increasing order according to the values of score function $s(x_i)$. Having obtained a binary sequence consisting of labels y_i, we applied PAVA to find the increasing sequence of them. The final isotonic calibrator g is a function mapping the increasing scores to the increasing sequence (i.e. probabilities). As the score increases, the object is more likely to be "1" in correlation with the increasing sequence.

Then the multi-probability prediction outputs for that the predicted label should be "1" is $\{p_0, p_1\}$, where $p_0 := g_0(s(x))$ and $p_1 := g_1(s(x))$. And for the reason that we need to predict the probability for the prediction label is correct, we should transform the bounds $\{p_0, p_1\}$ to $\{1 - p_1, 1 - p_0\}$ when the predicted label is "0".

3 Experimental Results

To compare our algorithm to SVM Venn Machine with homogeneous intervals and Venn-ABERS predictor, we used eight data sets from the real world which could be easily obtained from UCI Repository (http://archive.ics.uci.edu/ml/) except that SVMguide1 is obtained from the website of LibSVM [3]. The data sets we

Table 1. Main characteristics for each data set

Data Set	# of Objects	# of Features	# of Classes	Training Set Size	Testing Set Size
WBC	683	10	2	400	283
SVMguide1	7089	4	2	3089	4000
Splice	3175	60	2	1000	2175
Satimage	6435	36	6	4435	2000
Segment	2310	19	7	1500	810
DNA	3186	180	3	2000	1186
Wine	178	13	3	100	78
Vehicle	846	18	4	500	346

used in this paper could be divided into two parts based on their number of classes. The details of these data sets are summarised in Table 1.

3.1   Experimental Settings

For VM-SVM-KM, the number of clusters and the initial means, which are the two key features of k-means clustering, are often regarded as its biggest drawbacks. The number of clusters is an input parameter: an inappropriate choice of k may yield poor results. That is why, when performing k-means clustering, it is important to run diagnostic checks for determining the number of clusters in the data set. The choice of initial means might lead the convergence to a local minimum which may produce counterintuitive results. A good design of a combined decision function could make it easier to avoid these two drawbacks.

To have a more intuitive view of our combined decision function described in (5), we applied the algorithm to Satimage data set and plotted the histogram in Fig. 1, roughly representing the distribution of the decision values.

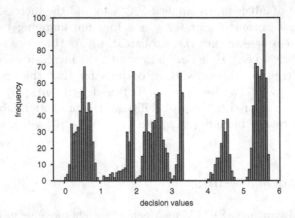

Fig. 1. Histogram of combined decision values for the Satimage data set

It can be seen obviously from the figure that there were 6 clusters in the data set, the exact number of the possible labels. The reason for this is that the decision function spreads out the values into $(0, k)$ by adding the most possible labels. Furthermore, each cluster i $(i = 1, 2, \ldots, k)$ is approximately within the range of $(i - 1, i)$, which means we could choose the initial means from each range to avoid the local minimum trap as much as possible and speed up the convergence process. We conducted k-means clustering to these decision values and calculated the 6 centroids: $0.63, 1.91, 2.64, 3.33, 4.57, 5.59$. The result seems to be a reasonable reflection of the histogram.

Then we could come to our decision that we set the number of clusters the same as the number of possible labels and we choose the initial means as $0.5, 1.5, \ldots, k - 0.5$ if the possible labels are $0, 1, \ldots, k - 1$.

Additionally, we need to notice that k-means clustering uses Euclidean distance as a metric and variance as a measure of cluster scatter, which makes it tend to produce equal-sized clusters. Since data is split halfway between cluster means, this can lead to suboptimal splits as some objects will be attributed to the incorrect cluster, especially for unbalanced data set as Satimage data set.

Except all the settings for the underlying algorithm, we need another setting for VA-SVM. It is the size of the calibration set. Having given careful consideration to both accuracy and narrowness of the bounds, we decided to take 30% of the whole data set as the calibration set, And the calibration set was stratified selected from the whole training set, which means the distribution of classes in the calibration set was the same as in the training set.

Although the size of proper training set in VA-SVM is smaller comparing to the size of training set in our algorithms, this is still a fair comparison because we use the same original training set for all algorithms, otherwise VA-SVM will need extra examples for probabilistic predictions. We also noticed that Venn-ABERS predictor is an inductive Venn predictor while Venn Machine is a transductive Venn predictor. The gap between inductive and transductive learning algorithms are not distinguishable in our offline setting. Because in offline setting, we use the fixed predictors to make predictions for testing set. Furthermore, in VA-SVM we repeat the computations of isotonic calibrators for each testing object which still involve all examples in calibration set.

3.2   Comparisons and Results

For binary cases, we applied VM-SVM-KM, VM-SVM-HI and VA-SVM to the data sets in the offline setting. While for multi-class cases, we only applied VM-SVM-KM and VM-SVM-HI to the data sets in both offline setting and online setting. Hence, there were three comparisons described as below. All the SVMs in these algorithms were using RBF kernel. Additionally, the parameters of SVM for each data set, including cost C and σ in RBF kernel, were determined by grid search on the training set and retained the same over corresponding algorithms respectively. The algorithms were compared in terms of their accuracies and probabilistic outputs in these data sets. In addition, we calculated the Brier scores (introduced in [2]) of the mean of the probabilistic bounds as evaluation for binary data sets.

The experimental results of VM-SVM-KM compared with VM-SVM-HI and VA-SVM are shown in Table 2.

Table 2. The offline accuracy and probability results on the binary data set

Data Set	Taxonomy	Accuracy	Prob. Outputs	Brier Score
WBC	VM-SVM-KM	97.53%	[86.34%,98.94%]	0.0325
	VM-SVM-HI	97.22%	[83.63%,98.70%]	0.0369
	VA-SVM	97.17%	[85.67%,95.97%]	0.0315
SVMguide1	VM-SVM-KM	96.93%	[91.27%,98.42%]	0.0362
	VM-SVM-HI	95.79%	[89.59%,98.97%]	0.0406
	VA-SVM	95.95%	[93.67%,96.29%]	0.0370
Splice	VM-SVM-KM	90.21%	[82.44%,96.07%]	0.0884
	VM-SVM-HI	89.52%	[80.15%,97.35%]	0.0939
	VA-SVM	89.15%	[83.40%,88.32%]	0.0878

The comparison results of our algorithm against VM-SVM-HI for all multi-class data sets in the offline setting are shown in Table 3.

Table 3. The offline accuracy and probability results on the multi-class data set

Data Set	Algorithm	Accuracy	Probabilistic Outputs
Satimage	VM-SVM-HI	84.18%	[75.48%,96.92%]
	VM-SVM-KM	86.56%	[81.18%,93.33%]
Segment	VM-SVM-HI	90.88%	[74.61%,96.68%]
	VM-SVM-KM	91.65%	[75.64%,95.60%]
DNA	VM-SVM-HI	94.34%	[81.39%,98.07%]
	VM-SVM-KM	96.65%	[87.25%,99.48%]
Wine	VM-SVM-HI	92.30%	[81.19%,97.11%]
	VM-SVM-KM	96.11%	[86.53%,98.47%]
Vehicle	VM-SVM-HI	67.63%	[56.42%,77.48%]
	VM-SVM-KM	69.15%	[60.65%,79.20%]

And the results for the online setting are shown in Table 4.

In order to giving a more intuitive comparison, we also give figures on online performance for Wine data set in Fig. 2.

Table 4. The online accuracy and probability results on the multi-class data set

Data Set	Algorithm	Accuracy	Probabilistic Outputs
Satimage	VM-SVM-HI	80.94%	[80.20%,81.69%]
	VM-SVM-KM	83.40%	[83.24%,83.86%]
Segment	VM-SVM-HI	88.40%	[88.92%,93.20%]
	VM-SVM-KM	89.96%	[90.11%,91.50%]
DNA	VM-SVM-HI	89.12%	[88.46%,89.86%]
	VM-SVM-KM	89.70%	[89.25%,90.48%]
Wine	VM-SVM-HI	91.53%	[87.57%,94.35%]
	VM-SVM-KM	93.22%	[91.67%,96.87%]
Vehicle	VM-SVM-HI	66.04%	[70.24%,72.17%]
	VM-SVM-KM	67.83%	[69.48%,71.02%]

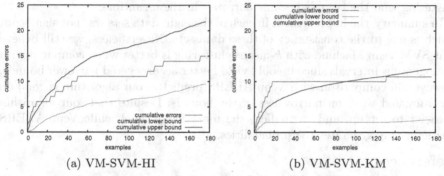

(a) VM-SVM-HI   (b) VM-SVM-KM

Fig. 2. Comparison of online performances for the Wine data set

4   Discussion and Conclusion

From the results shown in Table 2 which comparing our method to VM-SVM-HI and VA-SVM, we could draw the following conclusions.

First, VM-SVM-KM performed better in accuracy among these three data sets nevertheless the increases were small. Furthermore, VA-SVM used 30% of the training set as the calibration set which did not participate in the training of classifiers; hence it may lead to worse results. Second, the accuracies of VA-SVM slightly outnumbered the upper bound in WBC and Splice data sets, which could be due to the offline setting. Third, the probability bounds of VA-SVM were the narrowest while VM-SVM-HI had the widest bounds and VM-SVM-KM was in-between. This is the advantage of VA-SVM in view of our preference for narrow bounds. It is also backed by the Brier scores results: VA-SVM and VM-SVM-KM had close Brier scores while VM-SVM-HI had the worst results.

Another point is that VA-SVM does not calibrate their predicted label according to the probability, more specifically it is an algorithm that generates the probabilities from the scores only, while our algorithm gives predictions based on the highest likelihood. Except the improvement in accuracy, VM-SVM-KM is easy to configure because the number of clusters is the only input parameter of this algorithm which is equal to the number of classes.

From the results presented in Table 3 and Table 4 where the performance of VM-SVM-KM is compared with VM-SVM-HI in both offline and online setting, we can discover the following points.

First, it can be observed that all accuracies were within the probabilistic outputs in the offline setting, while in the online setting the accuracies exceeded the bounds in Segment and Vehicle data sets. Second, after implementing the k-means clustering, the accuracies are improved in both settings. However, in the offline setting, the improvements ranged from 0.8% to 3.8% depending on the data sets. In the online setting, the difference between these two algorithms became smaller, only 0.6% to 2.5%. Third, probability bounds become narrower after applying the k-means clustering, mostly benefiting from the rise of lower bounds. An intuitive comparison is shown in Fig. 2. The cumulative errors and cumulative error bounds in the figures all decreased after implementing k-means clustering, and the bounds became narrower in the meantime.

In summary, the improvement in each of the eight data sets was not significant which is due to the consistency of these data sets. Nevertheless, we still believe that SVM Venn Machine with k-means clustering is better when compared with homogeneous intervals since it could yield better accuracy and narrower bounds. However, in comparison with Venn-ABERS predictor, our algorithm is good on accuracy and weak on narrowness of the bounds. Despite that, our algorithm is easier to set up, and it predicts the most likely label while Venn-ABERS predictor only generates the probabilities.

References

1. Ayer, M., Brunk, H.D., Ewing, G.M., Reid, W.T., Silverman, E.: An empirical distribution function for sampling with incomplete information. Ann. Math. Statist. 26(4), 641–647 (1955)
2. Brier, G.W.: Verification of forecasts expressed in terms of probability. Monthly Weather Review 78(1), 1–3 (1950)
3. Chang, C.C., Lin, C.J.: Libsvm: a library for support vector machines. ACM Transactions on Intelligent Systems and Technology (TIST) 2(3), 27 (2011)
4. Forgy, E.W.: Cluster analysis of multivariate data: Efficiency vs interpretability of classifications. Biometrics 21, 768–769 (1965)
5. Lambrou, A., Papadopoulos, H., Nouretdinov, I., Gammerman, A.: Reliable probability estimates based on support vector machines for large multiclass datasets. In: Iliadis, L., Maglogiannis, I., Papadopoulos, H., Karatzas, K., Sioutas, S. (eds.) AIAI 2012, Part II. IFIP AICT, vol. 382, pp. 182–191. Springer, Heidelberg (2012)
6. Lloyd, S.P.: Least squares quantization in pcm. IEEE Transactions on Information Theory 28, 129–137 (1982)
7. Vovk, V.: Venn predictors and isotonic regression. CoRR abs/1211.0025 (2012)
8. Vovk, V., Gammerman, A., Shafer, G.: Algorithmic Learning in a Random World. Springer-Verlag New York, Inc., Secaucus (2005)
9. Zadrozny, B., Elkan, C.: Transforming classifier scores into accurate multiclass probability estimates. In: Proceedings of the Eighth ACM SIGKDD International Conference on Knowledge Discovery and Data Mining, pp. 694–699. ACM (2002)
10. Zhou, C., Nouretdinov, I., Luo, Z., Adamskiy, D., Randell, L., Coldham, N., Gammerman, A.: A comparison of venn machine with platt's method in probabilistic outputs. In: Iliadis, L., Maglogiannis, I., Papadopoulos, H. (eds.) EANN/AIAI 2011, Part II. IFIP AICT, vol. 364, pp. 483–490. Springer, Heidelberg (2011)

Efficiency Comparison of Unstable Transductive and Inductive Conformal Classifiers

Henrik Linusson[1], Ulf Johansson[1], Henrik Boström[2], and Tuve Löfström[1]

[1] School of Business and IT
University of Borås, Borås, Sweden
{henrik.linusson,ulf.johansson,tuve.lofstrom}@hb.se
[2] Dept. of Computer and Systems Sciences
Stockholm University, Kista, Sweden
henrik.bostrom@dsv.su.se

Abstract. In the conformal prediction literature, it appears axiomatic that transductive conformal classifiers possess a higher predictive efficiency than inductive conformal classifiers, however, this depends on whether or not the nonconformity function tends to overfit misclassified test examples. With the conformal prediction framework's increasing popularity, it thus becomes necessary to clarify the settings in which this claim holds true. In this paper, the efficiency of transductive conformal classifiers based on decision tree, random forest and support vector machine classification models is compared to the efficiency of corresponding inductive conformal classifiers. The results show that the efficiency of conformal classifiers based on standard decision trees or random forests is substantially improved when used in the inductive mode, while conformal classifiers based on support vector machines are more efficient in the transductive mode. In addition, an analysis is presented that discusses the effects of calibration set size on inductive conformal classifier efficiency.

1 Introduction

Conformal Prediction [1] is a machine learning framework for associating predictions for novel data with a measure of their *confidence*; whereas traditional machine learning algorithms produce *point predictions* — a single label $\hat{y}$ per test example — conformal predictors produce *prediction regions* — prediction sets $\hat{Y} \subseteq Y$ that, in the long run, contain the true labels for the test set with some predefined probability $1 - \epsilon$. Historically, such confidence predictions have relied on the Bayesian learning and Probably Approximately Correct (PAC) learning frameworks; however, the validity of Bayesian confidence predictions relies on an assumption of the *a priori* distribution, while PAC learning confidence measures apply to the entire model, and not the individual predictions [2]. In contrast, conformal predictors are able to produce confidence measures tailored for each separate prediction on novel data, and rely only on the assumption that the data is *exchangeable* — that the ordering of data points does not affect their joint

L. Iliadis et al. (Eds.): AIAI 2014 Workshops, IFIP AICT 437, pp. 261–270, 2014.

distributions — which is an even weaker assumption than the i.i.d. assumption typically required by traditional machine learning algorithms.

The method of constructing prediction regions using conformal predictors relies on measuring the strangeness — the *nonconformity* — of each data point, using some (arbitrary) real-valued function, called a *nonconformity function*, that measures the degree of strangeness of an example (x_i, y_i) in relation to a bag (multiset) of examples $Z = \{(x_1, y_1), ..., (x_i, y_i), ..., (x_k, y_k)\}$. For classification problems, this nonconformity measure is often based on the predictions of a traditional machine learning algorithm, called the *underlying model* of the conformal predictor. By comparing the nonconformity of a tentative classification $(x_{k+1}, \tilde{y})$ for a novel (test) input pattern x_{k+1} to the nonconformity scores of previously seen data, a conformal predictor can make inferences as to whether $\tilde{y}$ is likely to be a correct classification for x_{k+1}, and thus decide whether or not to include $\tilde{y}$ in the prediction region $\hat{Y}_{k+1}$.

There are two major categories of conformal predictors: transductive conformal predictors (TCP) [3,4] and inductive conformal predictors (ICP) [2,5]; both variants are based on the same principles, and place the same exchangability requirement on the data. Where they differ is in their usage of the training data, and their overall training scheme. For a novel input pattern x_{k+1} and every tentative prediction $\tilde{y} \in Y$, TCP measures the nonconformity of all examples in the bag $Z' = Z \cup \{(x_{k+1}, \tilde{y})\}$ relative to each other, meaning that the instance $(x_{k+1}, \tilde{y})$ is considered when calculating the nonconformity scores for the training set Z. Hence, for every new pattern x_{k+n} and every $\tilde{y}$, the underlying model needs to be retrained, and the nonconformity scores for the training data recomputed, rendering TCP computationally intensive. In ICP, only part of the data is used to train the underlying model (once), while the remaining data (a *calibration set*) is set aside to provide an unbiased estimate of the distribution of nonconformity scores.

The predictive *efficiency* — that is, the size of the prediction regions produced — of any conformal predictor is closely tied to the nonconformity function's ability to rank examples by order of strangeness. Moreover, as noted in several papers, ICP models typically suffer a small loss in predictive efficiency compared to corresponding TCP models due to the reduced number of training and calibration examples [1, 5–8]. However, as pointed out in [1], an *unstable* nonconformity function — one that is heavily influenced by an outlier example, i.e., an erroneously labled test instance $(x_{k+1}, \tilde{y})$ — can cause TCP models to become inefficient.

So, the choice between TCP and ICP is not so clear-cut: on the one hand, ICP will surely always produce its predictions faster than TCP, and TCP is often expected to have a higher predictive power than ICP. On the other hand, the efficiency of TCP relies on the chosen nonconformity function being *stable*, meaning that the underlying model does not train outlier examples into its learned rule [1]. When choosing a conformal prediction setup, a user should thus consider not only the trade-off between predictive speed and predictive power in

TCP and ICP, but also whether the chosen nonconformity function can be used effectively in TCP.

Decision trees and random forests are commonly used model types that are able to accommodate their training examples well, and, due to their ability to near-perfectly fit their training data, these model types should be expected to function better as underlying models in ICP than in TCP. To the best of our knowledge, however, this claim has not been investigated in existing literature.

In this study, the efficiency of TCP classifiers based on off-the-shelf decision trees and random forests, implemented by the scikit-learn [9] Python package, is compared to the efficiency of corresponding ICP classifiers; the results of this comparison are juxtaposed with an identical comparison of TCP and ICP classifiers based on stable support vector machines. To allow for a straightforward comparison of TCP and ICP classifiers, an analysis is also presented that discussess the effects of calibration set size on ICP classifier efficiency.

2   Background

Given some nonconformity scoring function $\Delta(Z, x_i, y_i) \to \mathbb{R}$, a conformal predictor can produce prediction regions that are *valid* in the sense that they contain the true target with some predefined probability $1 - \epsilon$. For classification problems, a common approach is to define a nonconformity function by combining a traditional machine learning algorithm and some error measure of the algorithm's predictions, e.g., the margin scoring function used in [10]:

$$\Delta(h, x_i, y_i) = P(y_i \mid h(x_i)) - \arg\max_{y' \neq y_i} P(y' \mid h(x_i)), \tag{1}$$

where h is a predictive model trained on some set Z, i.e., h represents a generalization of Z. Given the nature of classification problems in general, it is to be expected that a test pattern that deviates from the examples found in Z is likely to be misclassified by h, i.e., examples that are not conforming to Z are likely to be assigned large nonconformity scores.

When making a prediction using a conformal classifier, the nonconformity function is applied to a set of examples with known labels, resulting in a set of nonconformity scores $\alpha_1, ..., \alpha_k$ that represents a sample which is to be used for statistical inference (we here refer to this as the *nonconformity baseline*). The test pattern x_{k+1} is then assigned a tentative classification $(x_{k+1}, \tilde{y})$, where $\tilde{y} \in Y$, and a nonconformity score $\alpha_{k+1} = \Delta(Z, x_{k+1}, \tilde{y})$. Using a form of hypothesis testing (described in Sections 2.1 and 2.2), the conformal classifier attempts to reject the null hypothesis that $(x_{k+1}, \tilde{y})$ is conforming with Z, i.e., if the nonconformity score α_{k+1} is higher than for most examples in Z, as estimated by the nonconformity baseline, then $\tilde{y}$ is considered to be an unlikely classification for x_{k+1} and can be excluded from the final prediction set.

2.1   Transductive Conformal Predictors

A transductive conformal classifier uses the full training set Z to establish the nonconformity baseline. Due to the exchangeability assumption, this requires that the test pattern $(x_{k+1}, \tilde{y})$ is considered when calculating the nonconformity scores $\alpha_1, ..., \alpha_k$, i.e., if the nonconformity function Δ is based on an inductive model h, then $(x_{k+1}, \tilde{y})$ must be included in the training set for h. If it is not, then there is a possibility that h will have a larger bias towards its training examples than towards $(x_{k+1}, \tilde{y})$, meaning that the training examples and the test pattern might have different expected nonconformity values (which is a direct violation of the exchangeability assumption[1]). TCP thus requires that the underlying model h is trained $n\,|Y|$ times, where n is the number of test patterns, using the following scheme:

1. Assume a label $\tilde{y}$ and form the set $Z' = Z \cup \{(x_{k+1}, \tilde{y})\}$.
2. Use Z' to train a model h.
3. For each $(x_i, y_i) \in Z'$ calculate $\alpha_i = \Delta(h, x_i, y_i)$.
4. Calculate the p-value for $(x_{k+1}, \tilde{y})$ as

$$p(x_{k+1}, \tilde{y}) = \frac{|\{z_i \in Z' \mid \alpha_i \geq \alpha_{k+1}\}|}{|Z'|}. \tag{2}$$

5. If $p(x_{k+1}, \tilde{y}) > \epsilon$, include $\tilde{y}$ in the prediction region $\hat{Y}_{k+1}$.

2.2   Inductive Conformal Predictors

Inductive conformal predictors instead use only part of the training data to fit the underlying model h, setting aside a calibration set that is later used to establish the nonconformity baseline. Since the proper training set $Z^t \subset Z$ used to fit h and the calibration set $Z^c \subset Z$ are disjoint, the nonconformity scores $\alpha_1, ..., \alpha_{k-t}$ are exchangeable (unbiased) with the nonconformity score α_{k+1} without the need for including $(x_{k+1}, \tilde{y})$ in the training of h; thus, an ICP only needs to be trained once:

1. Divide Z into two disjoint subsets Z^t and Z^c.
2. Use Z^t to train a model h.
3. For each $(x_i, y_i) \in Z^c$, calculate $\alpha_i = \Delta(h, x_i, y_i)$.

For a novel test instance x_{k+1}:

1. Assume a label $\tilde{y}$ and calculate $\alpha_{k+1} = \Delta(h, x_{k+1}, \tilde{y})$.
2. Calculate the p-value for $(x_{k+1}, \tilde{y})$ as

$$p(x_{k+1}, \tilde{y}) = \frac{|\{z_i \in Z^c \mid \alpha_i \geq \alpha_{k+1}\}| + 1}{|Z^c| + 1}. \tag{3}$$

3. If $p(x_{k+1}, \tilde{y}) > \epsilon$, include $\tilde{y}$ in the prediction region $\hat{Y}_{k+1}$.

---

[1] If the model is able to better predict the correct output for the training examples than the test example, then the p-values for the true targets will no longer be uniformly distributed as required by the conformal prediction framework [1].

3   Method

The main point of interest of this study is the efficiency comparison of TCP and ICP classifiers based on decision tree, random forest and support vector machine models, but, to provide a straightforward comparison between the two, it is also necessary to find a suitable choice of calibration set size for the ICP classifiers. To the best of our knowledge, no thorough investigation has been published that discusses the effects of calibration set size on ICP classifier efficiency, and so the results presented in this paper contain first a discussion of ICP classifier setup, and second a comparison of ICP and TCP classifier efficiency.

To investigate what effect the calibration set size has on the efficiency of ICP classifiers using various types of underlying models, several ICP models were trained on five binary classification sets from the LIBSVM website [11] (a9a, covertype, cod-rna, ijcnn1 and w8a), using different amounts of training and calibration data. For each dataset, stratified random samples of $s = 500, 1000, ..., 4500$ examples were drawn, and for each s, several ICP models were applied to the same test set of 100 examples, each ICP model using $c = 100, 200, ..., s - 100$ calibration examples and $t = s - c$ training examples.

To compare the efficiency of the ICP and TCP variants, both types of classifiers were trained on 19 binary classification sets from the UCI repository [12]. The ICP models used a suitable calibration size as suggested by the results from the first experiment (20%). To limit the training time required for the TCP models, results from the larger datasets (kr-vs-kp, sick and spambase) were obtained using 1x10-fold cross-validation, while the results from the remaining sets were obtained using 5x10-fold cross-validation.

In both experiments, three different types of underlying models were used: decision trees (CART) [13], random forests (RF) [14] and support vector machines (SVM) [15]. The CART models used relative frequency probability estimates and no pruning, making them highly susceptible to overfitting noisy data. RF models (here consisting of 100 trees) are relatively robust to noise, but since the ensemble members are simple CART models, noisy examples are still fit into the learned rule to some extent. SVM models are, in contrast, stable to isolated noisy data points [16]. The scikit-learn library [9] for Python, using default settings, was used to train the underlying models. The margin nonconformity function (1) was used to construct the conformal classifiers, and predictions were made in the off-line (batch) mode.

4   Results

Figure 1 shows the relationships between training set size, calibration set size and efficiency (measured as AvgC — the average number of labels per prediction). The size of the calibration set is expressed as a portion of the full dataset, i.e., the point $(500, 1000)$ uses 500 of the total (1000) available examples as calibration data, and the remaining 500 examples as training data. The results are averaged over all five datasets and all five iterations. Clearly, the more data that is made available to the ICP classifier, the more efficient its predictions are, but only if the data is used sensibly. If more data is made available, and all newly available examples are added to the calibration set (any line $y = x+k$ in the six plots), the efficiency remains virtually unchanged at $\epsilon = 0.05$. If instead all newly available examples are added to the proper training set (any line parallel to the Y axis) the efficiency of the ICP model increases greatly, whereas, naturally then, the efficiency decreases if a larger calibration set is used while the total amount of available data remains unchanged (any line parallel to the X axis). The chosen (absolute) size of the calibration set thus has a small effect on ICP classifier efficiency, while the size of the proper training set is more strongly correlated with efficiency (cf. the correlations, r_t for training set size and r_c for calibration set size, listed below the plots).

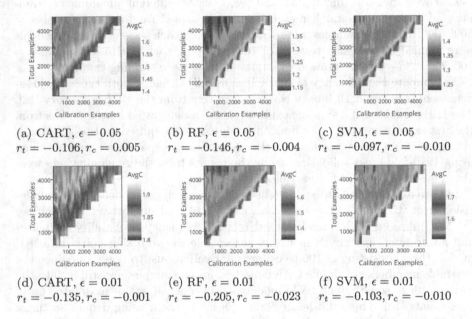

(a) CART, $\epsilon = 0.05$
$r_t = -0.106, r_c = 0.005$

(b) RF, $\epsilon = 0.05$
$r_t = -0.146, r_c = -0.004$

(c) SVM, $\epsilon = 0.05$
$r_t = -0.097, r_c = -0.010$

(d) CART, $\epsilon = 0.01$
$r_t = -0.135, r_c = -0.001$

(e) RF, $\epsilon = 0.01$
$r_t = -0.205, r_c = -0.023$

(f) SVM, $\epsilon = 0.01$
$r_t = -0.103, r_c = -0.010$

Fig. 1. ICP AvgC based on portion of examples used in the calibration set. The y-axis represents a growing dataset, while the x-axis represents a growing calibration set (relative to the full dataset). Blue areas represent the most efficient combinations of training and calibration set sizes, green and yellow areas are moderately efficient, while brown areas are the most inefficient.

As illustrated by Figures 1d, 1e and 1f, the size of the calibration set plays a larger role in determining efficiency when $\epsilon = 0.01$ than it does when $\epsilon = 0.05$, although the training set size is clearly the most important factor for maximizing efficiency at both significance levels. The best performing ICP classifiers are those using $15 - 30\%$ of the full training set as their calibration set. Notably, the performance of ICP classifiers using a calibration set containing less than 500 examples is quite poor when $\epsilon = 0.01$, regardless of training set size, suggesting that at least a few hundred examples should be used for calibration unless this leaves too few examples in the proper training set. The coverage (i.e. the 'empirical validity') of the conformal classifiers is not tabulated, however, all tested conformal classifiers show a coverage at or near the expected error rates on average. Note though, that the conformal classifiers using a very small calibration set (100 examples) displayed a larger variance in their coverage than the conformal classifiers using 500 calibration examples; in turn, using 500 calibration examples showed no substantial increase in variance compared to using a larger calibration set.

Table 1. AvgC of TCP and ICP based on CART, RF and SVM, $\epsilon = 0.05$

			CART		RF		SVM	
dataset	#f	#ex	ICP	TCP	ICP	TCP	ICP	TCP
balance-scale	4	577	1.571	1.891	1.037	1.044	0.957	0.951
breast-cancer	9	286	1.878	1.859	1.719	1.809	1.711	1.733
breast-w	10	699	1.259	1.520	0.973	0.976	0.987	0.993
credit-a	15	690	1.747	1.902	1.284	1.328	1.766	1.774
credit-g	20	1000	1.845	1.902	1.548	1.596	1.821	1.849
diabetes	20	769	1.842	1.896	1.541	1.601	1.822	1.861
haberman	3	306	1.849	1.843	1.649	1.773	1.680	1.607
heart-c	14	303	1.796	1.903	1.402	1.452	1.885	1.849
heart-h	14	264	1.808	1.898	1.402	1.460	1.838	1.836
heart-s	13	270	1.813	1.896	1.445	1.460	1.904	1.900
hepatitis	19	155	1.821	1.885	1.400	1.350	1.844	1.820
ionosphere	34	351	1.588	1.896	1.057	1.047	1.059	1.040
kr-vs-kp	36	3196	0.949	1.895	0.953	0.949	1.036	1.016
labor	16	57	1.719	1.909	1.400	1.115	1.525	1.039
liver-disorders	7	345	1.859	1.880	1.753	1.740	1.857	1.846
sonar	30	208	1.829	1.904	1.373	1.401	1.720	1.648
tic-tac-toe	9	958	1.255	1.906	0.960	0.963	1.161	1.045
sick	30	3772	0.964	1.886	0.954	0.950	1.062	1.036
spambase	57	4601	1.370	1.549	1.021	1.028	1.050	1.052
mean rank			**1.105**	1.895	1.316	1.684	1.684	1.316

Tables 1 and 2 show a comparison of the efficiency of ICP and TCP classifiers based on CART, random forest and SVM models. Note that the ranks are computed only within each TCP-ICP pair, i.e., the different underlying model types are not compared to each other in terms of efficiency ranks. Mean ranks in bold indicate significant differences at $\alpha = 0.05$ as reported by a two-tailed Wilcoxon signed-ranks test.

The stable SVM models do indeed appear to gain from being able to use the full dataset for both training and calibration when used in the TCP mode. At $\epsilon = 0.01$, the SVM-based TCP classifiers are significantly more efficient that SVM-based ICP classifiers; at $\epsilon = 0.05$ the SVM-TCP classifiers are more efficient on

Table 2. AvgC of TCP and ICP based on CART, RF and SVM, $\epsilon = 0.01$

dataset	#f	#ex	CART ICP	CART TCP	RF ICP	RF TCP	SVM ICP	SVM TCP
balance-scale	4	577	1.911	1.979	1.175	1.217	1.039	1.022
breast-cancer	9	286	1.977	1.976	1.931	1.965	1.940	1.972
breast-w	10	699	1.843	1.589	1.178	1.187	1.256	1.228
credit-a	15	690	1.951	1.982	1.806	1.810	1.934	1.936
credit-g	20	1000	1.969	1.981	1.838	1.871	1.959	1.972
diabetes	20	769	1.970	1.978	1.798	1.875	1.964	1.979
haberman	3	306	1.971	1.959	1.894	1.936	1.935	1.922
heart-c	14	303	1.966	1.981	1.800	1.815	1.982	1.976
heart-h	14	294	1.963	1.980	1.814	1.821	1.950	1.933
heart-s	13	270	1.953	1.979	1.837	1.827	1.984	1.979
hepatitis	19	155	1.960	1.983	1.832	1.758	1.976	1.966
ionosphere	34	351	1.917	1.974	1.589	1.454	1.543	1.330
kr-vs-kp	36	3196	1.067	1.980	1.008	1.000	1.203	1.258
labor	16	57	1.952	1.987	1.845	1.683	1.914	1.615
liver-disorders	7	345	1.978	1.959	1.945	1.929	1.969	1.948
sonar	30	208	1.968	1.980	1.804	1.672	1.922	1.872
tic-tac-toe	9	958	1.828	1.978	1.059	1.062	1.488	1.328
sick	30	3772	1.356	1.957	1.013	1.021	1.749	1.621
spambase	57	4601	1.875	1.673	1.272	1.623	1.352	1.344
mean rank			**1.263**	1.737	1.368	1.632	1.737	**1.263**

average, although the difference is not significant. It is evident, however, that the unpruned CART models — the most noise sensitive of the three model types — struggle with rejecting erroneous class labels in a TCP setting, likely due to the noisy test examples $(x_i, \tilde{y} \neq y_i)$ being fitted into the learned rules. TCP models based on CART are significantly less efficient than ICP-CART, both at $\epsilon = 0.05$ and $\epsilon = 0.01$. As expected, the RF models fare better in terms of TCP efficiency; the overfitting of the underlying CART models is counteracted by the smoothing effect of the ensemble constellation, however, not to such an extent that the TCP-RF models are more efficient than ICP-RF. ICP-RF is more efficient than TCP-RF on average, but the difference is not significant, neither at $\epsilon = 0.05$ nor at $\epsilon = 0.01$.

5   Related Work

As noted in [1], TCP classifiers using unstable nonconformity functions can be made more efficient through a slight modification of the transductive procedure. This modification involves calculating the nonconformity scores in a leave-one-out manner (LOOTCP), where the nonconformity of an example $(x_i, y_i) \in Z'$ is calculated from the set $Z' \setminus (x_i, y_i)$. Hence, in LOOTCP, the underlying model needs to be retrained not only for every test example and every tentative classification, but also for every training example. In principle, LOOTCP should indeed be expected to produce more efficient predictions than ICP, however, as LOOTCP increases the computational complexity of the (already computationally intensive) TCP classifier by a factor k, it is questionable whether it is truly applicable in practice when the nonconformity function requires training.

In [17], random forests are used to construct conformal classifiers run in the transductive mode. Rather than using the entire forest when calculating the nonconformity scores for calibration and test examples, the error of the out-of-bag prediction is used, i.e., to calculate the nonconformity score for a specific example, only ensemble members not trained on that particular example are considered (similar to LOOTCP). Although not explicitly stated by the authors, this effectively results in a random forest TCP that is not affected by the potential issues of strange examples being trained into the model.

In [18], a method for estimating the prediction certainty in decision trees, similar to conformal prediction, is proposed. The authors note that the default tree induction algorithm requires a modification for the predictions to be useful. Since the goal is to identify examples that are difficult to predict, the test instance (which is, as in TCP, included in the training data) is only allowed to affect node splitting to a limited extent, to avoid fitting strange examples into the models.

6 Concluding Remarks

This study shows that inductive conformal classifiers based on standard decision tree and random forest models can be more efficient than corresponding transductive conformal classifiers, while the opposite is true for support vector machines. This is contrary to the commonly accepted claim that transductive conformal predictors are by default more efficient than inductive conformal predictors. It has also been shown that to maximize the efficiency of inductive conformal classifiers, the calibration set should be kept small relative to the amount of available data $(15 - 30\%)$. At the same time, if the training set is large enough, it appears favourable to let the calibration set contain at least a few hundred examples.

For future work, we suggest that transductive and inductive conformal classifiers based on other types of classification models should be compared, to provide guidelines for designing conformal classification systems. Similarly, the efficiency of specialized transductive models, such as those proposed in [17], should be contrasted to the efficiency of standard transductive and inductive variants.

Acknowledgements. This work was supported by the Swedish Foundation for Strategic Research through the project High-Performance Data Mining for Drug Effect Detection (IIS11-0053) and the Knowledge Foundation through the project Big Data Analytics by Online Ensemble Learning (20120192).

References

1. Vovk, V., Gammerman, A., Shafer, G.: Algorithmic learning in a random world. Springer Verlag, DE (2006)
2. Papadopoulos, H.: Inductive conformal prediction: Theory and application to neural networks. Tools in Artificial Intelligence 18, 315–330 (2008)
3. Gammerman, A., Vovk, V., Vapnik, V.: Learning by transduction. In: Proceedings of the Fourteenth Conference on Uncertainty in Artificial Intelligence, pp. 148–155. Morgan Kaufmann Publishers Inc. (1998)
4. Saunders, C., Gammerman, A., Vovk, V.: Transduction with confidence and credibility. In: Proceedings of the Sixteenth International Joint Conference on Artificial Intelligence (IJCAI 1999), vol. 2, pp. 722–726 (1999)
5. Papadopoulos, H., Proedrou, K., Vovk, V., Gammerman, A.: Inductive confidence machines for regression. In: Elomaa, T., Mannila, H., Toivonen, H. (eds.) ECML 2002. LNCS (LNAI), vol. 2430, pp. 345–356. Springer, Heidelberg (2002)
6. Papadopoulos, H.: Inductive conformal prediction: Theory and application to neural networks. Tools in Artificial Intelligence 18(315-330), 2 (2008)
7. Papadopoulos, H., Vovk, V., Gammerman, A.: Conformal prediction with neural networks. In: 19th IEEE International Conference on Tools with Artificial Intelligence, ICTAI 2007, vol. 2, pp. 388–395. IEEE (2007)
8. Balasubramanian, V.N., Ho, S.S., Vovk, V.: Conformal prediction for reliable machine learning: theory, adaptations, and applications. Elsevier, Waltham (2013) (to appear)
9. Pedregosa, F., Varoquaux, G., Gramfort, A., Michel, V., Thirion, B., Grisel, O., Blondel, M., Prettenhofer, P., Weiss, R., Dubourg, V., et al.: Scikit-learn: Machine learning in python. The Journal of Machine Learning Research 12, 2825–2830 (2011)
10. Johansson, U., Boström, H., Löfström, T.: Conformal prediction using decision trees. In: IEEE International Conference on Data Mining (2013)
11. Chang, C.C., Lin, C.J.: Libsvm: a library for support vector machines. ACM Transactions on Intelligent Systems and Technology (TIST), Software, available at http://www.csie.ntu.edu.tw/~cjlin/libsvm
12. Bache, K., Lichman, M.: UCI machine learning repository (2013), http://archive.ics.uci.edu/ml
13. Breiman, L., Friedman, J., Stone, C.J., Olshen, R.A.: Classification and regression trees. CRC press (1984)
14. Breiman, L.: Random forests. Machine Learning 45(1), 5–32 (2001)
15. Cortes, C., Vapnik, V.: Support-vector networks. Machine Learning 20(3), 273–297 (1995)
16. Buciu, I., Kotropoulos, C., Pitas, I.: Demonstrating the stability of support vector machines for classification. Signal Processing 86(9), 2364–2380 (2006)
17. Devetyarov, D., Nouretdinov, I.: Prediction with confidence based on a random forest classifier. In: Papadopoulos, H., Andreou, A.S., Bramer, M. (eds.) AIAI 2010. IFIP AICT, vol. 339, pp. 37–44. Springer, Heidelberg (2010)
18. Costa, E.P., Verwer, S., Blockeel, H.: Estimating prediction certainty in decision trees. In: Tucker, A., Höppner, F., Siebes, A., Swift, S. (eds.) IDA 2013. LNCS, vol. 8207, pp. 138–149. Springer, Heidelberg (2013)

Anomaly Detection of Trajectories with Kernel Density Estimation by Conformal Prediction

James Smith[1], Ilia Nouretdinov[1], Rachel Craddock[2], Charles Offer[2],
and Alexander Gammerman[1]

[1] Computer Learning Research Center, Royal Holloway University of London
{James.Smith.2009,alex,ilia}@cs.rhul.ac.uk
[2] Thales UK
{firstname.lastname@uk.thalesgroup.com}

Abstract. This paper describes conformal prediction techniques for detecting anomalous trajectories in the maritime domain. The data used in experiments were obtained from Automatic Identification System (AIS) broadcasts – a system for tracking vessel locations. A dimensionality reduction package is used and a kernel density estimation function as a non-conformity measure has been applied to detect anomalies. We propose average p-value as an efficiency criteria for conformal anomaly detection. A comparison with a k-nearest neighbours non-conformity measure is presented and the results are discussed.

1 Introduction

Anomaly detection is a large area of research in machine learning and many interesting techniques have been developed to detect 'abnormal' behaviour of objects. The word 'anomaly' here is used in the sense that there are some patterns in the data that do not conform to typical behaviour. These non-conforming patterns are often called 'anomalies' or 'abnormalities' or 'outliers' [1]. Recently some new techniques known as conformal predictors (CP) have emerged which allow the detection of the non-conformal behaviour of objects using some measures of non-conformity [2,3]. This technique also has an advantage in delivering provably valid confidence measures under the exchangeability assumption that is usually weaker than those traditionally used.

Consider, for example, a set of moving objects (vessels, vehicles, planes, etc.) $z_1, z_2, \ldots$ and this movement might be normal (typical, conformal) or anomalous (atypical, non-conformal). We make an idealised assumption that $z_1, z_2, \ldots$ are from the same probability distribution P on the measurable feature space X independent from each other; however no further assumptions are made about P, which is completely unknown.

In this paper, the problem of anomaly detection in the maritime domain deals with trajectories of the ships to detect suspicious behaviours: a sudden change of direction, or speed, or anchoring, etc.

There has been previous research in applying conformal prediction to anomaly detection in the maritime surveillance domain [5]. Those methods focus on non-

L. Iliadis et al. (Eds.): AIAI 2014 Workshops, IFIP AICT 437, pp. 271–280, 2014.

conformity measures using nearest neighbours with Hausdorff distance or local densities of neighbourhoods with Local Outlier Factor [4,7].

In this paper we experiment with two different measures of non-conformity. In particular, the nearest neighbours non-conformity measure and the kernel density non-conformity measure have been used to detect anomalies. The data was obtained from Automatic Identification System (AIS) – a tracking system for vessels that is used to broadcast the location (retrieved by a GPS receiver onboard) of a vessel over radio-waves every few seconds.

The remaining part of this paper describes some details of conformal predictors including non-conformity measures, efficiency (performance) criteria, then a dimensionality reduction package T-SNE, the description of the data and the results with discussion. In particular, we propose average p-value as a level-independent criterion for assessing the efficiency.

2   Method

2.1   Conformal Prediction

Conformal prediction is a framework that allows making predictions with valid measures of confidence. Conformal Anomaly Detection (CAD) is an extension of Conformal Prediction that focuses on one-class (normal) in the unsupervised or semi-supervised setting [4].

Conformal Anomaly Detection

Input   : Non-Conformity Measure A, significance level ϵ, training objects $z_1, z_2, ..., z_{n-1}$ and new object z_n
Output: P-value p_n, boolean variable Anomaly

$D = \{z_1, ..., z_n\}$
for $i \leftarrow 1$ **to** n **do**
$\quad \lfloor \; \alpha_i \leftarrow A(D \setminus z_i, z_i)$
$\tau \leftarrow U(0, 1)$
$p_n \leftarrow \frac{|\{i : a_i > a_n\}| + \tau |\{i : a_i = a_n\}|}{n}$
if $p_n < \epsilon$ **then**
$\quad | \; Anomaly_j \leftarrow$ **true**
else
$\quad \lfloor \; Anomaly_j \leftarrow$ **false**

Basically, the method tests whether a *new object* z_n might be generated by the same distribution as the previous *(training)* objects $z_1, \ldots, z_{n-1}$. If produced *p-value* p_n is small, then the hypothesis of the new object's normalcy is likely to be rejected, so the abnormality is confirmed.

The *significance* level ϵ regulates the pre-determined level of confidence. According to the validity property[2], if all the data objects $z_1, \ldots, z_n$ are really generated by the same distribution, then probability that $p_n \leq \epsilon$ is at most ϵ. In the context of anomaly detection this means that if z_n is not an anomaly, it will be classified as anomaly with probability at most ϵ. This allows the false positive rate to be calibrated with a significance level parameter ϵ [7].

Another goal is efficiency: if z_n is an *anomaly*, we wish this to be captured by our test assigning a small p-value.

This performance depends on the selection of a *Non-Conformity measure (NCM)* denoted as A – that is a sort of information distance between an object and a set of the same type objects.

In this paper we use leave-one-out cross-validation to evaluate the performance. For each object a p-value is calculated using the rest of the objects as a training set. One advantage of using leave-one-out is independence on the order of data objects. Another is that it allows doing fair cross-validation using dimensionality reduction just once.

2.2 Performance Criterion

The validation of leave-one-out is done with *supervised anomaly detection* which has labelled anomalies and normal objects (from a *testing set*) where the correctness of output can be checked.

As mentioned above, the output (and the performance measure) typically depend on the significance level ϵ. Using a fixed ϵ, the more objects are classified as anomalies, the more sensitive the p-values as a test for randomness.

For supervised anomaly detection, to get an overall performance measure, independent of ϵ we adopt the well-known measure *receiver operating curve* (ROC) and use the– *area under ROC curve* (AUC) . For each value of ϵ we can produce two statistics: the percentage of normal objects classified as normal objects (that is close to $1 - \epsilon$ by validity), and the percentage of captured anomalies. AUC is the area under the corresponding ϵ-parametrized two-dimensional curve inside the square $[0, 1]^2$.

In [7], partial AUC (pAUC) is suggested for conformal anomaly detection, because it is important to minimise the number of false positives. pAUC only considers a subsection of the AUC, in particular false positive rate $\in [0, 0.01]$ and pAUC is normalised such that its outputs are $\in [0, 1]$.

Another important goal is to make the size of the *prediction set* as small as possible. The prediction set is the set of all the possible objects z_n from the feature space such that $p(z_n) \geq \epsilon$. Such kind of performance measure was investigated in the context of anomaly detection by Lei et al. [8].

We propose a new ϵ-independent version of this performance measure called *average p-value* (APV). Average p-value is the p-value of a potential new object, averaged over its location in the feature space. Its approximation can be calculated by using a finite grid of points uniformly spaced out. Every object in the grid will have a p-value calculated using a training set. The training set is fixed for each element of the grid. In the online setting it is possible to generate an APV after each iteration (using the set of normal examples as the training set). In this paper we choose to use all normal and abnormal objects in the training set to match the leave-one-out setting. We recommend using the min and max points from observed data to be used as the corners of the grid. A grid of g^d cells is generated where g is the grid resolution, and d is number of dimensions of the feature space. A p-value is generated for each cell using the center point of each cell as the object to be evaluated. In this paper we use $g = 100$ and $d = 2$ to give a grid of 100x100 cells.

An alternative setting is the *unsupervised anomaly detection* setting which is designed for when either the data is unlabelled or no examples of anomalous objects have been provided. It considers the whole feature space as an 'ideal' testing set and considers its training set $z_1, \ldots, z_{n-1}$ as normal. In this setting AUC, and pAUC are not applicable as they both require labels, however APV could be used as a criterion in this setting. In the supervised setting AUC is preferable to APV for measuring performance, but APV can still provide information on the efficiency of non-conformity measures.

2.3 Non-conformity Measures

In this work we consider two non-conformity measures: the first is based on Kernel Density Estimation (KDE) and another, for comparison, on Nearest Neighbours algorithm. Lei et al. [8] considered KDE as a conformity measure in the unsupervised setting.

We shall start with the Kernel Density Estimation (KDE) measure. It allows assessing non-conformity based on the density of data points. The normal objects usually are concentrated in a relatively small areas (high density areas or clusters) while anomalies will be outside of these clusters. This can be exploited by estimating a probability density function from empirical data set. A standard method to do this is to use kernel density estimation. It is a non-parametric technique that requires no knowledge of the underlying distribution.

We can interpret a density function as a measure of conformity – many similar type of data points will be located together; hence we can multiply it by minus one to convert it to a non-conformity measure for consistency.

Input   : Object z_i, Set of objects $z_1, z_2, \ldots z_n$ (note in this setup z_i is included in the set), bandwidth h, Kernel function K, number of dimensions d

Output: Non-conformity score A

$$A_i = -\left(\frac{1}{nh^d} \sum_{j=1}^{n} K\left(\frac{z_i - z_j}{h} \right) \right)$$

Kernel density estimators use the previous objects with a bandwidth parameter h that specifies the width of each object.

We will treat the bandwidth uniformly in each dimension, and fixed for each object. A kernel K is a symmetric function centred around each data point. In this work we use a Gaussian Kernel function:

$$K(u) = (2\pi)^{-d/2} e^{-\frac{1}{2} u^T u}$$

Lei et al. [8] have carried out work extending conformal prediction to produce minimal prediction regions with the use of kernel density estimators and initially proposed KDE as a conformity measure in the unsupervised setting. Their method is underpinned by utilizing a custom bandwidth estimator that minimises the Lebesgue measure of the prediction set in the space.

We have not applied any bandwidth estimators in this paper because we wish to compare KDE with another method that also has a parameter and test performance for the parameters against multiple performance criterion.

We also apply k-Nearest Neighbour (kNN) NCM [5]: d_{ij}^+ is the jth nearest distance to an object z_i from other objects.

Input   : Object z_i, Set of objects $z_1, z_2, ...z_n$, number of nearest
neighbours k

Output: Non-Conformity score A

$A_i = \sum_{j=1}^k d_{ij}^+$

The nearest neighbour non-conformity measure was found to be useful in detecting anomalies [5] and we shall use it to compare performance with the KDE NCM.

2.4 Dimensionality Reduction

The dimensionality of trajectory data is high ($4N$, where N is a number of points in a trajectory) and in order to apply kernel density estimation, we need to decrease the dimensionality of our data.

This is achieved by applying a package called T-SNE – a dimensionality reduction system. The t-Distributed Stochastic Neighbour Embedding (T-SNE) algorithm [9] is a non-deterministic and effective dimensionality reduction algorithm. It has been primarily used for visualisation but we use it to transform our data to lower-dimensional space to evaluate non-conformity measures.

In this particular application of T-SNE to trajectory data we replaced the Euclidean pairwise distance matrix with the Hausdorff distance matrix [4], but otherwise use the standard MATLAB implementation[1]. To remind the reader that the directional Hausdorff distance $\vec{H}(F, G)$ is the distance from set F to set G. $H(F, G)$ is the symmetrical Hausdorff distance. Hausdorff uses a distance metric $dist$ between the sets of points:

$$\vec{H}(F, G) = \max_{a \in F} \left\{ \min_{b \in G} \{ dist(a, b) \} \right\}$$

$$H(F, G) = \max \left\{ \vec{H}(F, G), \vec{H}(G, F) \right\}$$

3 Data

An object in our task is a trajectory that can be represented as a function of position over time. We convert the trajectories into a sequence of discrete $4D$ points $(x, y, x_{speed}, y_{speed})$ in a similar method to [4].

[1] http://homepage.tudelft.nl/19j49/t-SNE.html

The original broadcasts are interpolated at a sampling distance of 200m.

If a vessel leaves the observation area for a period of 10 minutes or more, or if the vessel is stationary for a period of 5 minutes or more we consider this as the end of a trajectory. Therefore a *trajectory* is a sequence of $4D$ Points and can have any length. The $4D$ points are normalised such that $x, y \in [0, 1]$ and $x_{speed}, y_{speed} \in [-1, 1]$.

The Portsmouth dataset we evaluate was collected from a single AIS receiver on the south coast of England, during July of 2012 for one week. We filtered the data such that it only contains AIS broadcasts that report their location in a specific area between the Isle of Wight and Portsmouth. This was done to ensure consistency as the further an AIS broadcast travels the more likely it is to be lost and the data becomes less reliable.

In this dataset we consider only passenger, tanker and cargo vessels to reflect a degree of 'regular' behaviour (going from A to B and back). We assume that this data does not contain anomalous behaviour. To add anomalies we artificially inserted two sources of anomalies data points. The first contains 22 search & rescue helicopter trajectories. The other source is 180 'artificial' anomalies: random walks that have been generated starting from a random position of a random observed normal vessel. They follow a random direction and speed and a new point is generated every 200m as it has been suggested in [5]. However, unlike in [4] we only consider the entire trajectory and do not calculate detection delay.

Instead of generating anomalous trajectories of 3km in length we are using different length of "artificial" anomalies. The composition of our 180 'artificial' trajectories is the following: 150 200m long, 20 400m long, 10 600m long, 10 800m long and 10 that are 1000m long. The aim is to diversify the difficulty by providing both easy and difficult anomalies to detect.

The dataset consists of 1124 normal trajectories with 202 anomalies added to it. All these trajectories can be seen in Fig 1.

Prior to applying conformal prediction we run the T-SNE algorithm to produce $2D$ representations of the trajectories.

4   Results

For measuring the performance of the non-conformity measures we use AUC as introduced in section 2.2. The partial AUC (pAUC) is also used to show performance for $fpr \in [0, 0.01]$, note that pAUC is normalised to be in the range [0,1]. The average p-value (APV) introduced in section 2.2 is calculated, recall the lower the APV the more efficient the classifier.

AUC and pAUC are our criteria for anomaly detection ability in the supervised setting and the average p-value in the unsupervised setting which doubles as a measure of efficiency. We compare both non-conformity measures for the best parameter values of AUC, pAUC and APV. The APV, AUC and pAUC for various parameter values of both NCMs can be found in the Table 1.

Table 2 was created to expand upon the k neighbours parameter as it is apparent that the highest AUC k-NN classifier was not in the initial parameter

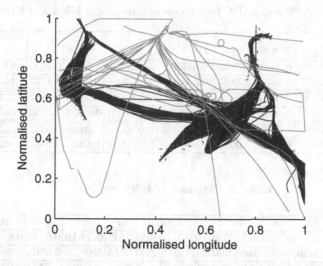

Fig. 1. Blue shows normal trajectories. Red shows the last 200m of the artificial anomalous trajectories. The green trajectories are the helicopters.

set. A rather important thing to note with testing leave-one-out is that anomalies are part of the training set, in practical applications ideally the training set would not contain anomalies. From the tables for all the parameters the highest AUC (supervised setting) are 0.7830 for KDE ($h = 3$) and 0.7616 for kNN ($k = 80$), it is clear that KDE has the higher AUC, and is therefore better at detecting anomalies across all ϵ in the leave-one-out setting. For both these parameters k-NN ($k = 80$) also has a larger APV 0.0638 against KDE ($h = 3$) 0.0606 which indicates that KDE is more efficient and offers better performance than k-NN when AUC is the criterion.

When we consider the most efficient APV (unsupervised setting) as a criterion k-NN's best parameter is $k = 7$ with APV of 0.0453 , whilst KDE's smallest APV is 0.0441 for $h = 1$.

The optimal result for the supervised problem requires more neighbours (k=80) than the unsupervised one (k=7) because most of the anomalies are close to each other (concentrating in a small area on Fig.2) which makes this problem harder. At the same time their influence on the unsupervised prediction is relatively small.

The pAUC Criterion in our leave-one-out setting may not be appropriate as the number of anomalies is far greater than a 1% composition of the dataset, but it is still a vital criterion for the purpose of minimising the false positive rate. KDE's best parameter by pAUC is $h = 2$ with a pAUC 0.484 and k-NN's best pAUC is with $k = 10$ with 0.484, however with these parameters $k = 10$ has a smaller APV and is thus more efficient. k-NN also achieves higher pAUC for more parameter values than KDE. This is quite apparent through with pAUC > 0.03 for $k = 7$ to $k = 20$, and for $k = 40$ to $k = 100$, where as for KDE only $h = \{2, 7, 8\}$ has pAUC above 0.03.

Table 1. AUC, APV and pAUC for various parameters of k-NN and KDE NCMs

k (k-NN) or h (KDE)	1	2	3	4	5	6	7	8	9	10
KDE AUC	0.6116	0.7620	**0.7830**	0.7455	0.6727	0.5932	0.5086	0.4406	0.3811	0.3518
k-NN AUC	0.2977	0.3051	0.3407	0.3611	0.3894	0.4066	0.4193	0.4323	0.4466	0.4618
KDE APV	**0.0441**	0.0519	0.0606	0.0694	0.0801	0.0936	0.1103	0.1307	0.1575	0.1941
k-NN APV	0.0490	0.0481	0.0469	0.0461	0.0458	0.0454	**0.0453**	0.0453	0.0455	0.0456
KDE pAUC	0.0082	**0.0484**	0.0285	0.0235	0.0270	0.0297	0.0415	0.0342	0.0250	0.0000
k-NN pAUC	0.0001	0.0099	0.0099	0.0099	0.0150	0.0276	0.0340	0.0381	0.0427	**0.0484**

Table 2. Extension of k-NN results

k (k-NN)	20	30	40	50	60	70	80	90	100	110
K-NN AUC	0.6157	0.7051	0.7257	0.7403	0.7519	0.7547	**0.7616**	0.6832	0.6301	0.6032
k-NN APV	0.0486	0.0519	0.0549	0.0574	0.0595	0.0615	0.0638	0.0705	0.0827	0.0954
k-NN pAUC	0.0304	0.0253	0.0327	0.0308	0.0400	0.0381	0.0384	0.0434	0.0375	0.0110

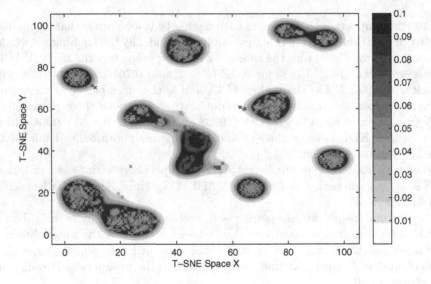

Fig. 2. Prediction sets for various parameters of ϵ in T-SNE space for KDE NCM (h=3). The labelled colours are for various values of ϵ, the red objects are anomalies, and the teal colour is used for the normal trajectories.

Figure 2 visualises the 'normal' class prediction sets of various ϵ for the KDE NCM. It is generated using a grid as the test set and all the objects from our dataset as the training set for each object in the grid.

5 Conclusions and Future Work

In this paper we applied conformal anomaly detection and studied applicable performance measures of efficiency. We have seen that it may be problematic to evaluate the performance of conformal anomaly detection directly because usually either the amount of labelled data for testing the accuracy is small or these data are not representative enough. Therefore we propose average p-value. Average p-value is an as a performance criterion that works in both the supervised and unsupervised settings and does not require labelled anomalies to evaluate performance. At the same time, it is independent on the significance level.

However, we applied some supervised criteria as well.

As examples of NCMs, we used two methods based on the idea of density approximation. One of them is nearest neighbours (k-NN) algorithm and the other is kernel density estimation (KDE) that considers an entire trajectory to the maritime surveillance domain. In addition, we reduced the dimensionality of our dataset to compare the different non-conformity measures.

In the leave-one-out supervised setting KDE NCM for our dataset in the supervised leave-one-out setting has higher AUC than the k-NN NCM. However for most anomaly detection applications performance at small false positive rates is more important. If small false positive rate (in the form of pAUC) is the primary criterion then k-NN NCM performs better than the KDE NCM.

Going to average p-value we see that KDE can lead to more efficient predictions with a smaller average p-value than k-NN, this indicates KDE NCM in the unsupervised setting with a good choice of parameter performs better with our dataset than the k-NN NCM.

Both KDE NCM and k-NN NCM performances for all criterion are dependent on the choice of parameter h and k respectively. We included some observations related to that.

For future work, it would be interesting to continue the work using other sources of data and to reach some explanation of the noticed effects. We also plan to apply various other NCMs in search of anomalous objects.

Acknowledgements. James Smith is very grateful for a PhD studentship jointly funded by Thales UK and Royal Holloway, University of London. This work is supported by EPSRC grant EP/K033344/1 ("Mining the Network Behaviour of Bots"); by the National Natural Science Foundation of China (No.61128003) grant; and by grant 'Development of New Venn Prediction Methods for Osteoporosis Risk Assessment' from the Cyprus Research Promotion Foundation. We are also grateful to Rikard Laxhammar, Vladimir Vovk, Christopher Watkins and Jiaxin Kou for useful discussions. AIS Data was provided by Thales UK.

References

1. Chandola, V., Banerjee, A., Kumar, V.: Anomaly Detection A Survey. ACM Computing Surveys (CSUR) (2009), dl.acm.org
2. Vovk, V., Gammerman, A., Shafer, G.: Algorithmic learning in a random world. Springer (2005)

3. Gammerman, A., Vovk, V.: Hedging predictions in machine learning. The Computer Journal 50(2), 151–163
4. Laxhammar, R., Falkman, G.: Sequential Conformal Anomaly Detection in Trajectories based on Hausdorff Distance. In: 2011 Proceedings of the 14th International Conference on Information Fusion (FUSION) (2011)
5. Laxhammar, R., Falkman, G.: Conformal prediction for distribution-independent anomaly detection in streaming vessel data. In: Proceedings of the First International Workshop on Novel Data Stream Pattern Mining Techniques, pp. 47–55. ACM (2010)
6. Laxhammar, R., Falkman, G.: Online Detection of Anomalous Sub-trajectories: A Sliding Window Approach Based on Conformal Anomaly Detection and Local Outlier Factor. In: Iliadis, L., Maglogiannis, I., Papadopoulos, H., Karatzas, K., Sioutas, S. (eds.) AIAI 2012, Part II. IFIP AICT, vol. 382, pp. 192–202. Springer, Heidelberg (2012)
7. Laxhammar, R.: Conformal anomaly detection: Detecting abnormal trajectories in surveillance applications. PhD Thesis, University of Skovde (2014)
8. Lei, J., Robins, J., Wasserman, L.: Distribution-Free Prediction Sets. Journal of the American Statistical Association 108(501), 278–287 (2013)
9. Van der Maaten, L., Hinton, G.: Visualizing Data using t-SNE. Journal of Machine Learning Research 9(11) (2008), http://homepage.tudelft.nl/19j49/t-SNE.html

Rule Extraction with Guaranteed Fidelity

Ulf Johansson[1,*], Rikard König[1], Henrik Linusson[1], Tuve Löfström[1],
and Henrik Boström[2]

[1] School of Business and IT
University of Borås, Sweden
{ulf.johansson,rikard.konig,henrik.linusson, tuve.lofstrom}@hb.se
[2] Department of Systems and Computer Sciences
Stockholm University, Sweden
henrik.bostrom@dsv.su.se

Abstract. This paper extends the conformal prediction framework to
rule extraction, making it possible to extract interpretable models from
opaque models in a setting where either the infidelity or the error rate
is bounded by a predefined significance level. Experimental results on 27
publicly available data sets show that all three setups evaluated produced
valid and rather efficient conformal predictors. The implication is that
augmenting rule extraction with conformal prediction allows extraction
of models where test set errors or test sets infidelities are guaranteed
to be lower than a chosen acceptable level. Clearly this is beneficial for
both typical rule extraction scenarios, i.e., either when the purpose is
to explain an existing opaque model, or when it is to build a predictive
model that must be interpretable.

Keywords: Rule extraction, Conformal prediction, Decision trees.

1 Introduction

When predictive models must be interpretable, most data miners will use decision trees like C4.5/C5.0 [1]. Unfortunately, decision trees are much weaker
in terms of predictive performance than opaque models like support vector machines, neural networks and ensembles. Opaque predictive models, on the other
hand, make it impossible to assess the model, or even to understand the reasoning behind individual predictions. This dilemma is often referred to as the
accuracy vs. comprehensibility trade-off.

One way of reducing this trade-off is to apply *rule extraction*, which is the
process of generating a transparent model based on a corresponding opaque
predictive model. Naturally, extracted models must be as good approximations
as possible of the opaque models. This criterion, called *fidelity*, is therefore a

* This work was supported by the Swedish Foundation for Strategic Research through
the project High-Performance Data Mining for Drug Effect Detection (IIS11-0053)
and the Knowledge Foundation through the project Big Data Analytics by Online
Ensemble Learning (20120192).

L. Iliadis et al. (Eds.): AIAI 2014 Workshops, IFIP AICT 437, pp. 281–290, 2014.
© IFIP International Federation for Information Processing 2014

key part of the optimization function in most rule extracting algorithms. For classification, the *infidelity rate* is the proportion of test instances where the extracted model outputs a different label than the opaque model. Similarly, the fidelity is the proportion of test instances where the two models agree. Unfortunately, when *black-box* rule extraction is used, i.e., when the rule extractor utilizes input-output patterns consisting of the original input vector and the corresponding prediction from the opaque model to learn the relationship represented by the opaque model, the result is often a too specific or too general model resulting in low fidelity on the test set, that is, the extracted model is actually a poor approximation of the opaque. Consequently, decision makers would like to have some guarantee *before* applying the extracted model to the test instances that the predictions will actually mimic the opaque.

In conformal prediction [2], prediction sets with a bounded error are produced, i.e., for classification, the probability of excluding the correct class label is guaranteed to be less than the predetermined significance level. The prediction sets can contain one, multiple or even zero class labels, so the price paid for the guaranteed error rate is that not all predictions are informative. In, inductive conformal prediction (ICP) [2], just one model is induced from the training data, and then used for predicting all test instances, but a separate data set (called the *calibration set*) must be used for calculating *conformity scores*.

The conformal prediction framework has been applied to several popular learning schemes, such as ANNs [3], kNN [4] and SVMs [5]. Until now, however, the guarantee provided by conformal prediction has always been related to the error rate. In this paper, we extend the conformal prediction framework to rule extraction, specifically introducing the possibility to bound the infidelity rate by a preset significance level.

2   Background

Rule extraction has been heavily investigated for ANNs, and the techniques have been applied mainly to ANN models; for an introduction and a good survey of traditional methods, see [6]. For ANN rule extraction, there are two fundamentally different extraction strategies, *decompositional* (*open-box* or *white-box*) and *pedagogical* (*black-box*). Decompositional approaches focus on extracting rules at the level of individual units within a trained ANN. Typically, the output of each hidden and output unit is first modeled as a consequent of their inputs, before the rules extracted at the individual unit level are aggregated to form the composite rule set for the ANN. Two classic open-box algorithms are RX [7] and Subset [8].

The core pedagogical idea is to view rule extraction as a learning task, where the target concept is the function originally learned by the opaque model. Black-box rule extraction is therefore an instance of predictive modeling, where each input-output pattern consists of the original input vector x_i and the corresponding prediction $f(x_i; \theta)$ from the opaque model. One typical and well-known black-box algorithm is TREPAN [9].

It must be noted that black-box rule extraction algorithms can be applied to any opaque model, including ensembles, and it can use any learning algorithm producing interpretable models as the actual rule extractor. An inherent problem for open-box methods, regarding both running time and comprehensibility, is the scalability. The potential size of a rule for a unit with n inputs each having k possible values is k^n, meaning that a straightforward search for possible rules is normally impossible for larger networks. Consequently, most modern rule extraction algorithms are black-box, see the more recent survey [10].

There is, however, one very important problem associated with black-box rule extraction. Even if the algorithm aims for maximizing fidelity in the learning phase, there is no guarantee that the extracted model will actually be faithful to the opaque model when applied to test set instances. Instead, since black-box rule extraction is just a special case of predictive modeling, the extracted models may very well overfit or underfit the training data, leading to poor fidelity on test data. The potentially low test set fidelity for black-box techniques stands in sharp contrast to open-box methods where the rules, at least in theory, should have perfect fidelity, even on the test set. Consequently, in situations where a very high fidelity is needed, open-box methods may be necessary; see e.g., [11]. Ideally though, we would like to have the best of both worlds, i.e., providing the efficiency and the freedom to use any type of opaque model present in black-box rule extractors, while guaranteeing test set fidelity. Again, the purpose of this paper is to show how the conformal prediction framework can be employed for achieving this.

An interesting discussion about the purpose of rule extraction is found in [12], where Zhou argues that rule extraction really should be seen as two very different tasks; rule extraction *for* neural networks and rule extraction *using* neural networks[1]. While the first task is solely aimed at understanding the inner workings of an opaque model, the second task is explicitly aimed at extracting a comprehensible model with higher accuracy than a comprehensible model created directly from the data set. More specifically, in rule extraction *for* opaque models, the purpose is most often to explain the reasoning behind individual predictions from an opaque model, i.e., the actual predictions are still made by the opaque model. In that situation, test set fidelity must be regarded as the most important criterion, since we use the extracted model to understand the opaque. In rule extraction *using* opaque models, the predictions are made by the extracted model, so it is used both as the predictive model and as a tool for understanding and analysis of the underlying relationship. In that situation, predictive performance is what matters, so the data miner must have reasons to believe that the extracted model will be more accurate than other comprehensible models induced directly from the data. The motivation for that rule extraction *using* opaque models may work is that even a highly accurate opaque model is a smoothed representation of the underlying relationship. In fact, train-

---

[1] Naturally this distinction is as relevant for rule extraction from any opaque model, not just from ANNs, so we use the terms rule extraction for or using *opaque models* instead.

ing instances misclassified by the opaque model are often atypical, i.e., learning such instances will reduce the generalization capability. Consequently, rule extraction is most often less prone to overfitting than standard induction, resulting in smaller and more general models.

2.1 Conformal Prediction

A key component in ICP is the conformity function, which produces a score for each instance-label pair. When classifying a test instance, scores are calculated for all possible class labels, and these scores are compared to scores obtained from a calibration set consisting of instances with known labels. Each class is assigned a probability that it does conform to the calibration set based on the fraction of calibration instances with a higher conformity score. For each test instance, the conformal predictor outputs a set of predictions with all class labels having a probability higher than some predetermined *significance level*. This prediction set may contain one, several, or even no class labels. Under very general assumptions, it can be guaranteed that the probability of excluding the true class label is bounded by the chosen significance level, independently of the conformity function used, for more details see [2].

In ICP, the conformity function A is normally defined relative to a trained model M:

$$A(\langle \bar{x}, c \rangle) = F(c, M(\bar{x})) \tag{1}$$

where $\bar{x}$ is a vector of feature values (representing the example to be classified), c is a class label, $M(\bar{x})$ returns the class probability distribution predicted by the model, and the function F returns a score calculated from the chosen class label and predicted class distribution.

Using a conformity function, a *p-value* for an example $\bar{x}$ and a class label c is calculated in the following way:

$$p_{\langle \bar{x}, c \rangle} = \frac{|\{s : s \in S \wedge A(s) \leq A(\langle \bar{x}, c \rangle)\}|}{|S|} \tag{2}$$

where S is the calibration set. The prediction for an example $\bar{x}$, where $\{c_1, \ldots, c_n\}$ are the possible class labels, is:

$$P(\bar{x}, \sigma) = \{c : c \in \{c_1, \ldots, c_n\} \wedge p_{\langle \bar{x}, c \rangle} > \sigma\} \tag{3}$$

where σ is a chosen significance level, e.g., 0.05.

3   Method

The purpose of this study is to extend the conformal prediction framework to rule extraction, and show how it can be used for both rule extraction *for* opaque models and rule extraction *using* opaque models. Since standard ICP is used, the difference between the scenarios is just how the calibration set is used. For the final modeling, all setups use J48 trees from the Weka workbench [13]. Here J48, which is the Weka implementation of C4.5, uses default settings, but pruning was turned off and Laplace smoothing was used for calculating the probability estimates. The three different setups evaluated are described below:

- **J48:** J48 trees built directly from the data. When used as a conformal predictor, the calibration set uses the true targets, i.e., the guarantee is that the error rate is bounded by the significance level.
- **RE-a:** Rule extraction *using* opaque models. Here, an opaque model is first trained, and then a J48 tree is built using original training data inputs, but with the predictions from the opaque model as targets. For the conformal prediction, the calibration set uses the true targets, so the guarantee is again that the error rate is bounded by the significance level.
- **RE-f:** Rule extraction *for* opaque models. The J48 model is trained identically to RE-a, but now the conformal predictor uses predictions from the opaque model as targets for the calibration. Consequently, the guarantee is that the infidelity rate will be lower than the significance level.

In the experimentation, bagged ensembles of 15 RBF networks were used as opaque models. With guaranteed validity, the most important criterion for comparing conformal predictors is *efficiency*. Since high efficiency roughly corresponds to a large number of singleton predictions, *OneC*, i.e., the proportion of predictions that include just one single class, is a natural choice. Similarly, *MultiC* and *ZeroC* are the proportions of predictions consisting of more than one class, and empty predictions, respectively. One way of aggregating these number is *AvgC*, which is the average number of classes in the predictions.

In this study, the well-known concept of *margin* was used as the conformity function. For an instance i with the true class Y, the higher the probability estimate for class Y the more conforming the instance, and the higher the other estimates the less conforming the instance. For the evaluation, 4-fold cross-validation is used. The training data was split 2:1; i.e., 50% of the available instances were used for training and 25% were used for calibration. The 27 data sets used are all publicly available from either the UCI repository [14] or the PROMISE Software Engineering Repository [15].

4   Results

Table 1 below shows the accuracy, AUC and size (total number of nodes) for the J48 models produced using either standard induction or rule extraction. As described in the introduction, the rule extraction is supposed to increase model accuracy, produce smaller models, or both. Comparing mean values and wins/ties/losses, the results show that the use of rule extraction actually produced models with higher accuracy. A standard sign test requires 19 wins for significance when $\alpha = 0.05$, so the difference is statistically significant at that level. Looking at models sizes, the extracted models are also significantly less complex. When comparing the ranking ability, however, the larger induced tree models obtained higher AUCs, on a majority of the data sets.

Table 1. Accuracy, AUC and Size

Data set	Accuracy Ind.	Ext.	AUC Ind.	Ext.	Size Ind.	Ext.	Data set	Accuracy Ind.	Ext.	AUC Ind.	Ext.	Size Ind.	Ext.
ar1	.909	.913	.457	.608	8.0	6.3	kc1	.839	.848	.680	.595	100.0	10.8
ar4	.817	.808	.664	.660	8.5	5.8	kc2	.827	.828	.785	.674	26.5	8.0
breast-w	.921	.928	.953	.945	20.8	15.0	kc3	.889	.903	.688	.629	25.5	4.8
colic	.705	.717	.713	.731	34.8	21.8	letter	.824	.800	.838	.824	20.0	23.3
credit-a	.712	.751	.771	.800	57.5	32.5	liver	.578	.620	.561	.610	22.3	23.5
credit-g	.683	.712	.620	.643	108.0	51.3	mw1	.901	.917	.679	.616	15.5	4.5
cylinder	.644	.634	.630	.638	63.3	50.3	sonar	.618	.680	.657	.733	18.8	14.0
diabetes	.691	.711	.690	.684	33.3	29.5	spect	.771	.793	.699	.731	25.0	11.8
heart-c	.719	.723	.754	.773	31.0	21.8	spectf	.744	.756	.718	.691	20.3	13.0
heart-h	.760	.786	.769	.804	28.5	14.5	tic-tac-toe	.770	.694	.775	.631	52.8	45.5
heart-s	.767	.748	.810	.784	31.3	17.3	vote	.899	.905	.933	.928	19.8	13.5
hepatitis	.781	.781	.746	.701	18.5	14.0	vowel	.786	.725	.804	.782	13.5	15.3
iono	.769	.789	.732	.750	13.3	13.8	**Mean**	**.761**	**.767**	**.719**	**.712**	**31.9**	**19.0**
jEdit4042	.642	.639	.669	.671	21.3	14.8	**Wins**	**8**	**19**	**15**	**12**	**4**	**23**
jEdit4243	.583	.589	.606	.599	23.5	15.8							

Turning to the results for conformal prediction, Fig 1 shows the behavior of extracted J48 trees as conformal predictors on the Iono data set, when using true targets for calibration. Since the conformal predictor is calibrated using true targets, it is the error and not the infidelity that is bounded by the significance level.

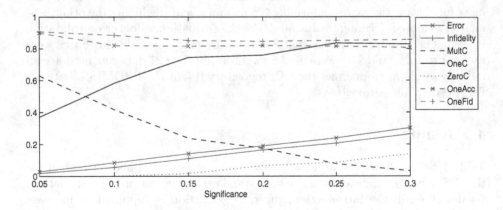

Fig. 1. Rule extraction *using* opaque model. Iono.

First of all, the conformal predictor is valid and well-calibrated, since the error rate is very close to the corresponding significance level. Analyzing the efficiency, the number of singleton predictions (OneC) starts at approximately

40% for $\epsilon = 0.05$, and then rises quickly to over 70% at $\epsilon = 0.15$. The number of multiple predictions (MultiC), i.e., predictions containing both classes, has the exact opposite behavior. The first empty predictions (ZeroC) appear at $\epsilon = 0.10$. Interestingly enough, OneAcc (the accuracy of the singleton predictions) is always higher than the accuracy of the underlying tree model (0.769), so singleton predictions from the conformal predictor could be trusted more than predictions from the original model. Finally, the fidelity of the singleton predictions (One-Fid) is very high, always over 80%. In fact, the infidelity rate is always lower than the error, indicating that the extracted conformal predictor is very faithful to the opaque model, even if this is not enforced by the conformal prediction framework in this setup.

Fig 2 below shows the behavior of extracted J48 trees as conformal predictors on the Iono data set, when using the ensemble predictions as targets for calibration.

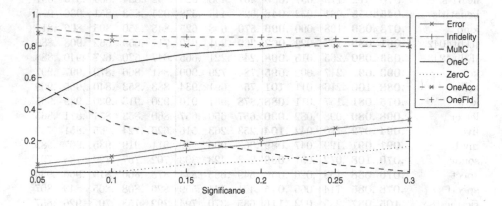

Fig. 2. Rule extraction *for* opaque model. Iono.

In this setup, it is the infidelity and not the error that is guaranteed, and indeed the actual infidelity rate is very close to the significance level. Here, singleton predictions are more common than for the other setup, i.e., it is easier to have high confidence in predictions about ensemble predictions than true targets. The error rate is slightly higher than the significance level, but interestingly enough both OneAcc and OneFid are comparable to the results for the previous setup.

Table 2 below shows detailed results for the three different conformal prediction setups, when the level of significance is $\epsilon = 0.10$. Investigating the errors and infidelities, it is obvious that the conformal prediction framework applies to both rule extraction scenarios, i.e., when the error rate or the infidelity rate must be lower than the significance level. On almost all data sets, the errors for J48 and RE-a are quite close to the significance level $\epsilon = 0.1$, indicating that the conformal predictors are valid and well-calibrated. Similarly, the infidelities for RE-f are also close to 0.1, on most data sets. Looking at the efficiency, measured

Table 2. Conformal prediction with $\epsilon = 0.1$. Bold numbers indicate criteria that are guaranteed by the conformal prediction framework.

	Error J48	RE-a	RE-f	Infidelity RE-a	RE-f	OneC J48	RE-a	RE-f	OneAcc J48	RE-a	RE-f	OneFid RE-a	RE-f
ar1	**.070**	**.070**	.124	.054	**.091**	.904	.821	.842	.919	.939	.930	.957	.950
ar4	**.043**	**.052**	.083	.042	**.056**	.369	.366	.649	.757	.870	.907	.881	.917
breast-w	**.090**	**.094**	.105	.095	**.102**	.936	.886	.852	.940	.944	.948	.946	.952
colic	**.098**	**.094**	.143	.063	**.086**	.538	.503	.652	.800	.797	.770	.870	.861
credit-a	**.094**	**.111**	.182	.055	**.089**	.556	.644	.818	.803	.818	.770	.908	.885
credit-g	**.104**	**.085**	.195	.024	**.090**	.440	.409	.745	.753	.780	.733	.935	.880
cylinder	**.097**	**.099**	.122	.073	**.093**	.317	.344	.427	.664	.695	.698	.764	.785
diabetes	**.083**	**.096**	.191	.044	**.098**	.415	.447	.723	.795	.774	.734	.888	.865
heart-c	**.073**	**.091**	.127	.045	**.084**	.407	.452	.586	.793	.787	.770	.879	.857
heart-h	**.083**	**.080**	.163	.034	**.075**	.509	.565	.858	.786	.845	.818	.931	.914
heart-s	**.070**	**.072**	.119	.037	**.059**	.461	.458	.625	.852	.834	.804	.915	.897
hepatitis	**.045**	**.055**	.081	.032	**.048**	.528	.445	.578	.925	.853	.851	.936	.914
iono	**.078**	**.089**	.108	.090	**.096**	.570	.608	.625	.845	.850	.805	.806	.813
jEdit4042	**.091**	**.089**	.197	.022	**.062**	.369	.291	.619	.717	.659	.662	.908	.880
jEdit4243	**.083**	**.080**	.265	.015	**.068**	.245	.221	.665	.647	.620	.627	.940	.888
kc1	**.093**	**.097**	.217	.002	**.095**	.784	.729	.900	.881	.866	.870	.997	.997
kc2	**.088**	**.100**	.216	.011	**.101**	.758	.691	.934	.883	.853	.840	.985	.969
kc3	**.075**	**.081**	.157	.011	**.088**	.878	.931	.916	.920	.913	.920	.989	.995
letter	**.098**	**.089**	.098	.082	**.090**	.657	.650	.657	.860	.853	.851	.871	.865
liver	**.094**	**.072**	.170	.042	**.104**	.253	.263	.516	.626	.724	.665	.851	.783
mw1	**.091**	**.091**	.129	.047	**.089**	.963	.990	.932	.911	.919	.935	.967	.983
sonar	**.070**	**.108**	.072	.091	**.089**	.233	.423	.303	.702	.737	.729	.788	.701
spect	**.070**	**.088**	.158	.026	**.056**	.549	.665	.844	.875	.866	.819	.960	.934
spectf	**.078**	**.085**	.114	.066	**.085**	.491	.444	.559	.826	.808	.805	.843	.807
tic-tac-toe	**.105**	**.087**	.235	.023	**.116**	.635	.370	.794	.792	.758	.707	.926	.857
vote	**.096**	**.079**	.091	.062	**.077**	.875	.845	.869	.923	.939	.934	.954	.946
vowel	**.083**	**.064**	.078	.100	**.097**	.581	.458	.378	.836	.854	.805	.777	.724
Mean	**.083**	**.085**	.146	.048	.085	.564	.553	.699	.816	.821	.804	.903	.882
Mean Rank	-	-	-	-	-	2.19	2.41	1.41	1.85	1.81	2.33	1.26	1.74

using the OneC metric, RE-f is clearly the most efficient conformal predictor. An interesting observation is that the errors for RE-f often are much higher than the corresponding significance level, thus indicating that the extracted model quite often is certain about the prediction from the ensemble, even when the ensemble prediction turns out to be wrong. This phenomenon is also obvious from the lower OneAcc exhibited by RE-f. Regarding infidelities and OneFid, it may be noted that RE-a turns out to be overly conservative. This actually results in a higher OneFid, compared to RE-f, but the explanation is the much fewer singleton predictions. Simply put, with a high demand on confidence in the selected singleton predictions, these tend to be predicted identically by the ensemble.

Table 3 below shows a summary, presenting averaged values and mean ranks over all data sets for three different significance levels. Included here is the metric AvgC, which is the average number of labels in the prediction sets. Since there are very few empty predictions at $\epsilon = 0.05$, OneC and AvgC will, for this significance level, produce the same ordering of the setups.

Table 3. Conformal prediction summary. Bold numbers indicate criteria that are guaranteed by the conformal prediction framework.

| | $\epsilon = 0.05$ | | | $\epsilon = 0.1$ | | | $\epsilon = 0.2$ | | |
	Ind	RE-a	RE-f	Ind	RE-a	RE-f	Ind	RE-a	RE-f
Error	**.034**	**.034**	.084	**.083**	**.085**	.146	**.184**	**.183**	.251
Infidelity	-	.018	**.035**	-	.046	**.084**	-	.124	**.190**
AvgC	1.66	1.70	1.46	1.43	1.44	1.26	1.15	1.15	1.01
Rank	2.11	2.70	1.19	2.30	2.48	1.22	2.48	2.33	1.19
OneC	.339	.297	.525	.564	.552	.701	.772	.778	.821
Rank	2.11	2.70	1.19	2.19	2.41	1.41	2.15	2.04	1.81
OneAcc	.772	.752	.778	.815	.819	.805	.794	.796	.794
Rank	1.78	1.89	2.33	1.85	1.89	2.26	2.07	1.93	2.00
OneFid	-	.824	.857	-	.906	.884	-	.878	.869
Rank	-	1.44	1.56	-	1.22	1.78	-	1.48	1.52

Even when analyzing all three significance levels, all conformal predictors seem to be valid and reasonably well-calibrated. Looking for instance at RE-a, the averaged errors over all data sets are 0.034 for $\epsilon = 0.05$, 0.084 for $\epsilon = 0.1$ and 0.183 for $\epsilon = 0.2$. Similarly, the averaged infidelities for RE-f are 0.035 for $\epsilon = 0.05$, 0.084 for $\epsilon = 0.1$ and 0.190 for $\epsilon = 0.2$.

Comparing efficiencies, RE-f is significantly more efficient, with regard to both OneC and AvgC, than the other two setups. J48 and RE-a have comparable efficiencies. Regarding OneAcc, J48 and RE-a are most often more accurate than RE-f. It must, however, be noted that RE-f has a fundamentally different purpose than RE-a and J48, so RE-a should only be compared directly to J48; they are both instances of, in Zhou's terminology, rule extraction *using* opaque models, while RE-f, is rule extraction *for* opaque models. Consequently the most important observation is that all setups have worked as intended, producing valid, well-calibrated and rather efficient conformal predictors for the two different rule extraction scenarios.

5   Concluding Remarks

In this paper, which should be regarded as a proof-of-concept, conformal prediction has been extended to rule extraction *for* opaque models and rule extraction *using* opaque models. The results show that conformal prediction enables extraction of efficient and comprehensible models, where either the error rate or the infidelity rate is guaranteed. This represents an important addition to the rule extraction tool-box, specifically addressing the problem with a potentially poor test set fidelity present in most black-box rule extractors.

For some reason rule extraction has not been extensively used on regression models, so the next step is to apply conformal prediction to this. We believe that the prediction intervals produced by conformal prediction regression will be a natural part of making extracted regression models accurate and comprehensible.

References

1. Quinlan, J.R.: C4.5: programs for machine learning. Morgan Kaufmann (1993)
2. Vovk, V., Gammerman, A., Shafer, G.: Algorithmic Learning in a Random World. Springer-Verlag New York, Inc. (2005)
3. Papadopoulos, H.: Inductive conformal prediction: Theory and application to neural networks. Tools in Artificial Intelligence 18, 315–330 (2008)
4. Nguyen, K., Luo, Z.: Conformal prediction for indoor localisation with fingerprinting method. In: Iliadis, L., Maglogiannis, I., Papadopoulos, H., Karatzas, K., Sioutas, S. (eds.) AIAI 2012, Part II. IFIP AICT, vol. 382, pp. 214–223. Springer, Heidelberg (2012)
5. Makili, L., Vega, J., Dormido-Canto, S., Pastor, I., Murari, A.: Computationally efficient svm multi-class image recognition with confidence measures. Fusion Engineering and Design 86(6), 1213–1216 (2011)
6. Andrews, R., Diederich, J., Tickle, A.B.: Survey and critique of techniques for extracting rules from trained artificial neural networks. Knowl.-Based Syst. 8(6), 373–389 (1995)
7. Rudy, H.L., Lu, H., Setiono, R., Liu, H.: Neurorule: A connectionist approach to data mining, 478–489 (1995)
8. Fu, L.: Rule learning by searching on adapted nets. In: AAAI, pp. 590–595 (1991)
9. Craven, M.W., Shavlik, J.W.: Extracting tree-structured representations of trained networks. In: Advances in Neural Information Processing Systems, pp. 24–30. MIT Press (1996)
10. Huysmans, J., Baesens, B., Vanthienen, J.: Using rule extraction to improve the comprehensibility of predictive models. FETEW Research Report KBI 0612, K. U. Leuven (2006)
11. Martens, D., Huysmans, J., Setiono, R., Vanthienen, J., Baesens, B.: Rule extraction from support vector machines: An overview of issues and application in credit scoring. In: Rule Extraction from Support Vector Machines, pp. 33–63 (2008)
12. Zhou, Z.H.: Rule extraction: using neural networks or for neural networks? J. Comput. Sci. Technol. 19(2), 249–253 (2004)
13. Witten, I.H., Frank, E.: Data Mining: Practical Machine Learning Tools and Techniques. Morgan Kaufmann (2005)
14. Asuncion, A., Newman, D.J.: UCI machine learning repository (2007)
15. Sayyad Shirabad, J., Menzies, T.: PROMISE Repository of Software Engineering Databases. School of Information Technology and Engineering, University of Ottawa, Canada (2005)

Conformal Prediction under Probabilistic Input

Ilia Nouretdinov

Computer Learning Research Centre,
Royal Holloway University of London

Abstract. In this paper we discuss a possible approach to confident prediction from data containing missing values presented in a probabilistic form. To achieve this we revise and generalize the notion of credibility known in the theory of conformal prediction.

1 Introduction

The task of machine learning is to predict a label for a new (or a testing) example x_{l+1} from a given training set of feature vectors $x_1, x_2, \ldots, x_l$ supplied with labels $y_1, y_2, \ldots, y_l$. The conformal prediction technique introduced in [1] and had many applications and extensions later. It allows to make a valid confident prediction.

Originally it was introduced for supervised machine learning problem with clear data structure. But in many practical problems data representation may be complex and combine multiple sorts of information. The conformal prediction was extended in previous works. In [6], it was *semi-supervised learning* when only some examples are presented with labels. In [7] training labels were available only for one of two classes. In [8] an unsupervised learning problem of *anomaly detection* was considered. Another kind of the task is Vapnik's Learning under privileged information [4] that can be interpreted as having missing values in testing examples. A conformal approach to it was made in the work [5].

The direction presented here is probabilistic representation of feature vectors or labels. Assume that there is kind of a priori distribution on features and/or labels. For example it is concentrated at one value when a feature is presented, it is uniform when it is completely missing, and other distributions are applicable when it is known partially or hypothetically. This means neither to try to exclude examples with missing values nor to fill them in a unique way.

An approach to this task is based on the notion of credibility that appears in the standard (supervised) conformal prediction. Unlike the confidence assigned to a likely hypothesis about the new example's label, the credibility answers the question whether any of these hypotheses is true at all. So the credibility is a characteristic of an unfinished data sequence, that includes a new example without its label. This can be naturally extended to the task when some part of training information is missing.

As an area of application needed for an illustration of the proposed method, we take LED data set from UCI repository [2], because a priori distribution on the values has a clear sense for these data.

L. Iliadis et al. (Eds.): AIAI 2014 Workshops, IFIP AICT 437, pp. 291–300, 2014.

2 Machine Learning Background

2.1 Conformal Prediction

Let us remind the properties of conformal prediction (in the case of classification) according to [1].

Assume that each data example z_i consists of x_i that is an m-dimensional vector $x_i = (x_{i1}, \ldots, x_{im})$ and a label y_i that is an element of a finite set Y.

Conformal predictor in supervised case assigns p-value (the value of a test for randomness) to a data sequence

$$p(y) = p((x_1, y_1), \ldots, (x_l, y_l); (x_{l+1}, y))$$

$$= \frac{card\{i = 1, \ldots, l+1 : \alpha_i \geq \alpha_{l+1}\}}{l+1}$$

where $x_1, \ldots, x_l \in X$ are feature vectors of training examples with known classifications $y_1, \ldots, y_l \in Y$, x_{l+1} is a new or testing example with a *hypothetical* label y, and

$$\alpha_i = A\left(\{(x_1, y_1), \ldots, (x_l, y_l), (x_{l+1}, y)\}, (x_i, y_i)\right)$$

for a *nonconformity measure A* that is a strangeness function of a set of labeled feature vectors and one of its elements.

The plan is to check each possible hypothesis about the label of a new example, and to the label of new example would conform the assumption of exchangeability, or with which label the example 'fits well' into the training set? The prediction set consists of satisfactory hypotheses y such that $p(y)$ exceeds a *significance level* γ. The calculations of prediction regions are based on a special function called *nonconformity measure (NCM)* that reflects how strange an example is with respect to others. Then p-value is assigned to each y.

There are two ways to present the results. One of them is the *prediction set*: a list of y which meet this confidence requirement $p(y) \geq \gamma$. The *validity* property implies that the probability of error is at most $1 - \gamma$ whenever the i.i.d. assumption is true. Here an error means true value of y_n being outside the prediction set.

Alternatively we can provide the prediction of a new label together with measures of its individual *confidence*. The correspondence between two types of output is that the confidence is the highest confidence level at which the prediction region consists of (at most) one value. In terms of p-values assigned to different labels, the confidence is a complement to 1 of the second highest p-value.

An individual prediction is also naturally completed with *credibility* that is the first highest p-value. If the credibility is low this means that any existing hypothesis about the label of the new object is unlikely. In other words, the new object itself is not credible enough as a continuation of the data sequence, and this could be said before its label is known. So it can be understood as dealing with an unknown testing label.

Our aim is to extend this idea, dealing other sort of incomplete information in analogous way.

2.2 Standard Credibility

In this work we call credibility a measure of conformity of an incomplete data sequence. Originally it was applied to the data sequences of the following type:

$$(x_1, y_1), \ldots, (x_l, y_l), x_{l+1}.$$

with y_{l+1} missing.

The credibility is obtained by maximization conformal p-values over all its possible completions:

$$p_{cred}((x_1, y_1), \ldots, (x_l, y_l), x_{l+1})$$

$$= \max_{y_{l+1} \in Y} p((x_1, y_1), \ldots, (x_l, y_l), (x_{l+1}, y_{l+1})).$$

The validity property of conformal prediction regions can be easily extended to the credibility. If a data sequence $(x_1, y_1), \ldots, (x_{l+1}, y_{l+1})$ is generated by $P = P_1^{l+1}$ where P_1 is a distribution on $X \times \{0, 1\}$, then

$$P\{p_{cred}((x_1, y_1), \ldots, (x_l, y_l), x_{l+1}) \leq \gamma\} \leq \gamma$$

for any $\gamma \in (0, 1)$.

In this form it was assumed that the incomplete sequence is obtained from the complete one by forgetting y_{l+1}. In other words it could be said that the incomplete sequence $(x_1, y_1), \ldots, (x_l, y_l), x_{l+1}$ is generated by $P^l \times P_X$ where P_X is the marginal distribution of the feature vector averaged over Y.

2.3 Extensions of Credibility

As we have seen, the standard credibility is the p-value (test for randomness) assigned to an incomplete sequence of examples. Incompleteness means there that the label of the last example is totally missing.

Sometimes a similar aproach can be applied to other kinds of missing values. A close problem is having an unknown feature (not a label) of a new (testing) example. This task is equivalent to learning under privileged (additional) information framework formulated in [4]. The conformal approach of this task was developed in the work [5]. An analogue of credibility was assigned to the sequence

$$(x_1, x_1^*, y_1), \ldots, (x_l, x_l^*, y_l), (x_{l+1}, y)$$

with x_{l+1}^* unknown. The feature x^* was called privileged bacause it is available for the training examples.

Next step might be related to missing values in training examples. But a straightforward approach to this task (maximizing p-value over possible fillings of the gap) is not effective because the conformal predictor concentrates on the conformity of the testing example without checking training examples for strangeness. Therefore we would like to consider missing values as distributions.

3   Conformal Approach for Probabilistic Input

For convenience of presentation, in this section we will start form the case of unclear information about binary labels y_i presented in a probabilistic form of a priori distribution. Then we will show how to apply it in a more general case.

3.1   Task and Assumptions

Suppose that $Y = \{0, 1\}$, but some information about $y_1, \ldots, y_l$ is missing. However, for each $i = 1, \ldots, l$ we know that p_i has a meaning of probability that $y_i = 1$. As for y_{l+1}, we assume that it is known as a hypothesis according to the conformal prediction procedure.

How to state this task in a well-defined way and what would be a proper analogue of the i.i.d. assumption in this case?

A mechanism should generate both the 'true' data sequence (including hidden values of y_i) and the 'visible' one (with probabilistic values p_i). This means that the triple (x_i, y_i, p_i) is generated simultaneously. But some agreement between p_i and y_i is also needed so that probabilistic values p_i make sense as probabilities.

To define this formally, assume that P_1 is a distribution on $X \times (0, 1)$ and Θ is the uniform distribution on $(0,1)$. First, $P = (P_1 \times \Theta)^{l+1}$ generates

$$(x_1, p_1, \theta_1), \ldots, (x_{l+1}, p_{l+1}, \theta_{l+1}).$$

Setting $y_i = 1$ if $p_i < \theta_i$ and $y_i = 0$ otherwise, we can also say that P^* generates a sequence of triples

$$(x_1, p_1, y_1), (x_2, p_2, y_2), \ldots, (x_{l+1}, p_{l+1}, y_{l+1})$$

where p_i is 'visible' label and y_i is the 'hidden' one, y_i is stochastically obtained from p_i.

3.2   Special Credibility

In order to make a conformal prediction of y_{l+1} for x_{l+1} we need to consider different hypotheses about it. When a hypothesis is chosen, we work with 'visible' labels $p_1, \ldots, p_l$ for training examples and for a 'hidden' value y_{l+1} for the new one. Thus the task is to assign a valid credibility value for a sequence $(x_1, p_1), \ldots, (x_l, p_l); (x_{l+1}, y_{l+1})$.

Fix a parameter $s > 0$ called *allowance* which is a trade-off between testing the hypothetical new label with respect to a version of the training data set, and testing the training data set with respect to a priori distribution on missing values.

Suppose that Q is the conditional distribution of $p((x_1, y_1), \ldots, (x_l, y_l), (x_{l+1}, y_{l+1}))$ given $(p_1, \ldots, p_l)$ and y_{l+1}, $q_0 = q_0(p_1, \ldots, p_l)$ is the smallest q such that

$$Q\{p > q | p_1, \ldots, p_l; y_{l+1}\} \leq s$$

and

$$p_{cred} = p_{cred}((x_1, p_1), \ldots, (x_l, p_l); (x_{l+1}, y_{l+1})) = q_0(p_1, \ldots, p_l; y_{l+1}) + s.$$

Proposition 1. *Assume tgat $(x_1, p_1), \ldots, (x_l, p_l)$ and $(x_{l+1}, y_{l+1}$ are generated by the mechanism described in Section 3.1 and p_{cred} us calculated as in Section 3.2, then*

$$P\{p_{cred} \le \gamma\} \le \gamma$$

for any $\gamma \in (0,1)$.

Proof: Recall that $p = p(y_1, \ldots, y_l) > q = q(p_1, \ldots, p_l)$ with probability at most s for any given $p_1, \ldots, p_l$. On the other hand, p is valid as a standard conformal predictor's output thus $p \le \gamma - s$ with overall probability at most $\gamma - s$. Therefore $\gamma - s < p < q$ with probability at least $1 - s - (\gamma - s) = 1 - \gamma$ and probability that $q + s < \gamma$ is bounded by γ. $\square$

3.3 Missing Values in Features

For convenience of presentation we earlier assumed that the labels $y_1, \ldots, y_l$ are given in probabilistic form, although this can be extended to the objects $x_1, \ldots, x_l$ as well.

So let us now assume that P^* generates (H_i, x_i, y_i) where 'visible' H_i is a distribution on X, while 'hidden' $x_i \in X$ is randomly generated by H_i.

If x_i is known clearly, this means that H_i is a distribution concentrated at one point. Otherwise H_i can be understood as an a priori distribution on its missing values. If X is discrete, then H_i can be presented in a vector form.

The extended credibility is defined by analogy. Suppose that Q is the conditional distribution of $p((x_1, y_1), \ldots, (x_l, y_l), (x_{l+1}, y_{l+1}))$ given $(H_1, \ldots, H_l, H_{l+1})$ and a fixed y_{l+1}, q_0 is the smallest q such that

$$Q\{p > q | H_1, \ldots, H_{l+1}; y_{l+1}\} \le s$$

and

$$p_{cred}((H_1, y_1), \ldots, (H_l, y_l), (H_{l+1}, y_{l+1})) = q_0 + s.$$

Obviously an analogue of Proposition 1 is also true in this case.

3.4 Efficient Approximation

To find q_0 exactly one has to know the condition distribution of p given 'visible' data. For the aims of computational efficiency this distribution can be replaced with an empirical one, using Monte-Carlo approximation. Let $H_1 \times H_2 \times \ldots \times H_{l+1}$ generate a large amount of vectors $(x_1, \ldots, x_{l+1})$ and calculate conformal p-value for each of them. Then we will get an empirical distribution of p that allows to estimate q_0 by sorting these p-values and taking one with corresponding rank. An example will be given in Section 4.3.

4 Experiments

For the experiments we use benchmark LED data sets generated by a program from the UCI repository[2].

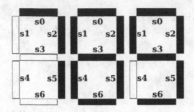

Fig. 1. Canonical images of 7,8,9 in LED data

LED means light emitting diode electronic display. A LED image of a digits has of 7 binary features (pixels). The task is to predict a digit from an image in the seven-segment display. Each of digits $0, 1, \ldots, 9$ has a canonical image that should normally represent it. Few examples are presented on Fig.1.

Assume now that visible displays can contain mistakes. Each pixel can occasionally show 'on' instead of 'off' or vice versa with probability p_0. For our example we assume that $p_0 = 0.1$ although normally it is much less. The data generating program first randomly selects a canonically represented digit then each of the attributes is inverted with a probability of noise p_0 and the noisy example is added to the data set.

In the work [3] the conformal approach was applied to LED data in its standard supervised form. Now we make some changes in the data statement. First, the probability p_0 itself is known for us. This means that all values in the training set are probabilistic ones. When we see that a pixel is 'on' this in fact means that it is on with probability $1 - p_0$ and 'off' with p_0, and vice versa. Second, in the testing examples there are no mistakes (as if $p_0 = 0$). The task is to classify a testing example with full information after training on the examples with probabilistic information. It is assumed that the canonical representations are not available for the learner, who has to make predictions based only on the examples with possible mistakes as they are presented in the data.

For experiments we generate some amount of LED digits. The number and distribution (frequency) of labels $(0,1,2,\ldots,9)$ is not restricted, we borrow it from well-known USPS (US Postal Service) benchmark data set in order to have imbalanced classes. Size of the classes is shown in Table 1.

Table 1. Size of different training classes

Class label (digit)	0	1	2	3	4	5	6	7	8	9	Total
Number of examples	359	264	198	166	200	160	170	147	166	177	2007

For a training example, given a label, we take its canonical LED image and make an error in each of the feature with probability γ. In the most of experiments $\gamma = 0.1$ unless stated another.

Testing examples are not probabilistic by the task, so in principle we can make predictions on $2^7 = 128$ possible images. This number includes 10 canonical images of digits.

Later we will consider two types of testing set. A proper one is generated with the same distribution on classes as the training set and therefore contains only canonical images. An auxiliary testing set contains all the possible images.

4.1 Nonconformity Measure

For convenience we use one of the simplest NCM that can be applied. As the space is discrete, NCM of an example with respect to the set of another ones is defined as the number of 'zero distance other class neighbors', i.e. number of examples in the set that have the same features but in fact belong to another class:

$$\alpha_i = A\left(\{(x_1, y_1), \ldots, (x_l, y_l), (x_{l+1}, y)\}, (x_i, y_i)\right) = card\{j : x_i = x_j, y_i \neq y_j\}.$$

4.2 Probabilistic Values of the Features

We apply our approach in its form mentioned in Section 3.3. All the features of X are binary and we assumed that the mistakes in features are done independently of each other. The connection between a 'hidden' vector

$$x = (x(1), \ldots, x(7)) \in X = \{0, 1\}^7$$

and a corresponding 'visible' distribution H is the following. H is a distribution on X such that:

- $x(1), \ldots, x(7)$ are H-independent on each other;
- for each $j = 1, \ldots, 7$, $H\{x(j) = 1\}$ is either $1 - \gamma$ or γ;
- a mistake $x(j)$ in a feature $x(j)$ is done with probability γ;
- if there is no mistake in $x(j)$ then $H\{x(j) = 1\} = 1 - \gamma$ if $x(j) = 1$, γ if $x(j) = 0$;
- if there is a mistake in $x(j)$ then $H\{x(j) = 1\} = \gamma$ if $x(j) = 1$, $1 - \gamma$ if $x(j) = 0$.

This means that in the 'visible' features vectors all the features are probabilistic. Each of the features is either *1 with probability* $1 - \gamma$, *0 with probability* γ or *1 with probability* γ, *0 with probability* $1 - \gamma$.

4.3 Other Details

Following 3.2 we set the 'allowance' coefficient to $s = 0.01$. Following the note 3.4 we avoid scanning all possible combinations by calculating p-values as Monte-Carlo approximations. the number of trials is 1000. Further we will see that this approximation does not affect validity properties.

Summarizing, there were 1000 trials (i.e. random filling of the missing values), and consider as the approximate credibility p_{cred} the 10-th largest of these p-values plus the allowance $s = 0.01$.

5   Results

Remind that p_{cred} finally is the p-value assigned to a new example (x_{l+1}, y_{l+1}).

To check the validity we wish to check what p-value is assigned to the true hypothesis about y_{l+1}. The correspoding p_{cred} is called p_{true}.

If y_{l+1} is unknown then each possible hypothesis about its value should be checked and assigned a p-value. As well as in the standard conformal predictor, the *prediction* is the hypothesis with the largest p-value and *confidence* in it is 1 minus the second largest p-value.

5.1   Validity

According to our problem statement, the validity is checked on testing examples that do not contain uncertainty and have the same distribution as the training examples *before* introducing mistakes. Therefore, each of the testing examples is one of ten digits ($y \in \{0, 1, \ldots, 9\}$) presented with its canonical image x. In order to satisfy i.i.d. assumption with training set, the distribution of ten types also corresponds to one from USPS data.

The corresponding validity plot is presented on Fig.2. It show that the probability of error (true value being outside the prediction set) does not exceed the selected significance level, for example:

$$P\{p_{true} \leq 0.16\} = 0.08;$$

$$P\{p_{true} \leq 0.27\} = 0.17.$$

The validity is satisfied with some excess. The same effect is known for the standard credibility and for LUPI due to involving incomplete information into the data.

5.2   Confidence

Recall that the testing set consists only of canonical images, so there are only 10 possible different configurations.

Individual confidences for them can be seen on Fig. 5.2 (boxed items), average value is 0.87. The smallest of these confidences is 0.79 assigned to the digit 7, because this digit is mixable with 1 (Hamming distance between them is the smallest) and relatively rare in the training set.

For comparison we also included confidence values that would be assigned to all 128 possible pixel combinations (auxiliary testing set) and they are much lower (0.18 in average).

The more indefinite the data are the smaller is the achieved level of confidence. For example, if we increase the probability of mistake from $p_0 = 0.1$ to $p_0 = 0.2$ then the figures of average confidence falls down to 0.55 and 0.11 respectively.

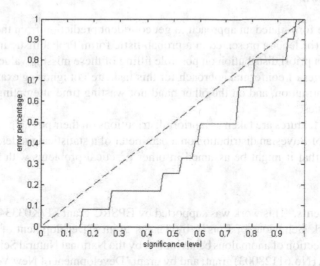

Fig. 2. Validity plot

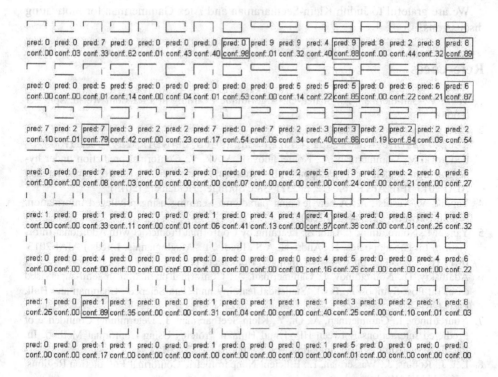

Fig. 3. Predictions for all possible pixel combinations. Predictions for the canonical images are put in boxes.

6 Conclusion

In this work we formulated an approach to get confident prediction from the data with missing values (or labels) presented in a probabilistic form. Probabilistic input means that there is an a priori distribution on possible filling of these missing values.

The advantages of conformal approach for this task are not ignoring examples with incomplete information, and on the other hand not wasting time attempting to restore the missing values.

The missing features are taken as a priori distributions on their possible values. This is an analogue of Bayesian distribution on a parameter of a statistical model. So we can expect as well that it might be assumed in other practical problems with incomplete information.

Acknowledgments. This work was supported by EPSRC grant EP/K033344/1 ("Mining the Network Behaviour of Bots"); by Thales grant ("Development of automated methods for detection of anomalous behaviour"); by the National Natural Science Foundation of China (No.61128003) grant; and by grant 'Development of New Venn Prediction Methods for Osteoporosis Risk Assessment' from the Cyprus Research Promotion Foundation.

We are grateful to Judith Klein-Seetharaman and Alex Gammerman for motivating discussions.

References

1. Vovk, V., Gammerman, A., Shafer, G.: Algorithmic Learning in a Random World. Springer (2005)
2. LED Display Domain Data Set,
   http://archive.ics.uci.edu/ml/datasets/LED+Display+Domain
3. Fedorova, V., Gammerman, A., Nouretdinov, I., Vovk, V.: Conformal prediction under hypergraphical models. In: Papadopoulos, H., Andreou, A.S., Iliadis, L., Maglogiannis, I. (eds.) AIAI 2013. IFIP AICT, vol. 412, pp. 371–383. Springer, Heidelberg (2013)
4. Vapnik, V., Vashist, A.: A new learning paradigm: Learning using privileged information. Neural Netwotks 22, 544–557 (2009)
5. Yang, M., Nouretdinov, I., Luo, Z.: Learning by Conformal Predictors with Additional Information. In: Papadopoulos, H., Andreou, A.S., Iliadis, L., Maglogiannis, I. (eds.) AIAI 2013. IFIP AICT, vol. 412, pp. 394–400. Springer, Heidelberg (2013)
6. Adamskiy, D., Nouretdinov, I., Gammerman, A.: Conformal prediction in semi-supervised case. In: Post-Symposium Book 'Statistical learning and Data Science'. Chapman and Hall, Paris (2011)
7. Nouretdinov, I., Gammerman, A., Qi, Y., Klein-Seetharaman, J.: Determining Confidence of Predicted Interactions Between HIV-1 and Human Proteins Using Conformal Method. In: Pacific Symposium on Biocomputing, vol. 17, pp. 311–322 (2012)
8. Lei, J., Robins, J., Wasserman, L.: Efficient Nonparametric Conformal Pre- diction Regions. arXiv:1111.1418v1

Rule-Based Impact Analysis
for Enterprise Business Intelligence

Kalle Tomingas[1], Tanel Tammet[1], and Margus Kliimask[2]

[1] Tallinn University of Technology, Ehitajate tee 5, Tallinn 19086 Estonia
[2] Eliko Competence Center, Teaduspargi 6/2, Tallinn 12618 Estonia

Abstract. We address several common problems in the field of Business Intelligence, Data Warehousing and Decision Support Systems: the complexity to manage, track and understand data lineage and system component dependencies in long series of data transformation chains. The paper presents practical methods to calculate meaningful data transformation and component dependency paths, based on program parsing, heuristic impact analysis, probabilistic rules and semantic technologies. Case studies are employed to explain further data aggregation and visualization of the results to address different planning and decision support problems for various user profiles like business users, managers, data stewards, system analysts, designers and developers.

Keywords: impact analysis, data lineage, data warehouse, rule-based reasoning, probabilistic reasoning, semantics.

1   Introduction

Developers and managers are facing similar Data Lineage (DL) and Impact Analysis (IA) problems in complex data integration (DI), business intelligence (BI) and Data Warehouse (DW) environments where the chains of data transformations are long and the complexity of structural changes is high. The management of data integration processes becomes unpredictable and the costs of changes can be very high due to the lack of information about data flows and internal relations of system components. The amount of different data flows and system component dependencies in a traditional data warehouse environment is large. Important contextual relations are coded into data transformation queries and programs (e.g. SQL queries, data loading scripts, open or closed DI system components etc.). Data lineage dependencies are spread between different systems and frequently exist only in program code or SQL queries. This leads to unmanageable complexity, lack of knowledge and a large amount of technical work with uncomfortable consequences like unpredictable results, wrong estimations, rigid administrative and development processes, high cost, lack of flexibility and lack of trust. We point out some of the most important and common questions for large DW environments (see Fig.1) which usually become a topic of research for system analysts and administrators:

L. Iliadis et al. (Eds.): AIAI 2014 Workshops, IFIP AICT 437, pp. 301–309, 2014.
© IFIP International Federation for Information Processing 2014

- Where does the data come or go to in a specific column, table, view or report?
- Which components (reports, queries, loadings and structures) are impacted when other components are changed?
- Which data, structure or report is used by whom and when?
- What is the cost of making changes?
- What will break when we change something?

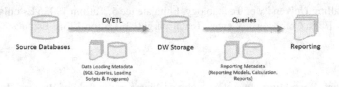

Fig. 1. General scheme of DW/BI data flows

The ability to find ad-hoc answers to many day-to-day questions determines not only the management capabilities and the cost of the system, but also the price and flexibility of making changes. The dynamics in business, environment and require-ments ensure that regular changes are a necessity for every living organization. Due to its reflective nature, the business intelligence is often the most fluid and unsteady part of enterprise information systems.

Obviously, the most promising way to tackle the challenges in such a rapidly grow-ing, changing and labor-intensive field is automation. We claim that efficient automa-tization in this particular field requires the use of semantic and probabilistic technolo-gies. Our goal is to aid the analysts with tools which can reduce several hard tasks from weeks to minutes, with better precision and smaller costs.

2   Related Work

Impact analysis, traceability and data lineage issues are not new. A good overview of the research activities of the last decade is presented in an article by Priebe et al. [9]. We can find various research approaches and published papers from the early 1990s with methodologies for software traceability [10]. The problem of data lineage tracing in data warehousing environments has been formally founded by Cui and Widom [2] and [3]. Our recent paper builds upon this theory by introducing the Abstract Map-ping representation of data transformations [14].

Other theoretical works for data lineage tracing can be found in [5] and [6]. Fan and Poulovassilis developed algorithms for deriving affected data items along the transformation pathway [5]. These approaches formalize a way to trace tuples (resp. attribute values) through rather complex transformations, given that the transforma-tions are known on a schema level. This assumption does not often hold in practice. Transformations may be documented in source-to-target matrices (specification li-neage) and implemented in ETL tools (implementation lineage).

Other practical works that based on conceptual models, ontologies and graphs for data quality and data lineage tracking can be found in [15] and [12]. De Santana proposes the integrated metadatada and the CWM metamodel based data lineage documentation approach [4]. The workflows and the manual annotations based solution proposed by Missier et al. [8].

Priebe et al. [9] concentrates on proper handling of specification lineage, a huge problem in large-scale DWH projects, especially in case different sources have to be consistently mapped to the same target. They propose a business information model (or conceptual business glossary) as the solution and a central mapping point to overcome those issues.

Our approach to Impact Analysis and Data Lineage differs from previous work in several aspects. Our aim is to merge technical data lineage [3] with semantic integration approaches [9], [11], using grammar based methods for metadata extraction from program texts and a probabilistic rule-based inference engine for weight calculations and reasoning approaches [7]. We also use the novel and powerful web based data flow and the graph visualization techniques with the multiple view approach [16] to deliver the extraction and the calculation of the result to the end-users.

3   System Architecture

We present a working Impact Analysis solution which can be adopted and implemented in an enterprise environment or provided as a service (SaaS) to manage organization information assets, analyze data flows and system component dependencies. The solution is modular, scalable and extendable. The core functions of our system architecture are built upon the following components presented in the Fig.2.

1. Scanners collect metadata from different systems that are part of DW data flows (DI/ETL processes, data structures, queries, reports etc.). We build scanners using our xml-based data transformation language and runtime engine XDTL [18].
2. The SQL parser is based on customized grammars, GoldParser parsing engine [1] and the Java-based XDTL engine.
3. The rule-based parse tree mapper extracts and collects meaningful expressions from the parsed text, using declared combinations of grammar rules and parsed text tokens.
4. The query resolver applies additional rules to expand and resolve all the variables, aliases, sub-query expressions and other SQL syntax structures which encode crucial information for data flow construction.
5. The expression weight calculator applies rules to calculate the meaning of data transformation, join and filter expressions for impact analysis and data flow construction.
6. The probabilistic rule-based reasoning engine propagates and aggregates weighted dependencies.
7. The directed and weighted sub-graph calculations, visualization and web based UI for data lineage and impact analysis applications.
8. The MMX open-schema relational database using PostgreSQL or Oracle for storing and sharing scanned, calculated and derived metadata [17].

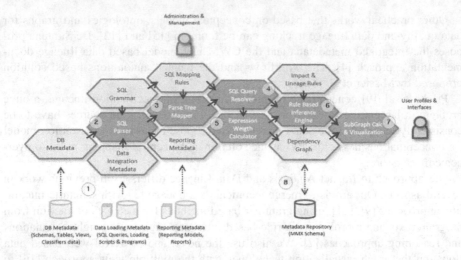

Fig. 2. Impact Analysis system architecture components

4   Query Parsing and Metadata Extraction

Scanners (No1 in Fig.2) collect metadata from external systems: database data dictionary structures, ETL system scripts and queries, reporting system query models and reports. All the structural information extracted is stored to the metadata database. The scanned objects and their properties are extracted and stored according to the meta-models we have designed: relational database, data integration, reporting and business terminology models. Meta-models contain ontological knowledge about the metadata and relations collected across different domains and models. The scanner technology (XDTL) and the open-schema metadata database (MMX) design have been described at a more detailed level in our previous work [13],[14].

In order to construct the data flows from the very beginning of the data sources (e.g. the accounting system) to the end points (e.g. the reporting system) we have to be able to find and connect both the identical and the related objects in different systems. In order to connect the objects we have to understand and extract the relations from the SQL queries (e.g. ETL tasks, database views, database procedures), scripts (e.g. loader utility scripts) and expressions (e.g. report structure) collected and stored by scanners. In order to understand the data transformation semantics encoded in the query language statements (e.g. insert, update, select and delete queries) and expressions, we have to involve external knowledge about the syntax and grammatical structure of the query language. Grammar-based parsing functionality is built into the scanner technology. A configurable "parse" command brings semi-structured text parsing and information extraction into the XDTL data integration environment. As the result of SQL parsing step (No2 in Fig.2) we get a large parse tree with every SQL query token assigned a special disambiguated meaning by the grammar.

In order to convert different texts into the tree structure, to reduce tokens and to convert the tree back to the meaningful expressions (depending on search goals), we use a declarative rule set presented in the Json format, combining token and grammar rules. A configurable grammar and a synchronized reduction rule set makes the XDTL parse command suitable for general purpose information extraction and captures the resource hungry computation steps into one single parse-and-map step with the flat table outcome. The Parse Tree Mapper (No3 in Fig.2) uses three different rule sets with more than eighty rules to map the parse tree to data transformation expressions.

After extraction and mapping of each SQL query statement into a series of expressions we execute the SQL Query Resolver (No4 in Fig.2) which contains a series of functions to resolve the SQL query structure:

- Solve source and target object aliases to full qualified object names;
- Solve sub-query aliases to context specific source and target object names;
- Solve sub-query expressions and identifiers to expand all the query level expressions and identifiers with fully qualified and functional ones;
- Solve syntactic dissymmetry in different data transformation expressions (e.g. insert statement column lists, select statement column lists and update statement assign list etc.);
- Extract quantitative metrics from data transformation, filter and join expressions to calculate expression weights (e.g. number of columns in expression, functions, predicates, string constants, number constants etc.).

5   Data Transformation Weight Calculation

Data structure transformations are parsed, extracted from queries and stored as formalized, declarative mappings in the system. To add additional quantitative measures to each column transformation or column usage in the join and filter conditions we evaluate each expression and calculate transformation and filter weights for those.

Expression Weight Calculation (No5 in Fig.2) is based on the idea that we can evaluate column data "transformation rate" and column data "filtering rate" using data structure and structure transformation information captured from SQL queries. Such a heuristic evaluation enables us to distinguish columns and structures used in the transformation expressions or in filtering conditions or both, and gives probabilistic weights to expressions without the need to understand the full semantics of each expression. We have defined two measures that we further use in our probabilistic rule system for deriving new facts:

Definition 1. A primitive data transformation operation $O(X,Y,M1,F1,W_t)$ is a data transformation between a source column X and a target column Y in a transformation set M (mapping or query) having expression similarity weight W_t and having conditions set F1.

Definition 2. Column transformation weight W_t is based on the similarity of each source column and column transformation expression: the calculated weight expresses the source column transfer rate or strength. The weight is calculated on scale [0,1] where 0 means that the source data are not transformed to target (e.g. constant assignment in a query) and 1 means that the source is directly copied to the target (no additional column transformations).

Definition 3. Column filter weight W_f is based on the similarity of each filter column in the filter expression and the calculated weight expresses the column filtering rate or strength. The weight is calculated on scale [0,1] where 0 means that the column is not used in the filter and 1 means that the column is directly used in the filter predicate (no additional expressions).

The general column weight W algorithm in each expression for W_t and W_f components is calculated as a column count ratio over all the expression component counts (e.g. column count, constant count, function count, predicate count):

```
W=IdCount/IdCount+FncCount+StrCount+NbrCount+PrdCount.
```

5.1   Rule System

The defined figures, operations and weights are used in combination with the declarative probabilistic inference rules to calculate the possible relations and dependencies between data structures and software components. Applying the rule system to the extracted query graphs we calculate and produce a full dependency graph that is used for data lineage and impact analysis.

The basic operations used in the rules for the dependency graph calculations are the following:

- The primitive data transformation is the elementary operation between the source column X and the target column Y in the query mapping id set M1 with the filter condition set F1 and the transformation weight W_t (see Definition 1). This is represented by the predicate `O(X,Y,M1,F1,Wt)`;
- The predicate `member(X,F1,Wt)` is used in the filter impact calculation rule to detect that the column X is a member of the filter id set F1 with the filter weight Wt;
- The predicate `disjoint(M1,M2)` is used in the impact aggregation rule to detect that two query mapping id sets M1 and M2 are disjoint. The disjointness condition is necessary for aggregating the data transformation relations and weights in case more than one path from different queries connects the same column pairs;
- The predicate `parent(X0,X1)` is used in the parent aggregation rule to detect that the table X0 is the owner or parent object of the column X1. This condition is necessary for aggregating all the column level relations and the weights between two tables for the table level impact relation;
- The function `union(M1,M2)` is used to calculate the impact relations over two disjoint query id sets M1 and M2: it returns the distinct id lists of two sets M1 and M2;
- The function `sum(W1,W2)` is used to calculate the aggregated impact relation weight in case the basic operations are disjoint, i.e. stem from independent queries. The weight calculation is based on non-mutually-exclusive event probabilities (two independent queries means there could be an overlap between two events) and is calculated as the sum of probabilities of W1 and W2: `sum(W1,W2)=(W1+W2)-(W1*W2)`;

- The function avg(W1,W2) is used to calculate the parent impact weight when the basic operations have the same parent structures. The weight is calculated as the arithmetic mean of W1 and W2: $\text{avg(W1,W2)} = \text{(W1+W2)/2}$;

The inference rules with the basic operations and the weighs for the dependency graph calculations are the following:

- The basic impact calculation rule for the operation O with no additional filters produces the impact predicate I:
  `O(X,Y,M1,F1,Wt) => I(X,Y,M1,F1,Wt);`
- The basic impact calculation rule for the operation O with a related filter condition produces the impact predicate I with multiplied weights:
  `O(X,Y,M1,F1,Wt)& member(X,F1,Wf) => I(X,Y,M1,F1,Wt*Wf);`
- The transitivity rule is used to calculate the sequences of the consecutive impact relations:
  `I(X,Y,M1,F1,W1) & I(Y,Z,M2,F2,W2) & disjoint(M1,M2) =>`
  `          I(X,Z,union(M1,M2),union(F1,F2),W1*W2);`
- The column aggregation rule is used when multiple different paths from the different queries connect the same columns: calculate the impact relations with aggregated query id-s and the aggregated weights:
  `I(X,Z,M1,F1,W1) & I(X,Z,M2,F2,W2) & disjoint(M1,M2) =>`
  `          I(X,Z,union(M1,M2),union(F1,F2),sum(W1,W2));`
- The parent aggregation rule is used when multiple different impact relations connect the column pairs of the same tables: calculate the table level impact relations with aggregated query id-s and aggregated weights:
  `I(X1,Y1,M1,F1,W1) & I(X2,Y2,M2,F2,W2) & parent(X0,X1)& par-`
  `ent(X0,X2) & parent(Y0,Y1) & parent(Y0,Y2) =>`
  `          I(X0,Y0,union(M1,M2),union(F1,F2),avg(W1,W2)).`

5.2   Dependency Score Calculation

We can use the derived dependency graph to solve different business tasks by calculating the selected component(s) lineage or impact over available layers and details chosen details. Business questions like: "What reports are using my data?", "Which components should be changed or tested?" or "What is the time and cost of change?" are converted to directed sub-graph navigation and calculation tasks. The following definitions add new quantitative measures to each component or node (e.g. table, view, column, etl task, report etc.) in the calculation. We use those measures in the UI to sort and select the right components for specific tasks.

Definition 4. Local Lineage Dependency % (LLD) is calculated as the ratio over the sum of the local source and target Lineage weights W_t:

`LLD = SUM(source(`W_t`)/source(`W_t`)+target(`W_t`)).`

Local Lineage Dependency 0 % means that there are no data sources detected for the object. Local Lineage Dependency 100 % means that there are no data consumers (targets) detected for the object. Local Lineage Dependency about 50 % means that there are equal numbers of weighted sources and consumers (targets) detected for the object.

Definition 5. Local Impact Dependency % (LID) is calculated as the ratio over the sum of local source and target impact weights $W(W_t, W_f)$:

```
LID=SUM(source(W) /source(W)+target(W)).
```

Local Impact Dependency 0 % means that there are no dependent data sources detected for the object. Local Dependency 100 % means that there are no dependent data consumers (targets) detected for the object. Local Impact Dependency about 50 % means that there are equal numbers of weighted dependent sources and consumers (targets) detected for the object.

Definition 6. Global Dependency Count (GDC) is the sum of all source and target Lineage and Impact relations counts: `GDC=GSC+GTC`.

The Global Dependency Count is a good differential metric that allows us to see clear distinctions in the dependencies of each object. We can take the GDC metric as a sort of "gravity" of the object that can be used to develop new rules, to infer the time and cost of changes of object(s) (e.g. database table, view, data loading programs or report).

6   Conclusions and Future Work

The previously described architecture and algorithms have been used to implement an integrated toolset dLineage (http://dlineage.com). Both the scanners and web-based tools of dLineage have been enhanced and tested in real-life projects and environments to support several popular DW database platforms (e.g. Oracle, Greenplum, Teradata, Vertica, PostgreSQL, MsSQL, Sybase), ETL tools (e.g. Pentaho, Oracle Data Integrator, SQL scripts and different data loading utilities) and BI tools (e.g. SAP Business Objects, Microstrategy). The dLineage dynamic visualization and graph navigation tools are implemented in Javascript using the d3.js graphics libraries.

We have tested our solution during two main case studies involving a thorough analysis of large international companies in the financial and the energy sectors. Both case studies analyzed thousands of database tables and views, tens of thousands of data loading scripts and BI reports. Those figures are far over the capacity limits of human analysts not assisted by the special tools and technologies.

We have presented several algorithms and techniques for quantitative impact analysis, data lineage and change management. The focus of these methods is on automated analysis of the semantics of data conversion systems followed by employing probabilistic rules for calculating chains and sums of impact estimations. The algorithms and techniques have been successfully employed in two large case studies, leading to practical data lineage and component dependency visualizations.

We are planning to continue this research by considering a more abstract, conceptual/business level in addition to the current physical/technical level of data representation.

Acknowledgments. This research has been supported by European Union through European Regional Development Fund.

References

1. Cook, D.: Gold parsing system (2010), http://www.goldparser.org
2. Cui, Y., Widom, J., Wiener, J.L.: Tracing the lineage of view data in a warehousing environment. ACM Transactions on Database Systems (TODS) 25(2), 179–227 (2000)
3. Cui, Y., Widom, J.: Lineage tracing for general data warehouse transformations. The VLDB Journal—The International Journal on Very Large Data Bases 12(1), 41–58 (2003)
4. de Santana, A.S., de Carvalho Moura, A.M.: Metadata to support transformations and data & metadata lineage in a warehousing environment. In: Kambayashi, Y., Mohania, M., Wöß, W. (eds.) DaWaK 2004. LNCS, vol. 3181, pp. 249–258. Springer, Heidelberg (2004)
5. Fan, H., Poulovassilis, A.: Using AutoMed metadata in data warehousing environments. In: Proceedings of the 6th ACM International Workshop on Data Warehousing and OLAP, pp. 86–93. ACM (November 2003)
6. Giorgini, P., Rizzi, S., Garzetti, M.: GRAnD: A goal-oriented approach to requirement analysis in data warehouses. Decision Support Systems 45(1), 4–21 (2008)
7. Luberg, A., Tammet, T., Järv, P.: Smart City: A Rule-based Tourist Recommendation System. In: Information and Communication Technologies in Tourism 2011, pp. 51–62. Springer Vienna (2011)
8. Missier, P., Belhajjame, K., Zhao, J., Roos, M., Goble, C.: Data Lineage Model for Taverna Workflows with Lightweight Annotation Requirements. In: Freire, J., Koop, D., Moreau, L. (eds.) IPAW 2008. LNCS, vol. 5272, pp. 17–30. Springer, Heidelberg (2008)
9. Priebe, T., Reisser, A., Hoang, D.T.A.: Reinventing the Wheel?! Why Harmonization and Reuse Fail in Complex Data Warehouse Environments and a Proposed Solution to the Problem (2011)
10. Ramesh, B., Jarke, M.: Toward reference models for requirements traceability. IEEE Transactions on Software Engineering 27(1), 58–93 (2001)
11. Reisser, A., Priebe, T.: Utilizing Semantic Web Technologies for Efficient Data Lineage and Impact Analyses in Data Warehouse Environments. In: Database and Expert Systems Application, DEXA 2009, pp. 59–63 (August 2009)
12. Skoutas, D., Simitsis, A.: Ontology-based conceptual design of ETL processes for both structured and semi-structured data. International Journal on Semantic Web and Information Systems (IJSWIS) 3(4), 1–24 (2007)
13. Tomingas, K., Kliimask, M., Tammet, T.: Mappings, Rules and Patterns in Template Based ETL Construction. In: The 11th International Baltic DB & IS2014 Conference (2014)
14. Tomingas, K., Kliimask, M., Tammet, T.: Data Integration Patterns for Data Warehouse Automation. In: The 18th East-European ADBIS 2014 Conference (2014)
15. Vassiliadis, P., Simitsis, A., Skiadopoulos, S.: Conceptual modeling for ETL processes. In: Proceedings of the 5th ACM International Workshop on Data Warehousing and OLAP, pp. 14–21. ACM (November 2002)
16. Wang Baldonado, M.Q., Woodruff, A., Kuchinsky, A.: Guidelines for using multiple views in information visualization. In: Proceedings of the Working Conference on Advanced Visual Interfaces, pp. 110–119. ACM (May 2000)
17. MMX Metadata Framework, http://mmxframework.org
18. XDTL Data Transformation Language, http://xdtl.org

Feature Extraction and Visual Feature Fusion for the Detection of Concurrent Prefix Hijacks*

Stavros Papadopoulos[1,2], Konstantinos Votis[2], Christos Alexakos[3], and Dimitrios Tzovaras[2]

[1] Department of Electrical and Electronic Engineering, Imperial College London, SW7 2AZ, London, UK
s.papadopoulos11@imperial.ac.uk

[2] Information Technologies Institute, Centre for Research and Technology Hellas, P.O. Box 361, 57001 Thermi-Thessaloniki, Greece
{spap,kvotis,tzovaras}@iti.gr

[3] Pattern Recognition Laboratory, Computer Engineering and Informatics, University of Patras, Patras, Greece
alexakos@ceid.upatras.gr

Abstract. This paper presents a method for visualizing and analyzing Multiple Origin Autonomous System (MOAS) incidents on Border Gateway Protocol (BGP), for the purpose of detecting concurrent prefix hijack. Concurrent prefix hijacks happen when an unauthorized network originates prefixes that belong to multiple other networks. Towards the goal of accurately identifying such events, multiple features are extracted from the BGP records and visualized using parallel coordinates enhanced with visual querying capabilities. The proposed visual queries enable the analyst to select a significant subset of the initial dataset for further analysis, based on the values of multiple features. This procedure allows for the efficient visual fusion of the proposed features and the accurate identification of prefix hijacks. Most of the previous approaches on BGP hijack detection depend on static methods in order to fuse the information from multiple features and identify anomalies. The proposed visual feature fusion, however, allows the human operator to incorporate his expert knowledge into the analysis, so as to dynamically investigate the observed events, and accurately identify anomalies. The efficiency of the proposed approach is demonstrated on state-of-the-art BGP events.

1 Introduction

Multiple Origin Autonomous System (MOAS) [1] conflicts occur when a particular prefix appears to originate from more than one AS (Autonomous System).

---

* This work has been partially supported by the European Commission through the project FP7-ICT-317888-NEMESYS funded by the 7th framework program. The opinions expressed in this paper are those of the authors and do not necessarily reflect the views of the European Commission.

L. Iliadis et al. (Eds.): AIAI 2014 Workshops, IFIP AICT 437, pp. 310–319, 2014.

Although MOAS incidents occur often for legitimate reasons (e.g. multi-homing, exchange point addresses etc[1]), they might also be the result of a fault or an attack, where a Border Gateway Protocol (BGP) router falsely originates prefixes belonging to other organizations. The detection of such faults and attacks has been the main focus of the research community in the last years. In this paper, the main focus is the detection of concurrent prefix hijacks, in which an unauthorized network originates prefixes that belong to multiple other networks, usually due to router misconfigurations.

The proposed methods for prefix hijack identification are separated into two categories: authentication and detection. Authentication schemes [2][3] depend on the a priori knowledge of the prefix owners, while detection schemes make no such assumption but instead analyze the routing behavior in order to identify suspicious events. The approach presented in this paper falls into the second category.

The detection schemes depend on some type of feature extraction and feature fusion in order to detect BGP prefix hijacks. Although there are many well known techniques for algorithmic fusion [4], the result of such approaches is static and it does not capture the dynamic nature of the modern multivariate systems and the corresponding alterations of the features' cross-correlations. For this reason, visualization techniques can be utilized for the purpose of performing the feature fusion, and allow the user to incorporate his experience and knowledge into the analysis.

In this respect, this paper introduces a novel visual feature fusion method for the combination of different BGP features, extracted from the observed MOAS incidents. The ultimate goal of the proposed approach is the detection of concurrent prefix hijacks, in which an unauthorized network originates prefixes that belong to multiple other networks. In order to achieve this, visual queries are applied on the parallel coordinates visualization in order to enable the analyst to focus on a significant subset of the data for further analysis, based on the values of multiple features.

It should be noted that the BGP exhibits continuous alterations, with over 20,000 BGP messages exchanged every 5 minutes (in 2014). All these messages are analyzed using the proposed approach in order to firstly identify MOAS incidents, and afterwards extract meaningful features that will enable the anomaly detection procedure. The proposed approach has been applied on data collected for over 3 months with out any problems. But in this paper and without loss of generality, the demonstration is carried out with respect to data corresponding to a time period of 5 days.

The rest of the paper is organized as follows: Section 2 presents the related work. Section 3 presents the feature extraction methodology, while the proposed visual feature fusion method is detailed in section 4. Use case analysis takes place in section 5. Finally, the paper concludes in section 6.

2 Related Work

The internet uses BGP as the defacto protocol for the exchange of routing information between Autonomous Systems (AS). An AS represents a collection of Internet Protocol prefixes under the control of one network operator. The distributed nature of BGP and the lack of security mechanisms, render it vulnerable to various types of attacks, as for example prefix hijacking or Man- In-the-Middle attacks [5]. Towards this end, the research community has focused its efforts on the development of methods that could enable the detection of such anomalies in the BGP infrastructure.

Specifically, Deshpande et al. [6] introduced multiple features that represent various aspects of the BGP update messages. A Generalized Likelihood Ratio (GLR) based hypothesis test is utilized onto each feature in order to detect time periods of instabilities. Majority voting is afterwards utilized in order to correlate the detected instabilities, and generate alerts for which more than half of the features indicate an anomaly. Unlike the proposed visual feature fusion method, the disadvantage of this approach is that the analyst is not able to change the parameters of the anomaly detection algorithm in order to get better detection rates.

Li et al. [7] propose the use of a signature based classifier, that is trained to recognize certain types of behaviors specific to BGP anomalies. The classifier is comprised of a collection of IF-THEN threshold based rules, which are applied onto multiple BGP features, extracted in different time windows. As with the aforementioned approach the anomaly results are also static, as the analyst is left out of the detection procedure.

Al-Rousan et al. [8] introduce multiple features that characterize the BGP activity. Feature selection methods are applied onto this set of features in order to select a specific subset that better characterizes BGP anomalies. Afterwards, Naive Bayes classifiers are used for the detection of BGP anomalies, including worm attacks and router misconfigurations. In comparison with the proposed approach the user is not able to change the parameters of the classification in order to find alternative or better results.

Zhang et al. [9] propose the use of signature and a statistical based methods for detecting anomalies in BGP. The signature based methods are comprised of a standard set of patterns in the BGP update messages, which are specific to BGP anomalies. The statistical based methods are applied onto multiple BGP features, in order to acquire an anomaly score for each one of them. A linear weighted sum of the scores of each feature is afterwards utilized, so as to fuse the features and acquire a single anomaly score. This procedure is repeated for each time window under investigation, in order to detect time periods of instability. The results of this work were utilized by Teoh et al. [10], and combined with visualization methods. Specifically, the anomaly scores of each feature, as well as the global anomaly score are visualized along the BGP update messages selected by the analyst. Using the feedback provided by the visualization, the analyst is able to adjust the parameters of the statistical and signature based methods and increase the detection accuracy. The adjustable parameters, however, do not refer to the feature fusion procedure but rather to the definitions of anomalous BGP events and the threshold of the fused anomaly metric. The approach of including the analyst into the analytical procedure is related to the present work, which also incorporates visual methods to aid the analysis procedure. But unlike [10], the present work utilizes the visualization to aid the visual feature fusion procedure.

3   MOAS Feature Extraction

This section presents the feature extraction methodology proposed in the context of this paper. The extracted features are able to quantify the degree of anomaly of each MOAS incident, while their fusion provides and complete view of the MOAS characteristics and enables the detection of concurrent prefix hijacks.

There are four features that are extracted from the MOAS events that occurred over a specific time period: 1)*Country MOAS Probability*, 2)*AS Relationship*, 3)*Ownership duration*, and 4)*MOAS duration*. Each one of them is described in detail in the paragraphs that follow.

The first feature is *Country MOAS Probability*, which represents the probability of occurrence of a MOAS event from the country of the new owner AS to the country of the old owner AS. This country based aggregation and analysis of the BGP activity, has been shown to capture the underlying geo-spatial coherence of the Internet routing information [11]. Although the probability of occurrence of a MOAS event between two ASes can be used directly for anomaly detection, the country aggregation takes into account all the MOAS events between the two countries, and thus holds additional semantic information regarding the geo-spatial distribution of the MOAS incidents. This fact leads to a more accurate identification of concurrent prefix hijacks, since most of them target countries that are usually not targeted in normal behavior.

The *AS Relationship* between the two ASes involved in the MOAS incident has been used as a heuristic for reducing the number of false positives in prefix hijack detection[12]. Under this consideration, it is also utilized in this context for the generation of a binary feature. This feature captures the AS pairs that have commercial relationships, and thus might use prefixes from the same set. The AS relationships where defined by Gao [13] and are comprised of Provider-To-Customer, Peer-To-Peer, and Sibling-To-Sibling relationships. If any type of relationship exists between two ASes, the *AS Relationship* feature has value 1, else it has value 0.

The *Ownership duration* feature captures the duration in hours that the new owner has been announcing the prefix involved in the MOAS incident. Since the new owner AS involved in the concurrent prefix hijacks is not the legitimate owner of the prefix, it is highly unlikely that it has been announcing the prefix for a long period. Thus, this feature captures this behavior and provides additional information for the accurate identification of concurrent prefix hijacks. It should be noted that absolute thresholds for anomaly detection using the aforementioned feature have been used in the literature [12], but in the proposed

Table 1. Overview of the MOAS features and metadata used for the identification of concurrent prefix hijacks

		Description
Features	Country MOAS Probability	The probability of a MOAS occurrence from the country of the Attacking AS to the country of the Victim AS
	AS Relationship	Captures the AS that have direct commercial relationships.
	Ownership duration	The duration in hours that the Attacking AS has been announcing the corresponding prefix
	MOAS duration	The duration in hours of the MOAS incident between the two ASes
Metadata	Attacking AS	The AS number of the new owner of the prefix
	Victim AS	The AS number of the old owner of the prefix
	Time	The exact date and time that the MOAS event occurred

approach combines the actual value of this feature with multiple other features, and allows the analyst to select appropriate thresholds dynamically, using visual queries.

The *MOAS duration* feature captures the duration in which each MOAS event is active. In the case of concurrent prefix hijacks, the MOAS incidents are highly unlikely to last for a long period. After the misconfiguration has been fixed, the prefixes return to their original owners. Although the MOAS duration can be a useful heuristic to distinguish between valid MOAS conflicts and invalid ones[1], such differentiation on this feature alone can not be accurate enough to be a solution prefix hijack identification. In this context, the visual feature fusion procedure proposed in the context of this paper combines the values of this feature with multiple other features, in order to provide a holistic view of the MOAS activity to the analyst.

In addition to the aforementioned features that characterize the degree of anomaly of each MOAS incident, the MOAS events are also characterized by metadata. These metadata provide additional semantic information about each event and are used for root cause analysis by the visualization. These metadata include the following:*Attacking AS*, i.e. the AS number of the new owner of the prefix, *Victim AS*, i.e. the AS number of the old owner of the prefix, and *Time*, i.e. the exact date and time that the MOAS event occurred.

An overview of the extracted features and the metadata that are utilized for visual feature fusion and prefix hijack detection is presented in Table 1.

4   Visual Feature Fusion Using Parallel Coordinates and Visual Queries

The proposed visual feature fusion scheme is detailed in this section. Specifically, the features and metadata describing each MOAS event, which were presented in section 3, are visualized using parallel coordinates. The parallel coordinates are enhanced with filtering capabilities in order to enable the analyst to detect significant MOAS incidents that might contain anomalies.

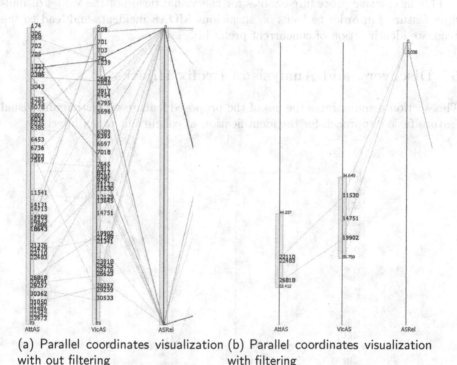

(a) Parallel coordinates visualization (b) Parallel coordinates visualization
with out filtering   with filtering

Fig. 1. The application of visual queries on the parallel coordinates utilized for visual feature fusion and the selection of significant data subsets

Specifically, bar sliders are attached onto each parallel coordinate that specify the allowed upper and lower bounds of each feature. These bar sliders can change size and position and are directly manipulated by the user of the visualization. This procedure allows the analyst to dynamically change the bounds of the features and as a result the MOAS events that are visualized. The direct feedback provided by the visualization with regards to the selection of MOAS events, enables the analyst to redefine the bounds, and find the best parameters in each case, according to the underling information.

Thus, for each feature and metadata f_i the bounds are defined as $B_i = \{f_i^{low}, f_i^{upper}\}$. Each MOAS event is characterized by its value in all the features and metadata. Furthermore, the following function is defined:

$$g(e_j) = \left\{ \begin{array}{cc} true, & if \; f_i^{low} \leq f_i(e_j) \leq f_i^{high}, \forall i \\ false, & else \end{array} \right\} \tag{1}$$

where $f_i(e_j)$ is the value of the j_{th} event e_j in the i_{th} feature f_i. If $g(e_j)$ is $true$ then the MOAS event e_j is visualized, while in the case that is $false$ it is not.

This interactive procedure enables for the visual fusion of the values of multiple features, in order to focus on suspicious MOAS incidents and lead to the accurate identification of concurrent prefix hijacks.

5   Discovery and Analysis of Prefix Hijacks

This section demonstrates the use of the proposed feature extraction and visual feature fusion approach for the identification of concurrent prefix hijacks.

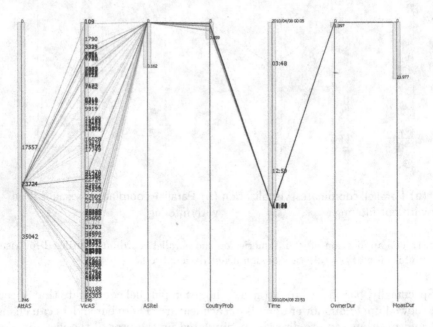

Fig. 2. Visual feature fusion for the detection of a router misconfiguration event which tool place on 08-April-2010[12]. The *AS Relationship* feature is set to 0 (no relationship exists), the *Country MOAS Probability* is set to be smaller than 6%, the *Ownership duration* feature is set to be smaller than 2 hours, and the *MOAS duration* feature is set to be smaller than 24 hours. After this filtering, AS-23724 is still involved in multiple MOAS events, occurring simultaneously (at around 16:00 GMT)

The BGP data were collected and analyzed from the RIPE routing monitoring service [14]. Specifically, the monitoring point of AS-3333 was utilized for the collection of the routing updates, and the analysis of the observed MOAS incidents.

It should be underlined that the monitoring period is larger than the period used for the visualization. The reason for this is that the values of the features are relevant to the total monitoring period, and small monitoring periods might not be enough for their accurate calculation. In both use cases examined in this section, the visualization period is set to one day, while the monitoring period is set to five days, two days before and two days after the visualization period.

The first use case involves a concurrent prefix hijack that occurred on 08-April-2010, around 16:00 GMT. In this incident, AS-23724 announced multiple prefixes it did not own, belonging to multiple other ASes[12]. The monitoring period is set from 06-April-2010 to 10-April-2010, while the visualization period is set to 08-April-2010. The result of the visual feature fusion is depicted in figure 2. The values of the *AS Relationship*, *Country MOAS Probability*, *Ownership duration*, and *MOAS duration* are set to be: 0 (no relationship exists), smaller than 6%, smaller than 2 hours, and smaller than 24 hours, respectively. After the application of these visual queries, there is a small set of MOAS incidents that remains. It is apparent that in the first parallel coordinate, which

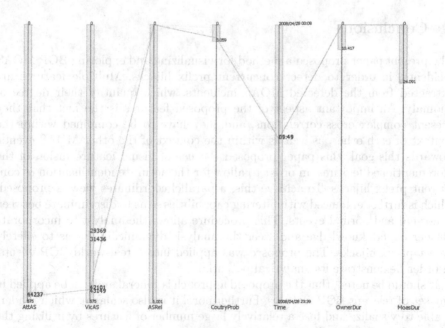

Fig. 3. Visual feature fusion for the detection of a router misconfiguration event which tool place on 28-April-2008[12]. The *Country MOAS Probability* is set to be smaller than 6%, the *Ownership duration* feature is set to be smaller than 10 hours, and the *MOAS duration* feature is set to be smaller than 24 hours. AS-44237 is involved in four suspicious MOAS events visible from the monitoring point of AS-3333.

represents the Attacking AS metadata, that AS-23724 is involved in multiple suspicious MOAS incidents, which last for a short period and have low probability of appearance. Furthermore, all these MOAS events are occurring almost simultaneously, around 16:00 GMT.

It should be noted that after the filtering procedure has taken place, there are two more MOAS incidents that remain. These incidents are not concurrent, since they do not concern multiple ASes. These events are short lived in the monitoring period and also have low probability of appearance. But due to the nature of BGP and the lack of ground truth, the answer to the legitimacy of these events is difficult and out of the scope of this paper, which focuses on the detection of concurrent prefix hijacks.

The second use case concerns a concurrent prefix hijack that occurred on 28-April-2008 by AS-44237[12]. The monitoring period is set from 26-April-2008 to 30-April-2008, while the visualization period is set to 28-April-2008. The result of the visual fusion is depicted in figure 3. The bounds of the features are set using visual queries to: *Country MOAS Probability* smaller than 6%, *Ownership duration* smaller than 10 hours, and *MOAS duration* feature smaller than 24 hours. It is apparent in this figure that AS-44237 is involved in four suspicious MOAS events, which are visible from the monitoring point of AS-3333. All these MOAS incidents occurred on 28-April-2008 around 9:49 GMT.

6   Conclusions

The present paper proposed a method for visualizing and exploring BGP MOAS incidents, in order to detect concurrent prefix hijacks. Multiple features are extracted from the detected MOAS incidents, which quantify their degree of anomaly. An important aspect of the proposed features is the fact that they present complex cross-correlations, and they have to be combined within the context of each other, as well as within the context of the other MOAS events. Towards this goal, this paper proposed the use of visual feature fusion of the aforementioned features, in order to allow for the accurate identification of concurrent prefix hijacks. To achieve this, a parallel coordinates view is proposed, which is further enhanced with filtering capabilities so as to discriminate between abnormal and normal events. This procedure allows the analyst to incorporate his/her expert knowledge and steer the analysis dynamically, so as to effetely detect prefix hijacks. The proposed was applied into a real-world BGP events, in order demonstrate its analytical potential.

It should be noted, that the proposed approach is general and can be applied in any set of relevant BGP features. Furthermore, it is also scalable, which renders it able to visualize and fuse a relatively large number of features by utilizing the power of parallel coordinates.

References

1. Zhao, X., Pei, D., Wang, L., Massey, D., Mankin, A., Wu, S.F., Zhang, L.: An Analysis of BGP Multiple Origin AS (MOAS) Conflicts. In: SIGCOMM Internet Measurement Workshop, p. 31. ACM Press (2001)
2. Zheng, C., Ji, L., Pei, D., Wang, J., Francis, P.: A light-weight distributed scheme for detecting ip prefix hijacks in real-time. ACM SIGCOMM Computer Communication Review 37(4), 277 (2007)
3. Lad, M., Massey, D., Pei, D., Wu, Y., Zhang, B., Zhang, L.: PHAS: A prefix hijack alert system. In: Proc. USENIX Security Symposium, vol. 2, pp. 153–166 (2006)
4. Chowdhury, P., Das, S., Samanta, S., Mangai, U.: A Survey of Decision Fusion and Feature Fusion Strategies for Pattern Classification. IETE Technical Review 27(4), 293–307 (2010)
5. Ballani, H., Francis, P., Zhang, X.: A study of prefix hijacking and interception in the internet. ACM SIGCOMM Computer Communication Review 37(4), 265 (2007)
6. Deshpande, S., Thottan, M., Ho, T.K., Sikdar, B.: An online mechanism for BGP instability detection and analysis. IEEE Transactions on Computers 58(11), 1470–1484 (2009)
7. Li, J., Dou, D., Wu, Z., Kim, S., Agarwal, V.: An Internet routing forensics framework for discovering rules of abnormal BGP events. ACM SIGCOMM Computer Communication Review 35(5), 55–66 (2005)
8. Al-Rousan, N.M., Haeri, S., Trajkovic, L.: Feature selection for classification of BGP anomalies using Bayesian models. In: ICMLC, pp. 140–147 (2012)
9. Zhang, K., Yen, A., Zhao, X., Massey, D., Wu, S.F., Zhang, L.: On detection of anomalous routing dynamics in BGP. In: Mitrou, N.M., Kontovasilis, K., Rouskas, G.N., Iliadis, I., Merakos, L. (eds.) NETWORKING 2004. LNCS, vol. 3042, pp. 259–270. Springer, Heidelberg (2004)
10. Teoh, S.T., Zhang, K., Tseng, S.-M., Ma, K.-L., Wu, S.F.: Combining visual and automated data mining for near-real-time anomaly detection and analysis in BGP. In: Proceedings of the 2004 ACM Workshop on Visualization and Data Mining for Computer Security, VizSECDMSEC 2004, p. 35 (2004)
11. Theodoridis, G., Tsigkas, O., Tzovaras, D.: A Novel Unsupervised Method for Securing BGP Against Routing Hijacks. In: Computer and Information Sciences III, pp. 21–29. Springer (2013)
12. Khare, V., Ju, Q., Zhang, B.: Concurrent prefix hijacks: Occurrence and impacts. In: Proceedings of the 2012 ACM Conference on Internet Measurement Conference, pp. 29–36. ACM (2012)
13. Gao, L.G.L.: On inferring autonomous system relationships in the Internet. IEEE/ACM Transactions on Networking 9(6), 733–745 (2001)
14. RIPE Network Coordination Centre, Routing Information Service project (RIS), http://www.ripe.net

Linking Data in the Insurance Sector: A Case Study

Katerina Kalou and Dimitrios Koutsomitropoulos

HPCLab, Computer Engineering and Informatics Dpt., University of Patas, Building B,
26500 Patras-Rio, Greece
{kaloukat,kotsomit}@hpclab.ceid.upatras.gr

Abstract. Enterprise data model (EDM) has been designated as a well-established approach in order to define a common way for communication and interoperability within an organization and across the industry. In recent years, business areas can also profit by the concept of Big Data, thus achieving effective management of massive amount of data. In this paper, we propose an example of how a data model can be enriched with the power of semantic web technologies, by converting it into an ontology. In this way, the data within the enterprise can be efficiently integrated into the Linked Open Data (LOD) cloud. The ultimate aim of this attempt is to set the infrastructure for exploring and linking Big Data in order to derive new value by exploiting Linked Data. We have selected to work on the insurance sector and apply the aforementioned ideas to the Property and Casualty data model, proposed by OMG.

Keywords: insurance, property, casualty, OMG, ontology, linked data, big data.

1   Introduction

An enterprise data model (EDM) captures the semantics of an organization for the purpose of communicating with the business community therefore, it can be considered a kind of ontology [3]. The components of an ontology, such as rules, classes, properties and individuals, can effectively reflect entities and activities described by the enterprise data model. Both an ontology and a data model can contribute to the meaning of data and to facilitate interoperability, integration, and adaptability of the informational systems that adopt them. Even though an EDM encapsulates the notion of ontology and shares common entities, it lacks the power of fine-grained expressiveness, reasoning and knowledge-based inference capabilities.

In recent years, Big Data [9] is used to describe all the new technologies that allow the manipulation of massive amounts of data in order to develop existing business areas or to drive new ones by improving on insight, decision-making and detecting sources of profit. Big Data can be thoroughly described by the following characteristics: Variety (the number of types of data), Velocity (the speed of data generation and processing), Volume (the amount of data) and Value (the knowledge).

In parallel of Big Data evolution, Linked Data [2], one of the most widely-used Semantic technologies, constitutes a means to evolve the Web into a global data space. Linked Data was introduced as a set of best practices for publishing and interlinking structured data on the Web by using RDF as the data model and HTTP as the

L. Iliadis et al. (Eds.): AIAI 2014 Workshops, IFIP AICT 437, pp. 320–329, 2014.

protocol. Even though Linked Data gains its success as a technology by fostering data publication in the WWW, it can also be utilized as an architectural style for integrating either applications or data within the Enterprise.

In the case of Big Data, the Linked Data principles can contribute effectively to the aspect of Variety [10] by simplifying data management. For example, data coming from unstructured databases can be linked to data in traditional OLTP and OLAP stores without changing existing schemas.

The interlinking of data following different models (OLAP/OLTP databases, documents, NoSQL etc.) allows more data to be linked and then queried more effectively. So, the context of data expands by achieving more insight as well as better data analysis. The usage of Linked Data in combination with Big Data supports and boosts the fourth V, the Value, too [10].

In this work, we make an attempt to gather all the aforementioned aspects, which can leverage the growth of an industry, into one unified approach. Beginning from an EDM, we move to a more semantically enhanced model, an ontology, thus developing the foundation for more efficient consumption of Big Data.

More precisely, we exploit the existing data model for Property and Casualty [6], inspired and proposed by the Object Management Group, in order to build an ontology. This ontology meets the needs of and could also be used as the basis for publishing Linked Data in the industry of insurance. The interlinking of data ensures more interoperability as well as sets the infrastructure for adopting Big Data in the Insurance sector.

The usage of an ontology to express the P&C Model as well as any other EDM in the industry may have the following benefits:

- Enhance the existing expressive power of the EDM
- Achieve interoperability and extendibility of higher level
- Support more efficiently the possibilities of scalability
- Accomplish knowledge inference and discovery by exploiting the reasoning power
- Bridge the gap between any other subsystems (different kinds of data storage, databases, documents etc.) in the industry without negating the benefits of conventional stores or NoSQL databases

The rest of this paper is organized as follows. In Section 2, we start by providing some broad definitions and discussing the concepts of OMG's Property and Casualty Model. Furthermore, in Section 3, we explain in detail the entire process of converting the terms of data model into ontology components. Next, Section 4 outlines an indicative application scenario in order to illustrate the ontology. Finally, Section 5 summarizes our conclusions.

2   OMG's Property and Casualty Model

The Object Management Group (OMG) [13] has developed a property-casualty insurance enterprise data model with the contribution of representatives from many carriers and using experts in each subject area. The P&C data model [6] addresses the data management needs of the Property and Casualty (P&C) insurance community. OMG's Model Driven Architecture principles and related standards have been utilized in order to construct the data model. Also, existing P&C industry standards (e.g., IBM's IAA) have been picked out as a source for the proposed P&C Business Glossary and associated models.

There are plenty of benefits for the organizations that decide to incorporate this data model into their procedures. P&C data model can provide significant value during a wide variety of warehousing, business intelligence, and data migration projects. In addition, these models can be leveraged to improve the use of data and in support of data governance within the enterprise.

Among the components of the P&C data model, there are the entities, attributes and relationships. Due to lack of space, only a snippet of all these is presented in this work. The main objects of the data model are the *Entities*. An Entity represents a person, organization, place, thing, or concept of interest to the enterprise. Next, an Entity can express a very broad and varied set of instance data, or it can be very specific. This is referred to as *levels of abstraction*. Some entities that have been used for the application scenario described in Section 4 are listed below.

- **Insurable Object:** An Insurable Object is an item which may be included or excluded from an insurance coverage or policy, either as the object for which possible damages of the object or loss of the object is insured, or as the object for which damages caused by the object are insured. Examples: residence, vehicle, class of employees.
- **Vehicle:** A Vehicle is an Insured Object that is a conveyance for transporting people and/or goods.
- **Person:** Person can be a human being, either alive or dead.
- **Organization:** An Organization is a Party that is a business concern or a group of individuals that are systematically bound by a common purpose. Organizations may be legal entities in their own right. Examples: commercial organizations such as limited companies, publicly quoted multinationals, subsidiaries.
- **Agreement:** Agreement is language that defines the terms and conditions of a legally binding contract among the identified parties, ordinarily leading to a contract. Examples; policy, reinsurance agreement, staff agreement.
- **Claim:** Claim is a request for indemnification by an insurance company for loss incurred from an insured peril or hazard.
- **Claim Amount:** Claim Amount is the money being paid or collected for settling a claim and paying the claimants, reinsurers, other insurers, and other interested parties. Claim amounts are classified by various attributes.
- **Claim Folder:** A Claim Folder is the physical file in which all claim documents are maintained.
- **Claim Document:** Claim Document is written information such as agreements, financial statements, offers, proposals, etc., to provide backup and depth to agreed-upon or discussed claim matters.

- **Claim Offer:** A Claim Offer is a proposal made by an insurer to a claimant or third party in order to settle a claim.
- **Party Role:** Party Role defines how a Party relates to another entity. Role is the definition of what a Person, Organization, or Grouping DOES whereas Relationship is about what a party IS.
- **Policy Coverage Detail:** Policy Coverage Detail defines the coverages included in an insurance policy (refer to Coverage definition). It is a key link in the data model among Policy, Claim, and Reinsurance.
- **Assessment:** Assessment is a formal evaluation of a particular person, organization, or thing that enables business decisions to be made. Examples: Credit Score, Underwriting Evaluation, Medical Condition.
- **Assessment Result:** An Assessment Result is the outcome of the assessment. There may be multiple results for one assessment.
- **Fraud Assessment:** Fraud Assessment is a subtype of Assessment Result that identifies whether fraud has occurred or is occurring, the extent and impact of the fraud, and to determine who is responsible for the fraud.

A *Relationship* should be a verb or a verb phrase that is always established between two entities, the 'Parent Entity' and the 'Child Entity'. The relationship is used in order to form a sentence between its two entities. Moreover, the type of relationship can be 'Identifying', 'Non-identifying' or 'Subtype'. Each relationship can be also characterized by the notion of cardinality. For example, the 'Subtype' relationship has always 'Zero-to-one' cardinality. Table 1 presents the majority of the relationships between the entities that are described above and utilized in the usage scenario of Section 4.

Table 1. P&C relationships definitions for the entities of Section 4

Parent Entity Name	Parent to Child Phrase	Relationship Cardinality	Child Entity Name
Insurable Object	is covered as defined in	One-to-Zero-One-or-More	Policy Coverage Detail
Insurable Object	involved in	Zero-or-One-to-Zero-One-or-More	Claim
Assessment	results in	Zero-or-One-to-Zero-One-or-More	Assessment Result
Claim	documented in	One-to-Zero-One-or-More	Claim Folder
Claim	settlement results in	One-to-Zero-One-or-More	Claim Amount
Claim Folder	contains	One-to-Zero-One-or-More	Claim Offer
Claim Folder	contains	One-to-Zero-One-or-More	Claim Folder Document
Claim Offer	results in	Zero-or-One-to-Zero-One-or-More	Claim Amount

Attributes are usually defined within an entity and is considered as a property or descriptor of this entity. An attribute is meaningless by itself. For example, date of birth comes from the context of the entity to which it is assigned, for example, date of birth of an employee. Every attribute in a data model is connected to a domain that provides for consistent names, data types, lengths, value sets, and validity rules. The main elements in order to define an attribute for an entity are the Attribute Name, Entity Name and Data Type. Table 2 presents some of the attributes for the entities *Insurable Object* and *Vehicle*.

Table 2. Attributes for the entities *Insurable Object* and *Vehicle*

Entity Name	Attribute Name	Data Type
Insurable Object	Insurable Object Identifier	INTEGER
	Insurable Object Type Code	VARCHAR(20)
	Geographic Location Identifier	INTEGER
Vehicle	Insurable Object Identifier	INTEGER
	Vehicle Model Year	NUMBER(4)
	Vehicle Model Name	VARCHAR(40)
	Vehicle Driving Wheel Quantity	INTEGER
	Vehicle Make Name	VARCHAR(40)
	Vehicle Identification Number	VARCHAR(100)

Figure 1 illustrates how *Insurable Object*, *Vehicle*, *Claim* and *Claim Folder* entities can be defined in terms of attribute and can be linked to each other via relationships in the context of P&C data model.

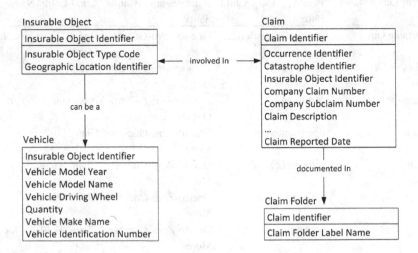

Fig. 1. A snippet of the P&C data model

3   P&C Model as Ontology

In this section, we describe how the Property and Casualty model can be expressed in terms of an ontology using the Web Ontology Language (OWL) [1]. Having the conceptual data model as a reference, we present how data entities representing most of P&C insurance business processes can be converted into ontology components. Note that the reverse is also feasible, i.e. the logical models can be generated from OWL ontologies. According to InfoSphere Data Architect guidelines [11], we can simply correlate logical data model's data types to OWL data types, logical data model's entities to ontology elements as well as logical data model's relationships to OWL object properties.

In the attempt to design our P&C ontology, we first convert all entities (see Section 2), included in the P&C Data Model, to ontology classes. In detail, the convention adopts the restrictions that are presented in Table 3.

Table 3. Logical data model to OWL object mappings

Logical Data model	OWL ontology elements
Entity	owl:Class
Entity - Name	rdf:about
Entity - Label	rdfs:label
Entity - Namespace	Namespace of owl:Class
Entity - Definition	rdfs:comment
Entity - Primary Key	owl:hasKey
Relationship	owl:ObjectProperty
Relationship – Name	rdf:about
Relationship – Label	rdfs:label
Relationship – Owner	rdfs:domain
Relationship – Namespace	Namespace of owl:ObjectProperty
Relationship – Child Entity	rdfs:domain
Relationship – Parent Entity	rdfs:range
Relationship – Annotation	rdfs:seeAlso
	rdfs:DefinedBy
	owl:FunctinalProperty
	owl:InverseFunctionalProperty
	owl:AnnotationProperty
Attribute	owl:DatatypeProperty
Attribute – Name	rdf:about
Attribute – Label	rdfs:label
Attribute – Namespace	rdfs:domain
Attribute – Data Type	rdfs:range
Attribute – Length/Precision	xsd:maxLength
	xsd:length
	xsd:totalDigits
Attribute – Scale	xsd:fractionDigits
Attribute – Primary Key	owl:hasKey
Attribute – Documentation	rdfs:comment
Attribure – Annotation	rdfs:seeAlso
	rdfs:isDefinedBy
	owl:AnnotationProperty

All 'Subtype' relationships (See section 2) have been utilized to form a class hierarchy. So, the 'Parent Entities' are translated into superclasses, while the 'Child Entities' into subclasses. Table 4 provides a listing of a part of 'Subtype' relationships included in the P&C Data Model which relates the entity 'Insurable Object' to any entity that can be characterized as 'Vehicle'.

Table 4. Logical data model 'can-be-a' relationship

Parent Entity Name	Parent to Child Phrase	Relationship Cardinality	Child Entity Name
Insurable Object			Vehicle
Vehicle			Automobile
Vehicle			Van
Vehicle			Trailer
Vehicle			Truck
Vehicle	can be a	Zero-to-One	Boat
Vehicle			Bus
Vehicle			RecreationVehicle
Vehicle			Watercraft
Vehicle			Motorcycle
Vehicle			ConstructionVehicle

Figure 2 depicts how the aforementioned relationships can be converted into components of the P&C ontology.

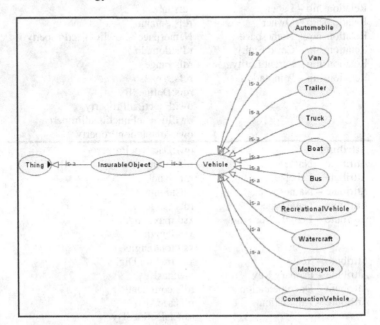

Fig. 2. The mapping of 'Insurable Object' entities in the P&C ontology

All the other types of relationships are represented as object properties in the proposed ontology. Table 3 gathers all the restrictions that must be followed for an effective conversion of these relationships. Two sample relationships, expressed in OWL, are listed below using RDF/XML syntax.

```
<owl:ObjectProperty
rdf:about="http://www.levantes.ceid.upatras.gr/OMG_Insurance#res
ultsIn">
        <rdfs:range rdf:resource="http://
        www.levantes.ceid.upatras.gr/OMG_Insurance #AssesmentRe-
        sult"/>
        <rdfs:domain rdf:resource="http://
        www.levantes.ceid.upatras.gr/OMG_Insurance #Assessment"/>
</owl:ObjectProperty>
<owl:ObjectProperty rdf:about="http://
www.levantes.ceid.upatras.gr/OMG_Insurance #documentedIn">
        <rdfs:domain rdf:resource="http://
        www.levantes.ceid.upatras.gr/OMG_Insurance#Claim"/>
        <rdfs:range rdf:resource="http://
        www.levantes.ceid.upatras.gr
        /OMG_Insurance#ClaimFolder"/>
</owl:ObjectProperty>
```

Each entity type will have one or more data attributes. Each attribute of the P&C model has been mapped to a datatype property in the ontology, according to the restrictions in Table 3.

4   Usage Scenario for Vehicle

A usage scenario (in natural language), relying on the OMG model vocabulary and the implemented ontology, is cited in this subsection. The scenario deals with the case of vehicle insurance and examines the procedure of a fraud assessment. All the nouns, which are in italics, reflect instances of ontology classes. Figure 3 illustrates more thoroughly the usage scenario. Since this is only a snapshot of the model's full range, further refinements are possible.

A vehicle (*Vehicle_1*) has an insurance (*Agreement_1*). This vehicle has been insured for a time interval A with B amount and X coverage (*PolicyCoverageDetail_1*). The *Person_1*, having the role of *Insurer* and being employed by the Organization, *Orgabnization_1* has insured the *Vehicle_1* via the *Agreement_1*. The *Person_2* is owner of the *Vehicle_1*. At the same time, the *Person_2* has also another role, the role of *Driver*.

An accident with significant damages occurred. *Person_2*, the owner of *Vehicle_1*, is involved in a claim (*Claim_1*). The *Claim_1* is documented in the folder (*Claim-Folder_1*) which contains two documents (*ClaimDocument_1* and *ClaimDocument_2*) as well as an offer (*ClaimOffer_1*). The *ClaimOffer_1* results in a specific amount, the *ClaimAmount_1*. An assessment (*Assesment_1*) has been done for the *Claim_1* and results in the *FraudAssesment_1*.

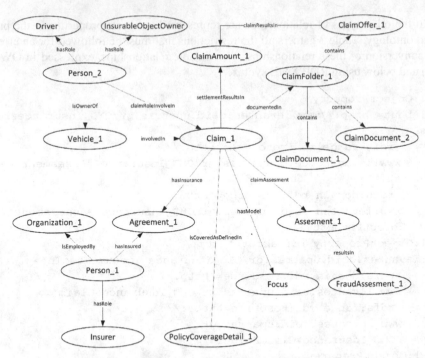

Fig. 3. Usage scenario for Insurance of Vehicle and Fraud Detection

Based on these, a series of business requirements can be represented in the form of logical rules or *axioms* that can be used to automatically infer possible added-value knowledge [5]. For example, a rule can be imposed stating that if a *Driver* has been *involved* in more than e.g. 5 insurance *claims* within a certain time interval, then he is susceptible to *FraudAssesment* to be performed on him.

5   Conclusion and Future Work

In this work, we have shown the feasibility and benefits of adopting a semantic approach towards Big Data manipulation in the insurance sector. To our knowledge, this is the first attempt to express the OMG P&C Model in the form of an OWL 2 ontology, so that reasoning-based conclusions can be enabled. As such, the added-value of adopting (even part of) the model at the insurance sector is leveraged, as it can be used towards automating standard legacy processes, such as fraud detection. Instead of manual resolution, an OWL 2 ontology can be used to highlight possible cases that need further investigation.

On the other hand, the specification of a relaxed ontology schema can serve as a scaffold for implementing Linked Data publishing and dissemination of insurance agreement and claim information. In this manner, the information flow between insurance companies, agents and insurers can be streamlined, by automatically transmitting messages through compatible services.

It is expected that contemporary insurance companies would have to deal with a great deal of such information, such resulting in an increased volume and velocity of data. It is our belief that Semantic Web technologies can stand up to these requirements, by establishing a scalable infrastructure around a semantic repository [12], [8] that secures performance without sacrificing the added-value inference capabilities inherently are equipped with.

References

1. Bechhofer, S., Van Harmelen, F., Hendler, J., Horrocks, I., Mc Guinness, D.L., Patel-Schneider, P.F., Stein, L.A.: OWL Web Ontology Language Reference, W3C Recommendation (2004), http://www.w3.org/TR/2004/REC-owl-ref-20040210/
2. Bizer, C., Heath, T., Berners-Lee, T.: Linked data - the story so far. Int. J. Semantic Web Inf. Syst. 5(3), 1–22 (2009)
3. Hay, C.H.: Data Modeling, RDF, & OWL - Part One: An Introduction To Ontologies (April 2006), http://TDAN.com
4. Heath, T., Bizer, B.: Linked Data: Evolving the Web into a Global Data Space, 1st edn. Synthesis Lectures on the Semantic Web: Theory and Technology, vol. 1, pp. 1–136. Morgan & Claypool (2011)
5. Hitzler, P., Parsia, B.: Ontologies and Rules. In: Staab, S., Studer, R. (eds.) Handbook on Ontologies, 2nd edn., pp. 111–132. Springer (2009)
6. Jenkins, W., Molnar, R., Wallman, B., Ford, T.: Property and Casualty Data Model Specification (2011)
7. Kim, H., Fox, M.S., Sengupta, A.: How To Build Enterprise Data Models To Achieve Compliance To Standards Or Regulatory Requirements (and share data). Journal of the Association for Information Systems 8(2), Article 5 (2007)
8. Koutsomitropoulos, A.D., Solomou, D.G., Pomonis, P., Aggelopoulos, P., Papatheodorou, T.S.: Developing Distributed Reasoning-Based Applications for the Semantic Web. In: Proc. of the 24th IEEE Int. Conference on Advanced Information and Networking (AINA 2010)-Int. Symposium on Mining and Web (MAW 2010), WAINA, pp. 593–598 (2010)
9. Manyika, J., Chui, M., Brown, B., Bughin, J., Dobbs, R., Roxburgh, C., Byers, A.H.: Big data: The next frontier for innovation, competition, and productivity. Technical report, Mc Kinsey (2011)
10. Mitchell, I., Wilson, M.: Linked Data: Connecting and exploiting big data. White paper. Fujitsu UK (2012)
11. Soares, S.: IBM InfoSphere: A Platform for Big Data Governance and Process Data Governance. MC Press Online, LLC (February 2013)
12. Solomou, D.G., Kalou, A.K., Koutsomitropoulos, A.D., Papatheodorou, T.S.: A Mashup Personalization Service based on Semantic Web Rules and Linked Data. In: Proc. of the 7th Int. Conference on Signal Image Technology and Internet Information Systems (SITIS 2011), pp. 89–96. IEEE (2011)
13. The Object Management Group (OMG). MDA Guide Version 1.0.1, omg/2003-06-01 edition (2003)

Enhanced Visual Analytics Services for the Optimal Planning of Renewable Energy Resources Installations

Votis Konstantinos[1], Yiannis Karras[2], Jörn Kohlhammer[3], Martin Steiger[3], Dimitrios Tzovaras[1], and Elias Gounopoulos[4]

[1] Information Technologies Institute, Centre for Research and Technology Hellas, 57001, Thessaloniki, Greece
{kvotis,dimitrios.tzovaras}@iti.gr
[2] Inaccess,
15125, Athens, Greece
jkarras@inaccess.com
[3] Fraunhofer IGD, Darmstadt, Germany
{joern.kohlhammer,Martin.Steiger}@igd.fraunhofer.de
[4] Technological Educational Institute of Western Macedonia (TEIWM),
51100, Grevena, Greece
elgounop@teikoz.gr

Abstract. This paper presents an integrated and novel service environment for real-time interactions between users, as well as enhanced visualization and decision support services over extremely large volumes of heterogeneous Renewable data sources. The integrated visual analytics methods, allow energy analysts to incorporate their expert knowledge into the analysis, so as to dynamically investigate the observed events and locations, and accurately identify the preferable results. The goal of visual analytics research is to turn the information overload into an opportunity by enabling decision-makers to examine this massive amount of information to make effective decisions.

1 Introduction

The appraisal of the technically and economically exploitable Renewable Energy Resources (RES) potential can be considered as one of the biggest problems worldwide, with enormous implications for the environment, economy and development for each country. Given a preferred location, energy planning stakeholders must typically evaluate energy demand and aim for the satisfaction of future needs, according to many constraints and factors. The availability of RES at a specific location depends on climate and weather conditions as well as on socioeconomic and landscape criteria and is highly variable. Thus, RES selection and installation, given a variety of exploitation proposals, is a highly complex task that requires the analysis of huge amounts of historic data.

As a consequence, a large number of parameters concerning the implementation of Renewable Energy Technologies must be evaluated by potential investors, in order to assess the technically and economically exploitable potential of RES in a wide geographical area. A comprehensive environmental assessment must consider the following subtopics; **(a)** a Technical Assessment to identify barriers to RES

L. Iliadis et al. (Eds.): AIAI 2014 Workshops, IFIP AICT 437, pp. 330–339, 2014.

deployment, such as inadequate electricity network infrastructure or natural obstacles, (b) a Land Use Analysis; assessing the changes in land use patterns for setting up RES Wind or solar energy plants, (c) a Social, Ecological and Environmental assessment (e.g. landscape alteration, visual change of rural life or noise) and its impact on the flora and fauna of the area, and (d) a Meteorological Assessment. This is fundamental because the selected assessment actions have to examine and synthesize thoroughly the current state, with respect to technical, land-use, environmental and socio-economic, issues at different levels of intervention. Decision makers must be able to identify a complete framework of performance indices for their planning alternatives. So, a prediction model for these assessment resources should be stochastic in nature to be able to account for the inherent variability. A majority of the existing tools for forecasting these renewable resources give a deterministic forecast, also known as a spot or a point forecast.

From these observations we derive the basic idea and goals of the proposed concept: An integrated service environment for real-time interaction (between human and machine), visualization and decision support by combining the best available data management and data analytics approaches with the best visualization technologies (i.e. innovative visual analytics [1]) over extremely large volumes of heterogeneous data sources. The goal of visual analytics research is to turn the information overload into an opportunity by enabling decision-makers to examine this massive amount of information to make effective decisions. Visual analytics enables people to find hidden relations and to turn the data into useful and defensible knowledge.

The motivation for the proposed framework is the need for energy planners from Greece and Germany to go beyond the analysis of conventional environmental and weather data, from well known databases, satellite images and existing systems, for understanding the relevant patterns in their energy plans within the German and Greek territories. Thus, there is a need for stakeholders to efficiently analyse and visualise an overlay of the historical data (energy, weather, environmental, energy consumption patterns and energy generation data, etc.) that were monitored and measured in real time from existing RES installations in a specific geographical location, along with all data from existing worldwide databases & tools (PVGIS [2], Poseidon, SoDA [3], etc.). Innovative, stochastic energy assessment modelling and optimization algorithms will be designed as part of services that allow end users to optimize their RES planning strategy through detailed simulations and innovative information visualisation exploration.

2   The State of the Art in this Area

2.1   Identification and Estimation of Potential Exploitable RES

Although at the present time they represent a tiny fraction of the world's energy use, their potential is sufficient to completely replace non-renewable sources in the long term. Greece's energy market is at the forefront of transformative changes that are attracting investors from around the globe and especially from Germany where a major energy plan (The Helios project) [4] for the utilization of solar energy in Greece, amounting to 20 billion Euros is under development. Moreover, Wind, or aeolic, power, that offers an alternative that is almost without limit, is driving growth in the renewable energy sector and represents a huge investment potential in Greece. So, with Greece emerging as the energy hub of Southeast Europe, and an aggressive drive for renewable sources to play a major role in the energy supply, the country is at the centre of significant growth opportunities.

The problem of the identification and estimation of potential exploitable RES in a specific geographical area and in a sustainable manner has been tackled in the last decade with the use of Geographical Information Systems and modelling simulation platforms focusing on numerical weather prediction models, and satellite images. The well known PVGIS (Photovoltaic Geographical Information System) tool [2], provides data (solar radiation data, PV module surface inclination and orientation and shadowing effect of the local terrain features (e.g. when the direct irradiation component is shadowed by the mountains)) for the analysis of the technical, environmental and socio-economic factors of solar PV electricity generation in Europe and Africa and supports systems for solar energy decision-making.

Another tentative tool that is being used by energy planners to address the problem is the SoDa Maps of Irradiation [3], Irradiance, and UV. Similarly, the SolarGIS [5] database and simulation tools provide data calculated using in-house developed algorithms that process satellite imagery, and atmospheric and geographical inputs in different locations. The Centre for Renewable Energy Sources (CRES) has also developed a GIS for assessing the technically and economically exploitable RES potential in Greece [6]. Similar initiatives have been also proposed for wind power exploitation (e.g. WindPro [7]) in Greece as well as in Europe [8]. Within the energy software simulation modelling domain, we can mention the RETscreen [9] tool and HOMER [10] which simulates and compares the performance of multiple energy systems (and can determine the optimal combination of power sources by comparing capital, operating costs and emissions over the life of the system).

Despite lots of interactive GIS and simulation tools and models that are now available to compute the required RES parameters in a given geographical area, the move towards the full exploitation of RES is nevertheless held back by the lack of innovative visualisation and simulation solutions which could improve the delivery of relevant information and end-user experience. With the new solutions the territorial RES potential could be evaluated, even for smaller areas, by enriching the data with big historical data from existing RES installations, so that energy planners could get a truly accurate picture of the long-term risks involved. This would be a fundamental shift from a model-based approach, to a data-driven approach which allows more accurate predictions, but requires services to manage and visualize large amounts of data. Right now there are large volumes of project- or module-level historical data (collected in real time and stored in an anonymised manner) that are being gathered from operating RES installations in Greece and Germany. After careful analysis, these could offer important insights into the true nature of RES operations and maintenance costs, failure rates, and energy generation.

2.2   Visual Analytics in the Area of Energy

Visual analytics [11, 12] is a novel research field that aims to provide visualisation based methods for human supported analysis of large, complex datasets. Visual analytics differs to classic information visualisation in that it attempts to take into account properties of human cognition, perception and sense making. Essentially, visual analytics aims to develop appropriate visualisation and interaction mechanisms that match the mental processes of humans. Moreover, visual analytics targets applications that present particular challenges, e.g. involve huge volumes of data or complex heterogeneous sources of information that are hard to analyse using automated mining methods. In essence, as it is often mentioned, the overarching vision of visual analytics [12] is to turn the information overload into an opportunity by making our way of processing data transparent for an analytic discourse.

Ultimately, visual analytics provides technology that combines the strengths of human and electronic data processing, drawing tools from the information visualisation, data mining and cognitive science communities. Visualisation becomes the medium of a semi-automated analytic process, where humans and machines cooperate in a manner which exploits their respective strengths for the most effective results. The user directs the analysis according to his or her specific goals. At the same time, the system provides both an effective means of interaction and data-mining tools for analysis steps, which can be automated.

Visual analytics has a very wide scope and relevant applications have appeared in many different areas. Successful applications have been developed in such diverse fields as environmental monitoring [13], software analytics [14], computer network security [15], electrical grids [16], etc. In general, research in the field of visual analytics has focused on specific applications. Work on general principles, e.g. on evaluation, sense making, perceptual and mental matching, is rather scarce compared to application-specific work.

Considering this, there are elements from applications that involve analysis of spatiotemporal data that could be used in order to benefit the goals of the proposed framework. A good review of the challenges, open issues and most important visual analytics systems that deal with analysis of spatiotemporal data can be found in [17]. In the following, some systems/tools, from which the proposed framework could transfer and further develop existing techniques, are presented.

The Visual analytics Law Enforcement Toolkit (VALET) [18] is a rich interactive environment that supports the exploration and analysis of complex spatiotemporal data. It utilises temporal plots and heatmaps in order to assist law enforcement analysts to detect patterns of criminal behaviour. A tool for powerful analysis of movement data is VisTracks [19]. VisTrack utilizes maps to represent movement data, but at the same time it offers a set of statistical analysis tools to assist in discovering patterns and correlations in data. An overview of issues related to visual analytics for decision support can be found in [20]. Other visual analytics tools for analysis of spatiotemporal data are presented in [21],where different aggregation mechanisms across the time and space dimensions are used in conjunction with heatmap visualisations in order to assist the analyst in detecting important spatiotemporal spots and in [22], where a choropleth map and a parallel coordinate plot allows the user to correlate spatiotemporal data with other sources of information.

3   The Introduced Framework

3.1   The Proposed EVRESIS System

Currently, energy planners and analysts are bombarded with enormous volumes of data that emanate from a variety of sources. So, vital decisions in RES installations can only be achieved through more efficient ways of processing and analysing terabytes of data. EVRESIS introduces a holistic RES assessment framework for future installations of RES assets in specific geographic areas that incorporates traditional well-known assessment models (numerical weather prediction (NWP) models [23], PVGIS data, models from advanced planning tools such as HELIOS 3D, etc.), critical RES operational processes and, consequently, the created micro and macro models that are mapped to appropriate geographic zones.

The main goal of the proposed framework is to tighten the trinity of data management – analytics – visualization through research and development of technologies to identify patterns, events and trends in massive amounts of data. This requires, the establishment of sophisticated and technology driven services for decision making and, hypothesis testing, close to the market. EVRESIS presents a unique opportunity for the establishment of a mutually beneficial cooperation scheme between Germany and Greece in the area of energy planning and production based on natural sources.

More specifically, the EVRESIS framework will address the following specific goals:

- To develop a scalable and technology agnostic cloud-based data collection and analysis service infrastructure, capable of dynamically incorporating a large number of heterogeneous input sources (meteorological, multi-sensorial, GIS, etc.) coming from existing RES installations at the global but also at the local level and by analysing them using novel data analytics methodologies. The aim is to build an inclusive and open service infrastructure that can be used in combination with the major global databases and existing information sources.

- To build relevant RES operational profiles (RES micro and macro operational models) for each type of installed RES generation asset in a specific geographic area, in order to match each zone with one and/or more models. The models will consider technical aspects (e.g., installed systems, size/capacity, produced energy, failures, etc.), environmental (weather data, location, GIS data, urban/industrialized area, CO_2 emissions, etc.), economic aspects (e.g., energy costs, installed costs, maintenance costs,), sociological data (e.g. noise, proximity to cultural monuments, etc.) and other short-term time-scaling uncertainties (production risks, climatic uncertainties; temperature, wind speed for wind turbines, solar radiation for solar cells). For the creation of the models we will take into account relevant dynamic and up-to-date data drawn from existing RES installations.

- To establish an Integrated exploitation Assessment RES Model that extends current energy assessment Models and 3D RES software design models (provided by established modeling software like HOMER, RETscreen, SAM [24], PVsyst [25], Helios 3D [26] etc,) that are being used by energy planners and designers by incorporating and integrating the implemented RES parameterized models coming from existing RES installations.

- To implement a Hypothesis-driven Data Investigation for short term and long Term Analysis of future RES installations. A fully-fledged visual analytics framework will offer alternative ways of exploiting large amounts of observed data in order to allow users to visualize and explore how this information changes through time in a spatio-temporal representation of multiple RES micro and macro models (visual, textual, time-series, graphs, vectors, etc) and in multiple visualization environments (Visual Analytics Techniques). The framework will provide extensive functionality to the end-users (planners, designers) and the ability to perform interactive hypothesis testing for evaluating different energy planning rules and options and visually representing the energy performance of different alternatives under real life operational conditions. This module will be open enough in order to be integrated with existing RES monitoring tools, like the insolar product of Inaccess.

- To develop and evaluate a Holistic Modeling and Simulation Framework, based on contingent and robust proactive-reactive energy planning approaches that take into account integrated RES assessment models.
- To create an "open database" of energy assets (instances for RES models from existing RES installations) with corresponding financial, technical, environmental parameters, for the Greek and German territories, that can be used for future analysis of RES market growth and RES improvement. Thus, end users may add their own RES installation and energy generation data, and then browse RES data entered by others, or view statistics derived from installation data.

3.2   The EVRESIS Conceptual Architecture

The overall EVRESIS conceptual architecture is graphically depicted in the following Figure 1.

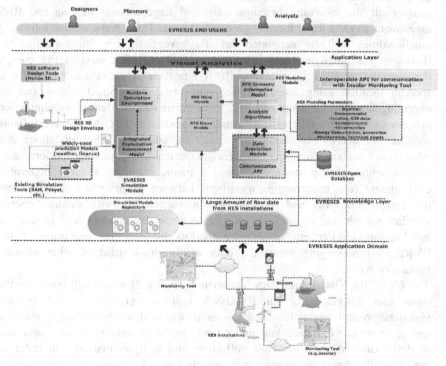

Fig. 1. The EVRESIS conceptual Architecture

The core technological components of the EVRESIS framework are:

- The **Data acquisition/collection module**, capable of dynamically incorporating a large number of heterogeneous input sources from existing RES installations. A specific API that will be implemented can support the future integration of dynamically updated additional databases of raw data coming from existing RES installations.

- The **RES Modeling Module Architecture.** The purpose of the RES modeling component will be to provide the necessary tools to access and analyse the data stemming from the data collection module in order to create generalized micro and macro patterns/models as well as specific business process models in a specific geographic area with improved accuracy that do not exist in the existing SoA paradigms. The design of the models will be as abstract as possible in order to enable their easy adaptation to a new plan of RES installation processes.

- The **EVRESIS Simulation & Decision Support Module Architecture.** The main purpose of the decision support & simulation system will be to provide the RES Designers and energy planners, in the early design phases, with the missing functionality (in terms of available simulation tools) addressing the simulated energy evaluation (via simulation runs) of the energy performance of RES installation designs prior to their realization, taking into account integrated RES assessment models. In this context, the system will be able to analyze all the incoming heterogeneous and large amount of data and RES assessment models for a focused geographic area. Thus, some of the following functionalities will be supported (a) To give designers the capability to represent via intelligent and innovative visual analytics services the energy performance of RES installations, (b) To provide a direct simulation for a specific location using integrated free maps (e.g. OpenStreet maps) and by executing appropriate integrated RES assessment models during the simulation process along with other existing common models and (c) to provide reliable and accurate energy assessment predictions for concrete RES installations that take into account multiple criteria (technical, operational, performance, environmental, socioeconomic, etc.) and dynamic activities within a preferable environment (e.g. uncertainties, transitions between geographic zones, etc) and (d) To provide tools and methods for interoperability with existing energy simulation tools and engines. A specific interoperable module will be able to involve the development of separate toolkits for seamless execution of the EVRESIS time-varying simulation data within the context of other existing simulation tools.

- The EVRESIS **Visual Analytics framework** (figure 2) which will comprise the main user interface for all EVRESIS end-user groups. It will provide visualization and interaction mechanisms to the analysts /modelers to abstract, query, retrieve, synthesize and understand the large amount of raw data and models extracted from the Data collection and analysis platform, in order to automatically detect spatio-temporal correlations of monitored parameters as well as for hypothesis testing and evaluation of different RES installation alternatives. Thereby, advanced visual interfaces will empower the users to interactively explore patterns in large-scale data sets by cross- linking the formerly disjoint data sources in a Visual Analytics Framework.

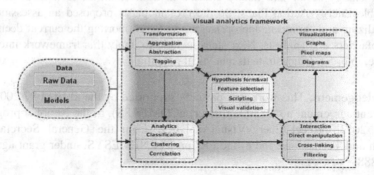

Fig. 2. The EVRESIS Visual Analytics Framework

The proposed system targets some the following user groups:

- Governmental authorities (cities, regions) to assess the economic and environmental impacts of RES planning scenarios, to optimise their local energy strategy to cost-effectively reduce CO2 emissions, to develop new energy plans in and/or identifying the most attractive and exploitable areas in their regions for RES planning.
- Private companies and industries involved in the energy sector and looking for a new business opportunity in RES, either as constructors or as consultants for such RES installations. The provided location analysis visualisation process can help them to identify alternative RES assets and examining candidate locations that are deemed to have fewer obstacles to the RES potential projects and that are more profitable at the initial stages of the development.
- RES installers and manufactures that have already installed their infrastructure in the Greek and German territories. The system will enable them to better design and configure their infrastructure that will result in upgraded products and services to their customers as well as to decreased operational and infrastructural costs and complexity.
- The general public and the individual citizens, who can have access to RES data entered by others with similar needs and capabilities, in order to compare data and identify the potential of RES within their location as well as the derived energy profit.

4   Conclusion and Planned Activities

In Berlin, on March 27 2013, the State Secretary at the Federal Environment Ministry in Germany signed a joint declaration of intent with the Greek Ministry of the Environment, Energy and Climate Change (YPEKA) and the EU Commission's

Task Force for Greece on strengthening cooperation in the field of renewable energies. This entails the future utilization of a platform like the proposed one to support integrated services and tools for the public authorities, the wind energy sector, the solar energy sector and finally citizens in both countries to help them in their Sustainable Energy Action Plans. So the present paper proposed an assessment tool for visualizing and exploring appropriate assets for supporting the current decisions of energy planners and analysts. Next steps include to apply this framework into a real-RES case, in order demonstrate its analytical potential.

Acknowledgement. This work was partially supported by the a) EU FP7/2007-2013, under grant agreement no. 288094 (project eCOMPASS), b) the National programme PAVET 2013, project EnergyVisualAnalytics, from the General Secretariat for Research and Technology and c) the FP7 project NEMESYS, under grant agreement no. 317888.

References

1. Visual Analytics for Large amount of data, http://en.wikipedia.org/wiki/Visual_analytics (accessed at April 2014)
2. European Photovoltaic Geographical Information System (PVGIS), http://re.jrc.ec.europa.eu/pvgis/apps4/pvest.php?lang=en&map=europe, (accessed at April 2014)
3. Solar Energy for professionals, Soda Maps of Irradiation, Irradiance, and UV, http://www.soda-is.com/eng/map/maps_for_free.htmlBruce (accessed at April 2014)
4. The Greece-Germany Helios project, http://www.project-helios.gr (accessed at April 2014)
5. Photovoltaic Softwares for simulation and decision support, http://photovoltaic-software.com/solar-irradiation-database.php (accessed at April 2014)
6. GIS for assessing the technically and economically exploitable RES potential in Greece, http://www.cres.gr/kape/projects_26_uk.htm, (accessed at April 2014)
7. WindPro energy simulation and decision support Tool, http://www.emd.dk/windpro/frontpage (accessed at April 2014)
8. Centre for Renewable Energy Sources (CRES) European projects for energy, http://www.cres.gr/kape/pdf/download/Black_Sea.pdf, (accessed at April 2014)
9. The RETScreen Clean Energy Project Analysis Software, http://en.wikipedia.org/wiki/RETScreen (accessed at April 2014)
10. HOMER- Energy Modeling Software for Hybrid Renewable Energy Systems, http://homerenergy.com/ (accessed at April 2014)
11. Keim, D.A., Andrienko, G., Fekete, J.-D., Görg, C., Kohlhammer, J., Melançon, G.: Visual Analytics: Definition, Process, and Challenges. In: Kerren, A., Stasko, J.T., Fekete, J.-D., North, C. (eds.) Information Visualization. LNCS, vol. 4950, pp. 154–175. Springer, Heidelberg (2008)
12. Thomas, J.J., Cook, K.A.: Illuminating the Path: The Research and Development Agenda for Visual Analytics. National Visualisation and Analytics Centre (2005)

13. Turdukulov, U., Blok, C.: Visual analytics to explore iceberg movement. In: Workshop on GeoVisualization of Dynamics, Movement and Change at the AGILE 2008 Conference (2008)
14. Reniers, D., Voinea, L., Ersoy, O., Telea, A.: A Visual Analytics Toolset for Program Structure, Metrics, and Evolution Comprehension. In: Proceedings of the 3rd Intl. Workshop on Academic Software Development Tools and Techniques (WASDeTT-3) (2010)
15. Mansmann, F., Keim, D.A., North, S.C., Rexroad, B., Sheleheda, D.: Visual Analysis of Network Traffic for Resource Planning, Interactive Monitoring, and Interpretation of Security Threats. IEEE Transactions on Visualization and Computer Graphics 13(6) (2007)
16. Wong, P.C., Schneider, K., Mackey, P., Foote, H., Chin, G., Guttromson, R., Thomas, J.: ANovel Visualization Technique for Electric Power Grid Analytics. IEEE Transactions on Visualization and Computer Graphics 15(3), 410–423 (2009)
17. Andrienko, G., et al.: Space, Time and Visual Analytics. International Journal of Geographical Information Science 24(10), 1577–1600 (2010); Geospatial Visual Analytics: Focus on Time Special Issue of the ICA Commission on GeoVisualization
18. Malik, A., Maciejewski, R., Collins, T.F., Ebert, D.S.: Visual Analytics Law Enforcement Technology. In: IEEE International Conference on Technologies for Homeland Security (2010)
19. Eick, S.G.: Geospatial visualization with VisTracks. Wiley Interdisciplinary Reviews: Computational Statistics 2(3), 272–286 (2010)
20. Kohlhammer, J., May, T., Hoffmann, M.: Visual Analytics for the Strategic Decision Making Process. NATO Science for Peace and Security Series C: Environmental Security, Part 5, 299–310 (2009)
21. Malik, A., Maciejewski, R., Collins, T.F., Ebert, D.S.: Visual Analytics Law Enforcement Technology. In: IEEE International Conference on Technologies for Homeland Security (2010)
22. Jern, M., Thygesen, L., Brezzi, M.: A web-enabled Geovisual Analytics tool applied to OECD Regional Data. In: Proceedings of Eurographics 2009, Munchen (March 2009)
23. Numerical_weather_prediction models, https://en.wikipedia.org/wiki/Numerical_weather_prediction (accessed at April 2014)
24. The System Advisor Model (SAM) to facilitate decision making for people involved in the renewable energy industry, https://sam.nrel.gov/ (accessed at April 2014)
25. PVsys Photovoltaic software, http://www.pvsyst.com/en/software (accessed at April 2014)
26. HELIOS 3D powerful tool for utility scale solar plants, http://www.helios3d.com/index.php/helios-3d.html (accessed at April 2014)

Integrating High Volume Financial Datasets to Achieve Profitable and Interpretable Short Term Trading with the FTSE100 Index

Thomas Amorgianiotis[1], Konstantinos Theofilatos[1], Sovan Mitra[2],
Efstratios F. Georgopoulos[3], and Spiros Likothanassis[1]

[1] Department of Computer Engineering and Informatics, University of Patras, Greece
{amorgianio,theofilk,likothan}@ceid.upatras.gr
[2] Glasgow Caledonian University, Glasgow, Scotland, UK
Sovan.Mitra@gcu.ac.uk
[3] Technological Educational Institute of Peloponnese, 24100, Kalamata, Greece
sfg@teikal.gr

Abstract. During the financial crisis of 2009 traditional models have failed to provide satisfactory results. Lately many techniques have been proposed to overcome the deficiencies of traditional models but most of them deal with the examined financial indices as they are cut off from the rest global market. However, many late studies are indicating that such dependencies exist. The enormous number of the potential financial time series which could be integrated to trade a single financial index enables the characterization of this problem as a "big data" problem and raises the need for advanced dimensionality reduction techniques which should additionally be interpretable in order to extract meaningful conclusions. In the present paper, ESVM-Fuzzy Inference Trader is introduced. This technique is based on the hybrid methodology ESVM Fuzzy Inference which combines genetic algorithms and some deterministic methods to extract interpretable fuzzy rules from SVM classification models.

The ESVM-Fuzzy Inference Trader was applied to the task of modeling and trading the FTSE100 index using a plethora of inputs including the closing prices of various European indexes. Its experimental results were compared with a state of the art hybrid technique which combines genetic algorithm with Multilayer Perceptron Neural Networks and indicated the superiority of ESVM-Fuzzy Inference Trader. Moreover, the proposed method extracted a compact set of fuzzy trading rules which among others can be utilized to describe the dependencies between other financial indices and FTSE100 index.

Keywords: Trading Strategies, Financial Forecasting, FTSE100, Genetic Algorithms, Support Vector Machines, Fuzzy logic.

1   Introduction

Modeling and trading financial indices remains nowadays a very challenging and demanding problem for the scientific community because of their complexity and

L. Iliadis et al. (Eds.): AIAI 2014 Workshops, IFIP AICT 437, pp. 340–349, 2014.

their dynamic and noisy nature. The FTSE 100 index trades futures contracts were utilized as a case study. A future contract is a contract between two parties and its cash settlement is determined by calculating the difference between the traded price and the closing price of the index on the expiration day of the contract. Most traditional methods and more complex machine learning ones have recently failed to capture the complexity and the nonlinearities that exist in financial time series during the latest crisis period. Certain disadvantages have been identified on traditional modelling and trading methods including the difficulties in tuning the parameters of the algorithm, the disability of linear methods to provide good prediction results, the overfitting problem and the fact that modelling and trading are most of the times considered as different problems. Lately, several machine learning methods have been proposed to solve these problems [1]. Despite the encouraging results of new hybrid methodologies their performance could be further enhanced.

Considering modelling and trading of the FTSE100 index, many researchers have been occupied with this problem. These methods are either based on simple traditional methods such as ARMA [2], or on machine learning techniques. The machine learning applications range from simple neural network techniques, such as Higher Order Neural Networks (HONN) [3] to more elaborate techniques such as the hybrid method combining Artificial Bee Colony Algorithm with Recurrent Neural Networks [4] and the hybrid methodologies which combine Genetic Algorithms with Support Vector Machines (SVM) [5].

Despite the promising methods of the aforementioned techniques most of them deal with the FTSE100 index as being independent and cut off from the global stock market. However, the reality is very different. As expected several articles indicated dependencies between the FTSE100 index and several other financial indices [6, 7] and these dependencies have not yet been studied thoroughly. Thus, in order to achieve optimal prediction and trading results using the FTSE100 index, several inputs from other financial indices should be utilized alongside with the traditional autoregressive and technical indicator inputs. The size of the universe of financial indices which could possibly be dependent with FTSE100 is so high that makes this problem relevant to the "big data" [8] topic which has gained the attention of the scientific community lately.

In the present paper, we have created an integrated dataset for modelling and daily trading with the FTSE100 index. This dataset, includes inputs and technical indicators from a variety of financial time-series including VIX, S&P500, DAXX, Euro Stoxx 50, and EURGBP exchange rate. The major problems for utilizing such integrated datasets for modelling and trading are: a) the difficulty to solve effectively the dimensionality reduction problem and b) the difficulty to achieve interpretability for extracting useful knowledge about the uncovered dependencies.

In the present study, we propose the ESVM Fuzzy Inference Trader to solve both these problems. This method is based on a Support Vector Machine methodology to achieve high performance predictions and it proposes an advanced technique to extract a meaningful compact set of fuzzy prediction rules. Moreover, a simple genetic algorithm finds the optimal feature subsets which should be used as inputs and optimizes the parameters of the overall modelling procedure. The final extracted interpretable fuzzy rules presents extremely high statistical and trading performance surpassing another machine learning technique (the hybrid combination of genetic algorithms with Multilayer Perceptron Neural Networks) and enabled for the extraction of important conclusions about the dependencies of FTSE100 and other examined financial indices.

2   Dataset Description

The dataset deployed in the present study incorporates information from 5 financial indexes and one exchange rate.

The basic financial index examined in the present paper is the FTSE100 index. The FTSE 100 index is a share index of the 100 companies listed in the London Stock Exchange having highest market capitalization. It is traded by futures contracts that are cash settled upon maturity of the contract with the value of the index fluctuating on a daily basis. The cash settlement of this index is simply determined by calculating the difference between the traded price and the closing price of the index on the expiration day of the contract.

The FTSE 100 daily time series is non-normal (Jarque-Bera statistics confirms this at the 99% confidence interval), containing slight skewness and high kurtosis. In the present study arithmetic returns were used and they are estimated using the following procedure: Given the price level P_1, P_2,…,P_t, the arithmetic return at time t is formed by:

$$R_t = \frac{P_t - P_{t-1}}{P_{t-1}} \tag{1}$$

Forecasting and trading the FTSE100 was attempted using an extended universe of inputs (50 inputs) containing both autoregressive inputs and technical indicators, such as moving averages of the FTSE 100 index, the VIX index, the S&P500 index, the DAX index, the Euro Stoxx 50 index and the Euro-GBP exchange rate. Specifically the technical indicators which were used by our model are for the FTSE 100 index the growth rate of the previous day, the maximum reduction rate of previous day, the seven previous days, the 5,15,30,50 and 252 days moving averages, the yesterday volume of transactions, the 5,15,30,50 and 252 days moving averages of volume of transactions, the highest close price for the past 30&252 days, the lowest close price for the past 30 and 252 days. The inputs computed for the VIX index, the S&P500 index, the DAX index, the Euro Stoxx 50 index and the Euro-GBP exchange rate are the 5,15,30,50 and 252 days moving average of their closing price.

VIX is a measure of the implied volatility of S&P 500 index options and it is used as an advanced volatility indicator which is able to measure the risk of the market. The S&P 500 consists of the 500 large companies having common stock listed on the NYSE or NASDAQ. The DAX index consists of the 30 major German companies trading on the Frankfurt stock Exchange. The Euro Stoxx 50 is made up of fifty of the largest and most liquid stocks of Eurozone. It was first introduced on February 26, 1988. Historical data of the five examined indexes were downloaded from the finance yahoo site (http://finance.yahoo.com/) and the Euro-GBP exchange rate was down-loaded from the global-view site (http://www.global-view.com/forex-trading-tools/forex-history/).

Our dataset consists of a three-year in-sample dataset (02/01/2009 - 31/12/2011) and a two-year out-of-sample dataset (02/01/2012 - 31/12/2013). Specifically the in-sample dataset consists of 757 rows and 50 columns and the out-of-sample dataset consists of 502 rows and 50 columns. Each row represents a day and each column the 49 input features and the output. This dataset includes inputs and technical indicators from a variety of financial time-series including VIX, S&P500, DAXX, Euro Stoxx 50, and EUR-GBP exchange rate. The ESVM fuzzy inference trader tries to find the dependencies between the above financial indices and the FTSE100. FTSE100 index can also be dependent on an enormous number of indices. From this point of view, the problem of modeling and forecasting the FTSE100 can be seen as a big data problem. The ESVM fuzzy inference trader can find these dependencies and eliminate the financial time series that seem to have no dependencies with the FTSE100 index and this is how we deal with the big data problem. This is done by the feature selection technique of the proposed evolutionary algorithm which is explained at the next section. Due to the market's non stationary nature and its continuing, we deployed a sliding window approach. In particular, the out of sample dataset (testing dataset) was divided in three-month sub-datasets and in order to forecast and trade each individual sub-dataset the algorithms were trained using as in-sample dataset (training dataset) the past three years.

3 Method Description

3.1 GAMLP Methodology

The first machine learning methodology which was applied in the present study is a hybrid combination of Genetic Algorithms and Multi-Layer Perceptron Neural Networks. It has been successfully applied in forecasting and trading the DJIE index [9] and the FTSE100 index [10]. This technique deploys an adaptive evolutionary algorithm to optimize the inputs, the structure and the parameters of MLP NNs. At this work three layered MLP NNs were utilized. The first is the input layer, the second is the hidden layer and the third is the output layer. The fitness function which was utilized by the evolutionary algorithm is described below:

$$Fitness = annualized\ return - MSE - 0.001 * \#selected\_inputs \qquad (2)$$

where #selected_inputs is the number of inputs selected by every iteration of the evolutionary algorithm.

This fitness function enables the extraction of profitable trading strategies while at the same time selecting only the most compact feature subset to be used as inputs.

The deployed evolutionary algorithm utilized an adaptive mutation rate operator which decreases the mutation probability as the evolutionary process progresses. Moreover, when stagnation effect is diagnosed the mutation probability is increased to escape local optimal solutions. The adaptive mutation rate, enables the overall algorithm to present advanced exploration properties in its initial steps and gradually reinforcing its exploitation properties.

3.2   ESVM Fuzzy Inference Trader

Support Vector Machines (SVM) is a method which belongs to the category of supervised learning models and it is nowadays considered as state-of-the-art in the field of machine learning. SVM attempts to locate the optimal hyperplane which can classify the points at categories and presents the largest distance between the points of different categories. SVMs do not require prior knowledge of the problem and they are general purpose classifiers. Non linear SVMs use a kernel function to project the input space to a space of higher dimensions in hope of transforming non-linear separable problems to linear separable ones. Through the years, there have been proposed several kernel functions. At this work we used the radial basis function because it is the most commonly used kernel function in non-linear classification problems [11]. The main limitations of the SVM models are their black-box nature which restricts its interpretability, and the dependence of their performance to the selection of their inputs and parameters.

To overcome the disadvantages of the SVMs we combined them with a Genetic algorithm and a fuzzy rule based system. The genetic algorithm is used for the feature selection problem in an embedded framework and for the optimization of the SVM's parameters. Fuzzy logic is more close to the linguistic language and domain experts can interpret fuzzy rule-based classifiers more easily. Fuzzy logic has the inherit advantage of hiding the imprecise knowledge through fuzziness. Their main disadvantage is overfitting when the dataset includes non linear data.

The ESVM Fuzzy Inference Trader approach deploys an advanced technique to extract interpretable fuzzy rules from SVM classification models.

The problem of extracting fuzzy rules from trained classification SVMs models has been of interest to many researchers through the last decade. Our model is based at the work of [12]. The proposed methodology initially deploys the technique which was proposed by [13]. This is an algorithm to implicitly transform the SVM classification model to a set of Support Vector Fuzzy Inference (SVFI) rules which are proved to be equivalent to the initial classification model.

SVFI rules are in the following general form:

Rule k: if P_1^k **and** P_2^k **and**... P_N^k then c_k

where $P_i^k, i = 1, ..., N$ are   fuzzy   clauses,   having   the   form   where x_i is $CloseToSV(k, i)$. These fuzzy clauses examine the membership of the i-th input value in the i-th fuzzy set of the k-th support vector. The support vectors are the training samples that are selected by the SVM algorithm to define the final classification hyperplane. A Gaussian function of the form $\mu_i^k(x_i) = \exp\left(-\frac{1}{2}\left(\frac{x_i^k - x_i}{\sigma_\kappa}\right)^2\right)$, estimates the membership function by quantifying the distance of the inputs component x_i from the value x_i^k of the ith component of SV_k. The parameters σ_k are real constant numbers ($\sigma_k \in R$).

Because SVFI rules are large and do not include linguistic clauses they are hard to be interpreted by domain experts. The ESVM Fuzzy Inference Trader applies a series of filtering and merging rules to obtain a more compact set of rules. In particular, fuzzy clauses with membership values less than a predefined threshold (β) and even whole fuzzy rules whose total membership value is less than another predefined threshold (δ) are discarded. As a second step, in the produced more compact set of rules *CloseToSV(k, i)* clauses are replaced with linguistic clauses (low, medium, high) which are represented as Gaussian functions with centers placed in the minimum, median and maximum value of each feature.

To fine tune the whole procedure of training SVMs and extracting interpretable fuzzy rules from them, the ESVM Fuzzy Inference Trader algorithm deploys a simple Genetic Algorithm. The GA is used to select the optimal feature subset, and find the optimal values for the regularization parameter C of the SVM, the parameter gamma for the RBF Kernel function, the threshold β for discarding a fuzzy clause, the threshold δ for discarding a fuzzy rule and the variations of the linguistic clauses' Gaussian function (low, medium, high).

The chromosome of the deployed GA consisted of input genes and parameter genes. The input genes take values 0 or 1 and force the classifier to use a specific feature as input if the feature value for this input is 1. The parameter genes are represented as double variables. The basic variation operators which were applied are the one point crossover and mutation operators. The crossover probability was set to 90% as this is the most important operator in GAs. For the mutation operator we used the binary mutation operator for the feature genes and the Gaussian mutation for the other genes as they are real valued genes. The binary mutation randomly alters a gene value from 0 to 1 and opposite. The Gaussian mutation operator adds a random number in a randomly selected gene. This random number is taken from the Gaussian distribution using as center the zero value and as width the interval of allowed values for this gene divided by 10.

The fitness function which guides the evolutionary process of ESVM Fuzzy Inference System is described below:

$$fitness = g1 * annualized_{return} + g2 * geometric_{mean} + g3 * \\ inter\_(sv\_ratio) + g4 * sel\_(feat\_ratio) \qquad (3)$$

where *annualized_return* is the annualized return of the extracted trading strategy, geometric mean is the geometric mean of sensitivity and specificity metrics, $inter\_(sv\_ratio) = 1 - \frac{\#Interpretable\_Rules}{\#SVFI\_Rules}$ and $sel\_(feat\_ratio) = 1 - \frac{\#Selected\_Features}{\#Total\_Features}$.

The parameters g1, g2, g3 and g4 were set to 2, 1, 0.5 and 0.25 respectively to reflect the importance of each goal. In particular the most important goal is to extract profitable trading strategies and the second most important goal is to achieve high classification performance. Secondary goals are the reduced number of fuzzy rules and selected inputs in order to increase the interpretability of the final set of fuzzy rules. The population of the genetic algorithm was set to 30 and the generations to 500.

The ESVM Fuzzy Inference Trading System also calculates a regression prediction for every data sample by calculating its distance from the classification hyperplane.

4   Experimental Results

In this section we present the results of the studied models in the problem of trading the FTSE100 index. The trading strategy for the examined models is simple and identical for both of them: go or stay long when the forecast return is above zero and go or stay short when the forecast return is below zero. Additionally, a confirmation filter was used which disables changing positions when the forecasted value is below the threshold 0.001.

Because of the stochastic nature of the proposed methodologies a simple run is not enough to measure their performance. This is the reason why ten runs where executed and the mean results are presented in the next table. When using the proposed trading strategy traders could utilize the execution which provided the highest in sample performance or use a more elaborate technique to combine the independent traders as proposed in [14].

Table 1. Trading Results

	GA-MLP	ESVM Fuzzy Inference Trader
Annualized Return (excluding costs)	15,41%	25,70%
Positions Taken (annualized)	54	40
Transaction Costs	5,40%	3,95%
Annualized Return (including costs)	10,01%	21,75%
Annualized Volatility	12,71%	9,75%
Information Ratio (including costs)	0,79	2,23
Maximum Drawdown	-12,88	-12,78%
Correct Directional Change	53,43%	59,34%

From Table 3 it is easily observed that both methodologies presented highly profitable trading results even when transaction costs were considered. This is mainly attributed to the extended universe of candidate inputs which was utilized and the sliding window approach which enables the model to be recalculated every three months. Furthermore, both deployed machine learning methods utilize an advanced feature selection mechanism which enables them to ignore redundant feature and extract more compact and simple prediction models.

The ESVM Fuzzy Inference Trader, clearly outperformed GAMLP approach in all examined metrics [15]. The main reason for this fact is the advanced generalization properties that the SVM provides to the overall methodology of ESVM Fuzzy Inference Trader. Moreover, the prediction model of ESVM Fuzzy Inference Trader could be further analyzed by studying the extracted interpretable fuzzy rules sets. The ten most highly rated, in terms of aggregated weight, rules which came up during the sliding window runs of ESVM Fuzzy Inference Trader are provided in Table 2. From this table it is shown that traditional autoregressive inputs where not so significant in the prediction models which were extracted from the ESVM Fuzzy Inference Trader. Important role seem to play as expected the different moving averages technical indicators. Moreover, other not traditional inputs such as past volumes of transactions, highest value of the indexes, lowest values of the indexes and external inputs such as

VIX, DAX, S&P500 and EURGBP related inputs seems to be present in the final prediction models. With a closer examination of the extracted fuzzy rules low values of the moving averages of VIX past values seem to be an indicator for positive predictions. This is explained from the fact that low volatility in the market which is indicated by low values of the VIX index may have positive effects on the FTSE100 index. It is noteworthy that both DAX and S&P500 present negative effects on the FTSE100 outcome. In particular the extracted fuzzy rules predict positive values when negative values for the moving averages of these indexes are observed. EURGBP has similar action on the extracted fuzzy rules as when high values are observed for its moving averages the fuzzy rules decide to predict negative values.

Table 2. Ten most significant extracted fuzzy rules

a/a	Fuzzy Rule	Aggrgated Weight - Signifiance
1	IF 15_days_moving_average_of_volume_of_transactions = Low & year-ly_moving_average_of_volume_of_transactions = High & high-est_close_price_for_the_past_30_days = High & lowest_close_price_for_the_past_252_days = Low & vix_index_moving_5_days_average = Low & vix_moving_15_days_average = Low & vix_moving_30_days_average = Low & dax_moving_252_days_average = Low THEN class = Positive Prediction	1.948
2	IF maximum_growth_rate_of_the_previous_day = Low & yesterday_volume_of_transactions = Low & 5_days_moving_average_of_volume_of_transactions = Low & 15_days_moving_average_of_volume_of_transactions = Low & high_close_price_for_the_past_30_days = High THEN class = Negative Prediction	1.333
3	IF vix_index_moving_5_days_average = Low & eur_gbp_moving_252_days_average = High THEN class = Negative Prediction	1.207
4	IF maximum_growth_rate_of_the_previous_day = Low & moving_average_50_days_of_FTSE100 = High & dax_moving_30_days_average = Low & sp500_moving_30_days_average = Low & sp500_moving_50_days_average = Low THEN class = Positive Prediction	1.125
5	IF vix_index_moving_5_days_average = Low & vix_moving_15_days_average = Low & vix_moving_30_days_average = Low THEN class = Positive Prediction	1.087
6	IF maximum_growth_rate_of_the_previous_day = Low & yesterday_volume_of_transactions = Low & 15_days_moving_average_of_volume_of_transactions = Low & high-est_close_price_for_the_past_30_days = High THEN class = Negative Prediction	1.077
7	IF maximum_growth_rate_of_the_previous_day = Low THEN class = Positive Prediction	1.008
8	IF maximum_growth_rate_of_the_previous_day = Low & vix_index_moving_5_days_average = High & st_50_moving_30_days_average = Low THEN class = Positive Prediction	1.000
9	IF maximum_growth_rate_of_the_previous_day = Low THEN class = Negative Prediction	0.986
10	IF maximum_growth_rate_of_the_previous_day = Low & seven_previous_day_return = Low & vix_index_moving_5_days_average = High & st_50_moving_30_days_average = Low THEN class = Positive Prediction	0.934

5   Conclusions

In the present paper, the problem of modelling and daily trading of the FTSE100 index was approached for the first time with as a big data problem. In specific an extended universe of 50 potential inputs were integrated to a single dataset. These inputs were related to simple autoregressive inputs, and technical indicators of the FTSE100 index and several other financial time-series.

The proposed novel machine learning algorithm for solving the aforementioned problem was the ESVM Fuzzy Inference Trader. It is a hybrid methodology which combines genetic algorithms, Support Vector Machines, Fuzzy systems and some deterministic techniques to extract a compact and interpretable set of linguistic fuzzy prediction rules. The extracted fuzzy rules presented extremely high statistical and trading performance and outperformed one other machine learning approach which was utilized even when transactions costs were taken into account. Further analysis of the extracted interpretable fuzzy rules revealed that traditional autoregressive inputs are not significant for trading the FTSE100 index. Moreover, other technical indicators, such as moving averages, of FTSE100 and of other examined time series are the most important inputs for forecasting and trading the FTSE100 on a daily bases. Considering the role of the other financial time series which were studies, both the moving averages of VIX, S&P500, DAX and EURGBP were found to be inversely related with the FTSE100 index.

The present study could be further extended using additional time series for generating even more potential inputs. The ESVM Fuzzy Inference system is selecting only the more relevant inputs and its performance is not affected by the number of the initial potential features used. For this reason, the incorporation of even more inputs will only affect positively the performance of the overall method and will additionally reveal even more dependencies of the FTSE100 with the rest of the global market. Moreover, The proposed algorithmic technique can be applied to extract intraday trading strategies which are even closer to the big data problem definition.

References

[1] Dunis, C., Likothanassis, S., Karathanasopoulos, A., Sermpinis, G., Theofilatos, K.: Computational Intelligent Techniques for Trading and Investment. Advances in Experimental and Computable Economics. Taylor and Francis (2013) ISBN: 978-0-415-63680-3

[2] Chen, C.C., Tsay, W.J.: A Markov regime-switching ARMA approach for hedg-ing stock indices. Journal of Futures Markets 31(2), 165–191 (2011)

[3] Sermpinis, G., Laws, J., Dunis, C.L.: Modelling and trading the realised volatility of the FTSE100 futures with higher order neural networks. European Journal of Finance 19(3), 165–179 (2011)

[4] Hsieha, T.J., Hsiaob, H.F., Yeh, W.C.: Forecasting stock markets using wavelet transforms and recurrent neural networks: An integrated system based on artificial bee colony algorithm. Applied Soft Computing 11(2), 2510–2525 (2011)

[5] Dunis, C., Likothanassis, S., Karathanasopoulos, A., Sermpinis, G., Theofilatos, K.: A hybrid genetic algorithm–support vector machine approach in the task of forecasting and trading. Journal of Asset Management 14, 52–71 (2012)

[6] Doman, M., Doman, R.: Dependencies between Stock Markets During the Period Including Late-2000s Financial Crisis. Procedia Economics and Finance 1, 108–117 (2012)

[7] Graham, M., Kiviaho, J., Nikkinen, J.: Short term and long-term dependencies of the S&P500 index and commodity prices 13(4), 583–592 (2013)

[8] Boyd, D., Crawford, K.: Critical Questions for Big Data. Information, Communication & Society 15(5), 662–679 (2012)

[9] Theofilatos, K., Karathanasopoulos, A., Sermpinis, G., Amorgianiotis, T., Georgopoulos, E., Likothanassis, S.: Modelling and Trading the DJIA Financial Index Using Neural Networks Optimized with Adaptive Evolutionary Algorithms. In: Jayne, C., Yue, S., Iliadis, L. (eds.) EANN 2012. CCIS, vol. 311, pp. 453–462. Springer, Heidelberg (2012)

[10] Theofilatos, K., Amorgianiotis, T., Karathanasopoulos, A., Sermpinis, G., Geor-gopoulos, E., Likothanassis, S.: Advanced short-term forecasting and trading deploy-ing neural networks optimized with an adaptive evolutionary algorithm. In: Computational Intelligence Techniques for Trading and Investment, pp. 133–145. Routledge (2014)

[11] Theofilatos, K., Pylarinos, D., Likothanassis, S., Melidis, D., Siderakis, K., Thalassinakis, E., Mavroudi, S.: A Hybrid Sup-port Vector Fuzzy Inference System for the Classification of Leakage Current Waveforms Portraying Discharges. Electric Power Components and Systems 42(2), 180–189 (2014)

[12] Papadimitriou, S., Terzidis, K.: Efficient and interpretable fuzzy classifiers from data with support vector learning. Intelligent Data Analysis 9(6), 527–550 (2005)

[13] Chen, Y., Wang, J.: Support vector learning for fuzzy rule based Classifica-tion. IEEE Transactions on Fuzzy Systems 11(6), 716–728 (2003)

[14] Pendaraki, K., Spanoudakis, N.: Constructing Portfolios Using Argumentation Based Decision Making and Performance Persistence of Mutual Funds. In: the 2nd International Symposium & 24th National Conference on Operational Research (EEEE 2013), pp. 24–29 (2013)

[15] Lindemann, A., Dunis, C.L., Lisboa, P.: Probability distributions, trading strategies and leverage: an application of Gaussian mixture models. J. Forecast 23, 559–585 (2004)

Author Index

Printed in the United States
By Bookmasters